CHECKPOINT
AND BEYOND

ASPIRE
SUCCEED
PROGRESS

Complete
Mathematics
for Cambridge
Secondary 1

2

Deborah Barton

OXFORD

OXFORD
UNIVERSITY PRESS

Great Clarendon Street, Oxford OX2 6DP

Oxford University Press is a department of the University of Oxford.
It furthers the University's objective of excellence in research,
scholarship, and education by publishing worldwide in

Oxford   New York

Auckland   Cape Town   Dar es Salaam   Hong Kong   Karachi
Kuala Lumpur   Madrid   Melbourne   Mexico City   Nairobi
New Delhi   Shanghai   Taipei   Toronto

With offices in

Argentina   Austria   Brazil   Chile   Czech Republic   France   Greece
Guatemala   Hungary   Italy   Japan   Poland   Portugal   Singapore
South Korea   Switzerland   Thailand   Turkey   Ukraine   Vietnam

Oxford is a registered trade mark of Oxford University Press
in the UK and in certain other countries

British Library Cataloguing in Publication Data

Data available

ISBN 978-0-19-913707-7

20 19 18 17 16 15 14 13 12 11

Printed in China by Leo Paper Products Ltd

Paper used in the production of this book is a natural, recyclable product
made from wood grown in sustainable forests. The manufacturing process
conforms to environmental regulations of the country of origin.

Acknowledgements

® IGCSE is the registered trademark of Cambridge International
Examinations.

The publisher would like to thank the following copyright holders for their kind permission to
reproduce photographs:

P7: U.S. Air Force; **P25:** Jorg Hackemann/Shutterstock.com; **P31:** © Mark Weiss/Corbis; **P31:** Feng Yu/
Shutterstock; **P35:** MalDix/Shutterstock; **P36:** Jim Hughes/Shutterstock; **P36:** Daniel Allan/Getty
Images; **P60:** Lynee Sladsky/AP Photo; **P61:** Aaron Amat/Shutterstock; **P61:** OUP; **P61:** OUP;
P61: OUP; **P62:** danymages/Istock; **P63:** wang song/Shutterstock; **P63:** Winai Tepsuttinun/Shutterstock;
P63: Alex Staroseltsev/Shutterstock; **P63:** Rouzes/Istock; **P70:** James Marshall/Corbis; **P77:** Skylines/
Shutterstock; **P78:** Cretolamna/Shutterstock; **P81:** Syahreil Hafiz/Shutterstock.com; **P82:** Petekarici/Istock;
P101: FogStock LLC/SuperStock; **P116:** hatman12/Istock; **P128:** Peter Parks/AFP IMAGES; **P144:** Corbis/
SuperStock; **P158:** © Sally and Richard Greenhill/Alamy; **P169:** Ugurhan/Istock; **P179:** Catherine Yeulet/
Istock; **P185:** Filip Fuxa/Shutterstock; **P185:** Pal Teravagimov/Shutterstock; **P202:** DAJ/Getty; **P213:** Science
Photo Library; **P233:** Anton Gvozdikov/Shutterstock; **P242:** Andersphoto/Shutterstock; **P251:** Elena Larina/
Shutterstock; **P255:** Bhathaway/Shutterstock; **P257:** Nattika/Shutterstock; **P260:** Donald Joski/Shutterstock;
P285: Dmitry Naumov/Shutterstock; **P291:** Nikolai Pozdeev/Shutterstock; **P303:** ID1974/Shutterstock.

Front Cover Image courtesy of Illustrart/Shutterstock.

Illustrations by Ian West and Q2A Media.

Contents

Contents

About this book

This book follows the Cambridge Secondary 1 Mathematics curriculum framework for Cambridge International Examinations in preparation for the Cambridge Checkpoint assessments. It has been written by a highly experienced teacher, examiner and author.

This book is part of a series of nine books. There are three student textbooks, each covering stages 7, 8 and 9 and three homework books written to closely match the textbooks, as well as a teacher book for each stage.

The books are carefully balanced between all the content areas in the curriculum framework: number, algebra, geometry, measure, data handling and problem solving. Some of the questions in the exercises and the investigations within the book are underpinned by the final framework area: problem solving – providing a structure for the application of mathematical skills.

Features of the book:

- **Objectives** – taken from the Cambridge Secondary 1 curriculum framework.
- **What's the point?** – providing rationale for inclusion of topics in a real-world setting.
- **Chapter Check in** – to assess whether the student has the required prior knowledge.
- **Notes and worked examples** – in a clear style using accessible English and culturally appropriate material.
- **Exercises** – carefully designed to gradually increase in difficulty, providing plenty of practice and techniques.
- **Considerable variation in question style** – encouraging deeper thinking and learning, including open questions.
- **Comprehensive practice** – plenty of initial practice questions followed by varied questions for stretch, challenge, cross-over between topics and links to the real world with questions set in context.
- **Extension questions** – providing stretch and challenge for students:
  - questions with a box e.g. $\boxed{1}$ provide challenge for the average student
  - questions with a filled box e.g. $\blacksquare$ provide extra challenge for more able students.

- **Technology boxes** – direct students to websites for review material, fun games and challenges to enhance learning.
- **Investigation and Game boxes** – providing extra fun, challenge and interest.
- **Full colour with modern artwork** – to engage students and maintain their interest.
- **Consolidation examples and exercises** – providing review material on the chapter.
- **Summary and Check out** – providing a quick review of the chapter's key points aiding revision and enabling you to assess progress.
- **Review exercises** – provided every six chapters with mixed questions covering all chapters.
- **Bonus chapter** – the work from Chapter 19 is not in the Cambridge Secondary 1 curriculum framework. It is in the Cambridge IGCSE® curriculum framework and is included to stretch and challenge more able students.

A note from the author:

If you don't already love maths as much as I do, I hope that after working through this book you will enjoy it more. Maths is more than just learning concepts and applying them. It isn't just about right and wrong answers. It is a wonderful subject full of challenges, puzzles and beautiful proofs. Studying maths develops your analysis and problem-solving skills and improves your logical thinking – all important skills in the workplace.

Be a responsible learner – if you don't understand something, ask or look it up. Be determined and courageous. Keep trying without giving up when things go wrong. No-one needs to be 'bad at maths'. Anyone can improve with hard work and practice in just the same way sports men and women improve their skills through training. If you are finding work too easy, say. Look for challenges, then maths will never be boring.

Most of all, enjoy the book. Do the 'training', enjoy the challenges and have fun!

Deborah Barton

® IGCSE is the registered trademark of Cambridge International Examinations.

1 Number and calculation 1

Objectives

- Add, subtract, multiply and divide integers.
- Calculate squares, positive and negative square roots, cubes and cube roots; use notation in the form $\sqrt{49}$ and $\sqrt[3]{64}$ and index notation for positive integer powers.
- Identify and use multiples, factors, common factors, highest common factors, lowest common multiples and

  primes; write a number in terms of its prime factors, e.g. $500 = 2^2 \times 5^3$.
- Recall squares to 20^2, cubes to 5^3 and corresponding roots.
- Use the order of operations, including brackets, with more complex calculations.
- Use known facts to derive new facts, e.g. given $20 \times 38 = 760$, work out 21×38.

What's the point?

The alternator in an aircraft is used to generate electrical power. To check that the alternator is working properly the pilot looks at an instrument, called an ammeter, that measures the flow of current. When the ammeter shows a negative flow of current it means that the battery is discharging and the alternator is not supplying power to the system. The aeroplane is in trouble!

Before you start

You should know ...

1 Negative numbers are smaller than zero.

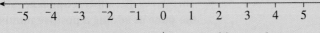

Numbers to the right of another number are always larger.
For example:
  1 is larger than $^-3$
  $^-2$ is larger than $^-4$

Check in

1 Write down the smaller of these number pairs:

  a $^-1, 5$

  b $3, {}^-2$

  c $^-2, {}^-3$

  d $^-1, 0$

  e $^-6, {}^-4$

  f $^-3, 1$

2 Square numbers come from squaring an integer.
Symbols are used for squaring and square roots.
For example:
$3^2 = 9$ 9 is a square number, it is the square of 3
$\sqrt{25} = 5$ 5 is the square root of 25

3 The volume of a cuboid is $w \times l \times h$
For example:
What is the volume of a cuboid measuring 3 cm by 4 cm by 10 cm?
Volume = $3 \times 4 \times 10 = 120 \, \text{cm}^3$

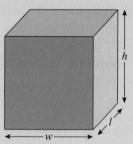

4 How the rules of arithmetic work.
For example:
$2 + 3 \times 5 = 17$

> Multiplication and division are completed before addition and subtraction.

2 Work out:
a 4^2
b $\sqrt{36}$
c 1^2
d $\sqrt{144}$

3 Find the volume of these cuboids.
a A cuboid measuring 2 cm by 4 cm by 11 cm.
b A cuboid measuring 7 cm by 1 cm by 3 cm.
c A cube with side length 2 cm.

4 Work out:
a $1 + 4 \times 6$
b $3 + 2 \times 5$
c $100 - 15 \times 5$

1.1 Adding and subtracting integers

Numbers such as $1, 2, 3, \ldots$ are positive whole numbers. Numbers such as $^-1, {}^-2, {}^-3, \ldots$ are negative whole numbers. The set of numbers that contains both positive and negative whole numbers and 0 is called the set of **integers**.

The symbol for the set of integers is usually written using a 'double' capital letter $\mathbb{Z}$, which stands for the German word for 'numbers', *Zahlen*.

One way to see how to add and subtract integers is to work these operations using number lines.

Look at these number lines:

$6 + {}^-4 = 2$

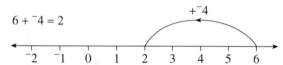

is the same as

$6 - 4 = 2$

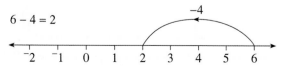

- The number lines show that **adding a negative number is the same as subtracting a positive number**.

Some people learn 'mix means minus', that is, a mixture of a plus sign and a minus sign together results in a minus sign.

Look at the pattern below:

$3 - 2 = 1$
$3 - 1 = 2$
$3 - 0 = 3$
$3 - {}^-1 = \square$

What do you think $3 - {}^-1$ should be? In the pattern the answers are increasing by one, so $3 - {}^-1 = 4$.

- This shows that **subtracting a negative number is the same as adding a positive number**.

Some people learn 'two minuses make a plus', that is, two minus signs together result in a plus sign.

EXAMPLE 1

Work out:

a $4 - {}^-2$ **b** $^-3 - {}^-5 - 4$

....

a $4 - {}^-2 = 4 + 2$

        $= 6$

b $^-3 - {}^-5 - 4 = {}^-3 + 5 - 4$

               $= 2 - 4$

               $= {}^-2$

Exercise 1A

1   **a**   Draw diagrams to show $5 - 7 = {}^-2$ and $5 + {}^-7 = {}^-2$.

    b $+{}^-3$ is the same as -3 Use this idea to write down the answers to:
$5 + {}^-3 = \square$, $3 + {}^-3 = \square$, $1 + {}^-3 = \square$

2   Write down the answers to:
    a $5 + {}^-2$ **b** $6 + {}^-3$ **c** $8 + {}^-10$
    d $4 + {}^-4$ **e** $1 + {}^-2$ **f** $2 - {}^-1$

3   Find the answer to:
    a $0 + {}^-3$ **b** $^-2 + {}^-3$
    c $^-1 + {}^-6$ **d** $^-8 - {}^-2$
    e $^-11 - {}^-1$ **f** $^-1 + {}^-1$

4   Try these questions:
    a $4 - 8$ **b** $4 + {}^-2$ **c** $3 - {}^-3$
    d $3 + {}^-6$ **e** $^-8 + 6$ **f** $^-8 - 8$
    g $^-8 + 10$ **h** $6 + {}^-3$

5   Work out:
    a $6 + {}^-6$ **b** $6 + {}^-7$ **c** $^-2 - 5$
    d $^-2 + {}^-5$ **e** $^-3 - {}^-6$ **f** $^-4 + {}^-3$

6   The commutative law can be used to add negative numbers:
$6 + {}^-4 = {}^-4 + 6 = 2$

    Find the answer to:
    a $3 + {}^-5$ **b** $4 + {}^-2$ **c** $8 + {}^-7$
    d $8 + {}^-10$ **e** $14 + {}^-6$ **f** $5 + {}^-5$

7   Work out:
    a $5 - {}^-2$ **b** $11 - {}^-1$
    c $4 - {}^-9$ **d** $^-3 - {}^-7$
    e $^-6 - {}^-1$ **f** $^-13 - {}^-8$
    g $^-17 - {}^-20$ **h** $^-16 - {}^-27$

8   Work out:
    a $6 + {}^-2 + 4$ **b** $3 + {}^-4 + 2$
    c $8 + 2 + {}^-6$ **d** $4 + 10 + {}^-3$

    e $^-4 + {}^-3 + 7$ **f** $^-4 + 6 + {}^-2$
    g $4 + {}^-4 - {}^-6$ **h** $7 + {}^-5 - {}^-3$
    i $7 - {}^-7 - 7$ **j** $9 - {}^-3 + {}^-1$
    k $^-4 + {}^-3 - {}^-6$ **l** $^-12 - {}^-15 - 16$

9   Steve saves his money with the local bank. He keeps a record of his savings. If he puts in $20 he writes down 20. When he takes out $20 he writes down $^-20$. He starts with $100. Here is a record of his entries for six weeks:

      $30, {}^-20, 20, {}^-60, {}^-30, 70$

    Find out how much he has in the bank at the end of the six weeks.

10   On Thursday afternoon the temperature fell by 3°C. On Friday it had risen by 5°C. If Thursday morning's temperature was $^-3$°C, what was the temperature on Friday?

1.2 Multiplying negative numbers

To multiply by a negative number remember that multiplication is just repeated addition.

For example,
    $3 \times 4 = 3 + 3 + 3 + 3 = 12$
so
    $^-3 \times 4 = {}^-3 + {}^-3 + {}^-3 + {}^-3$
            $= {}^-3 - 3 - 3 - 3$
            $= {}^-12$

Multiplication is **commutative**, that is,
    $3 \times 4 = 4 \times 3$
so
    $^-3 \times 4 = 4 \times {}^-3 = {}^-12$

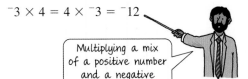

Multiplying a mix of a positive number and a negative number means minus.

Exercise 1B

1   Copy and complete:
    a $^-2 \times 4 = {}^-2 + {}^-2 + {}^-2 + {}^-2 = \square$
    b $^-4 \times 5 = \ldots$
    c $^-6 \times 2 = \ldots$

2   Without writing down the addition, find the answer to:
    a $^-4 \times 3$ **b** $^-8 \times 2$ **c** $^-6 \times 3$
    d $^-5 \times 4$ **e** $^-1 \times 6$ **f** $^-7 \times 8$

3 Copy and complete.
The first one has been done for you.

a $2 \times {}^-3 = {}^-3 \times 2 = {}^-6$
b $4 \times {}^-5 = \ldots$
c $6 \times {}^-2 = \ldots$
d $5 \times {}^-6 = \ldots$
e $2 \times {}^-1 = \ldots$

4 Find the answer to:

a ${}^-6 \times 4$ **b** ${}^-7 \times 3$ **c** $2 \times {}^-2$
d ${}^-1 \times 8$ **e** $8 \times {}^-2$ **f** $10 \times {}^-10$
g $4 \times {}^-7$ **h** ${}^-2 \times 9$ **i** $9 \times {}^-3$

5 Find the answer to:

a ${}^-9 \times 6$ **b** ${}^-8 \times 4$ **c** ${}^-10 \times 4$
d $6 \times {}^-4$ **e** $12 \times {}^-2$ **f** $3 \times {}^-11$
g $3 \times {}^-1$ **h** ${}^-1 \times 9$ **i** $4 \times {}^-9$

Multiplication tables

You can show your times tables on a graph.

The diagram shows a graph of the 2-times table.

Notice all the points lie in a straight line.

Graph of 2-times table

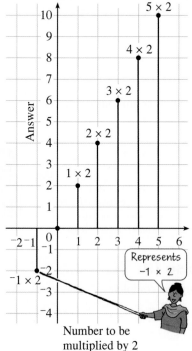

Exercise 1C

1 a Draw a graph of the 3-times table for multiplying numbers from ${}^-3$ to 4. The answer axis will need to show numbers from ${}^-9$ to 12.
  b Do all the answers lie on a straight line?
  c If you extended the line, what would be the answers to ${}^-4 \times 3$ and ${}^-5 \times 3$?

2 a Draw a graph of the 4-times table for multiplying numbers from ${}^-4$ to 3.
  b If you extended the line, what answers would you find for ${}^-5 \times 4$ and ${}^-7 \times 4$?

3 a Write down the answers to:
   i $3 \times {}^-2$ **ii** $4 \times {}^-2$ **iii** $5 \times {}^-2$
  b Copy the graph. Complete it with your answers to part **a**.

Graph of ${}^-2$-times table

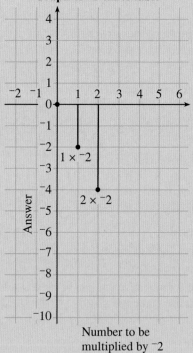

c Lay a ruler along the points. What answer does the graph give for ${}^-1 \times {}^-2$?
d Extend the graph to find the answer to ${}^-2 \times {}^-2$.

4 What does this graph show?

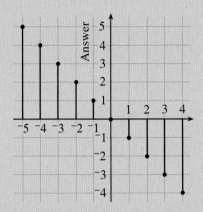

5 Use the graph in Question **4** to write down the answer to:
  a 4 × ⁻1
  b ⁻4 × ⁻1
  c ⁻2 × ⁻1

6 Look at the graph of the ⁻3-times table,
  a What is 0 × ⁻3?
  b What answers does the graph suggest for ⁻1 × ⁻3, ⁻2 × ⁻3, ⁻3 × ⁻3?

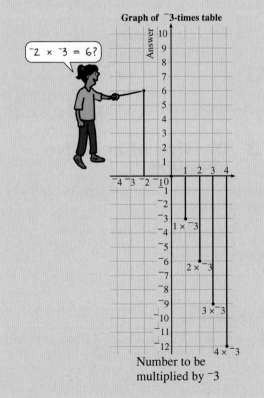

Graph of ⁻3-times table

⁻2 × ⁻3 = 6?

Number to be
multiplied by ⁻3

7 Copy this table. Match up the multiplications and answers using arrows.

⁻3 × ⁻2		⁻6
3 × ⁻2		6
⁻3 × 2		6
3 × 2		⁻6

8 Copy and complete the multiplication table. Use your graphs of the ⁻1-times, ⁻2-times and ⁻3-times tables to help you. Follow the number patterns up the columns to complete the upper left-hand corner of the table, where negative numbers are multiplied together.

×	⁻5	⁻4	⁻3	⁻2	⁻1	0	1	2	3	4	5
⁻5											
⁻4											
⁻3											
⁻2											
⁻1											
0											
1											
2								⁻2			8
3										6	
4											
5											

3 × 2 = 6

Remember:
- When you multiply a positive number by a negative number the answer is negative (mix means minus).
- When you multiply two negative numbers together the answer is positive (two minuses make a plus).

EXAMPLE 2

Work out:

a $3 \times {}^-5$ **b** ${}^-3 \times {}^-5$

....

a positive $\times$ negative = negative:
$3 \times {}^-5 = {}^-15$

b negative $\times$ negative = positive:
${}^-3 \times {}^-5 = 15$

The same rules hold when you multiply more than two numbers.

EXAMPLE 3

Find ${}^-3 \times 4 \times {}^-2$

....

$$
\begin{aligned}
{}^-3 \times 4 \times {}^-2 &= ({}^-3 \times 4) \times {}^-2 \\
&= {}^-12 \times {}^-2 \\
&= 24
\end{aligned}
$$

or
$$
\begin{aligned}
{}^-3 \times 4 \times {}^-2 &= {}^-3 \times (4 \times {}^-2) \\
&= {}^-3 \times {}^-8 \\
&= 24
\end{aligned}
$$

Exercise 1D

1 Write down the answers.

a 4×5 **b** ${}^-4 \times 5$ **c** $4 \times {}^-5$
d ${}^-4 \times {}^-5$ **e** $2 \times {}^-5$ **f** ${}^-4 \times {}^-3$
g 2×4 **h** ${}^-2 \times {}^-4$ **i** ${}^-2 \times 4$

2 Write down the answers.

a ${}^-6 \times {}^-8$ **b** $6 \times {}^-8$
c ${}^-2 \times 5$ **d** ${}^-7 \times {}^-5$
e 14×5 **f** ${}^-5 \times {}^-14$
g $100 \times {}^-1$ **h** $37 \times {}^-2$
i ${}^-1 \times {}^-1$ **j** ${}^-3 \times {}^-3$
k ${}^-20 \times {}^-20$ **l** $20 \times {}^-20$

3 Use the multiplication table in Question **8** of Exercise 1C to help find the missing numbers:

a $3 \times \square = 12$ **b** $3 \times \square = {}^-12$
c ${}^-3 \times \square = {}^-12$ **d** ${}^-3 \times \square = 12$
e $5 \times \square = {}^-10$ **f** $\square \times {}^-2 = 10$
g $\square \times {}^-4 = 8$ **h** $\square \times {}^-4 = {}^-8$
i ${}^-5 \times \square = 15$ **j** $\square \times 4 = {}^-16$

4 Find the answer to:

a ${}^-1 \times 3 \times {}^-2$ **b** ${}^-6 \times {}^-4 \times 2$
c $4 \times {}^-1 \times 2$ **d** ${}^-8 \times {}^-3 \times {}^-2$
e ${}^-5 \times {}^-2 \times 2$ **f** $8 \times 2 \times {}^-1$
g ${}^-8 \times {}^-1 \times {}^-3$ **h** $4 \times {}^-3 \times {}^-9$

5 Find the answer to:

a $6 \times 2 \times {}^-1 \times {}^-3$
b $8 \times {}^-3 \times {}^-2 \times 2$
c $1 \times {}^-2 \times {}^-3 \times 4$
d ${}^-1 \times {}^-2 \times {}^-4 \times {}^-8$
e $5 \times {}^-4 \times {}^-2 \times {}^-3$
f ${}^-1 \times {}^-4 \times 4 \times 5$

6 Find the missing number:

a $3 \times \square \times {}^-2 = 12$
b $4 \times \square \times 4 = 64$
c ${}^-3 \times \square \times {}^-3 = {}^-27$
d ${}^-1 \times \square \times 8 = {}^-16$
e $2 \times \square \times {}^-6 = 72$
f ${}^-3 \times 4 \times \square = 72$
g $\square \times {}^-3 \times {}^-4 = 72$
h ${}^-9 \times \square \times {}^-8 = {}^-72$

7 Find the answer to:

a ${}^-1 \times {}^-1$
b ${}^-1 \times {}^-1 \times {}^-1$
c ${}^-1 \times {}^-1 \times {}^-1 \times {}^-1$
d ${}^-1 \times {}^-1 \times {}^-1 \times {}^-1 \times {}^-1$

8 Is the answer positive or negative, when you multiply together
a three negative numbers
b four negative numbers
c five negative numbers?

9 Find a rule that tells you whether the answer will be positive or negative when you multiply several negative numbers together.

1.3 Division of negative numbers

Multiplication and division of positive numbers are connected like this:

$$3 \times 4 = 12 \qquad \text{so} \qquad 12 \div 4 = 3$$

For negative numbers:

$${}^-3 \times 4 = {}^-12 \qquad \text{so} \qquad {}^-12 \div 4 = {}^-3$$

Exercise 1E

1 Use the multiplication table in Question **8** of Exercise 1C to find the missing number in:

a $\square \times 3 = 6$ **b** $4 \times \square = 12$
c ${}^-4 \times \square = {}^-20$ **d** $2 \times \square = {}^-8$
e $3 \times \square = {}^-12$ **f** $\square \times 5 = {}^-15$

2 Use your answers for Question **1** to write down the answers to:

a $6 \div 3$ **b** $12 \div 4$

c $^{-}20 \div {}^{-}4$ **d** $^{-}8 \div 2$

e $^{-}12 \div 3$ **f** $^{-}15 \div 5$

3 Rewrite as a multiplication, using □.

a $18 \div 3 = \square$ **b** $^{-}8 \div 2 = \square$

c $12 \div {}^{-}3 = \square$ **d** $9 \div {}^{-}3 = \square$

e $\dfrac{16}{{}^{-}4} = \square$ **f** $\dfrac{{}^{-}4}{{}^{-}1} = \square$

4 Find the missing number in each part of Question **3**.

5 How can you tell whether the answer to a division will be a positive or a negative number?

6 Copy this table. Match each division with its answer using an arrow.

$\dfrac{15}{{}^{-}3}$	5
$\dfrac{15}{3}$	5
$\dfrac{{}^{-}15}{3}$	$^{-}5$
$\dfrac{{}^{-}15}{{}^{-}3}$	$^{-}5$

7 Multiplication and division of positive and negative numbers follow a pattern. Copy and complete these tables.

Second number

×	Positive	Negative
Positive	Positive	
Negative		

First number

Second number

÷	Positive	Negative
Positive		
Negative		

First number

8 Find the value of:

a $\dfrac{20}{{}^{-}2}$ **b** $\dfrac{8}{{}^{-}4}$ **c** $\dfrac{{}^{-}6}{3}$

d $\dfrac{{}^{-}100}{50}$ **e** $\dfrac{{}^{-}21}{3}$ **f** $\dfrac{{}^{-}21}{7}$

g $\dfrac{{}^{-}16}{{}^{-}4}$ **h** $\dfrac{{}^{-}4}{{}^{-}4}$ **i** $\dfrac{{}^{-}1}{1}$

9 Find the answer to:

a $\dfrac{{}^{-}4 \times {}^{-}9}{{}^{-}3}$ **b** $\dfrac{{}^{-}21 \times {}^{-}2}{{}^{-}7}$

c $\dfrac{{}^{-}70}{10} + \dfrac{{}^{-}35}{{}^{-}7}$ **d** $\dfrac{6}{{}^{-}2} - \dfrac{{}^{-}16}{4}$

10 a What is the square of $^{-}3$?

b Give two possible values of $\sqrt{9}$.

c Look at the multiplication table in Question **8** of Exercise 1C. Can you find a number whose square is $^{-}9$? Explain your answer.

1.4 Squares, cubes, roots and indices

Squares and square roots

Square numbers come from squaring integers (multiplying a whole number by itself).

1, 4 and 9 are the first three square numbers, which come from 1×1, 2×2 and 3×3. There is a short way to write these using **indices**. For example, the fourth square number is $4 \times 4 = 16$ and this can be written as $4^2 = 16$. Any number can be squared. Sometimes you may want to use your calculator.

EXAMPLE 4

Find the square of 2.7

....

$2.7 \times 2.7 = 2.7^2$

Key into your calculator $\boxed{2}$ $\boxed{\cdot}$ $\boxed{7}$ $\boxed{x^2}$ $\boxed{=}$

$2.7^2 = 7.29$

Note that 7.29 is not a square number because 2.7 is not an integer.

Finding the **square root** of a number is the inverse of squaring a number.

$\sqrt{49}$ means $? \times ? = 49$ (where both numbers $?$ are the same)

$7 \times 7 = 49$ so $\sqrt{49} = 7$

You can find the square root of a decimal using your calculator.

EXAMPLE 5

Find the square root of 11.56

....

To do $\sqrt{11.56}$, key into your calculator

$\boxed{\sqrt{}}\ \boxed{1}\ \boxed{1}\ \boxed{\cdot}\ \boxed{5}\ \boxed{6}\ \boxed{=}$

$\sqrt{11.56} = 3.4$

On some calculators you may need to key this in differently, for example as 11.56, then $\sqrt{}$. Check how yours works.

Squares and square roots are useful for working with squared shapes to find areas and side lengths.

EXAMPLE 6

a Find the area of a square with side length 15 cm.

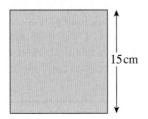

15 cm

b Find the side length of a square with area 104.04 m².

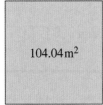

104.04 m²

....

a Area of square is $15^2 = 225$ cm²
b Side length of square is $\sqrt{104.04} = 10.2$ m

From your work on negative numbers you know that $^-3 \times ^-3 = 9$ as well as $3 \times 3 = 9$. So if you want to work out the square root of 9 there are two possible answers: $\pm\sqrt{9} = 3$ or $^-3$. If you are working with side lengths of squares it does not make sense to write down the negative square root as you can't have a negative side length. In some cases, though, the negative square root is very important – you will learn about this in your Cambridge IGCSE® course. If a positive and negative square root are needed, $\pm\sqrt{}$ is used.

Exercise 1F

1 Work out the square or square root. (Where there are two possible answers, write both.)

    a 9^2 **b** $\pm\sqrt{121}$

    c 14^2 **d** $\pm\sqrt{0}$

    e 13^2 **f** $\pm\sqrt{64}$

2 Find the square or square root. You may use a calculator. (Where there are two possible answers, write both.)

    a 3.6^2 **b** $\pm\sqrt{16.81}$

    c 0.32^2 **d** $\pm\sqrt{0.16}$

    e 11.4^2 **f** $\pm\sqrt{0.64}$

3 Find the square or square root. Round the answers to 1 decimal place. In parts **b** and **d**, give positive and negative roots.

    a 1.13^2 **b** $\pm\sqrt{6.4}$

    c 3.1^2 **d** $\pm\sqrt{2.5}$

4 What is the side length of a square with area
    a 7.84 cm²   **b** 60.84 m²   **c** 156.25 mm²?

5 What is the area of a square with side length
    a 3.2 m   **b** 6.9 mm   **c** 1.7 cm?

6 To work out the square root of a fraction you find the square roots of the top and bottom separately:

$$\sqrt{\frac{9}{25}} = \frac{\sqrt{9}}{\sqrt{25}} = \frac{3}{5}$$

Using this method, work out:

    a $\sqrt{\frac{16}{49}}$ **b** $\sqrt{\frac{81}{100}}$ **c** $\sqrt{\frac{25}{64}}$

7 To work out $\sqrt{0.25}$ without a calculator you can write the decimal as a fraction.
(The numbers in the fraction should be square numbers.)

So $\sqrt{0.25} = \sqrt{\frac{25}{100}}$, then find the square roots of the top and bottom separately:

$$\sqrt{\frac{25}{100}} = \frac{\sqrt{25}}{\sqrt{100}} = \frac{5}{10} = 0.5$$

Using this method, work out:

    a $\sqrt{0.36}$ **b** $\sqrt{0.81}$ **c** $\sqrt{1.44}$

8 Can you think of a way of working out $\sqrt{\frac{32}{50}}$ without a calculator?

Cubes and cube roots

Cube numbers come from cubing integers (multiplying a whole number by itself and then multiplying by itself again). 1, 8 and 27 are the first three cube numbers, which come from $1 \times 1 \times 1$, $2 \times 2 \times 2$ and $3 \times 3 \times 3$. There is a short way to write these using **indices**. For example, the fourth cube number is $4 \times 4 \times 4 = 64$ and this can be written as $4^3 = 64$. Any number can be cubed. Sometimes you may want to use your calculator.

EXAMPLE 7

Find the cube of 3.1

....

$3.1 \times 3.1 \times 3.1 = 3.1^3$

Key into your calculator:

$\boxed{3}\ \boxed{\cdot}\ \boxed{1}\ \boxed{x^3}\ \boxed{=}$

$3.1^3 = 29.791$

Note that 29.791 is not a cube number because 3.1 is not an integer.

Some calculators do not have the x^3 button. You may need to use a $\boxed{y^x}$ key or a $\boxed{x^y}$ key. This allows a number to be raised to any power. You need to type in the power you wish to find after pressing this key. For example, to find 3.1^3, key in $\boxed{3}\ \boxed{\cdot}\ \boxed{1}\ \boxed{x^y}\ \boxed{3}\ \boxed{=}$.

Find out how your calculator works.

Finding the **cube root** of a number is the inverse of cubing a number. We use the square root symbol with a small, raised '3' in front to show we need to find the cube root instead of the square root.

$\sqrt[3]{125}$ means $? \times ? \times ? = 125$ (where all numbers ? are the same)

$5 \times 5 \times 5 = 125$ so $\sqrt[3]{125} = 5$

You can find the cube root of a decimal using your calculator.

EXAMPLE 8

Find the cube root of 1.728.

....

To do $\sqrt[3]{1.728}$, key into your calculator

$\boxed{\sqrt[3]{}}\ \boxed{1}\ \boxed{\cdot}\ \boxed{7}\ \boxed{2}\ \boxed{8}\ \boxed{=}$

$\sqrt[3]{1.728} = 1.2$

To get the $\sqrt[3]{}$ function you may need to press the $\boxed{\text{SHIFT}}$ or $\boxed{\text{2nd F}}$ button first. Some calculators have a $\boxed{\sqrt[x]{}}$ button which allows any root to be taken. You will need to tell the calculator to find a cube root by keying in '3' before you press this button. Find out how your calculator works.

Cubes and cube roots are useful for finding volumes and side lengths of cubes.

EXAMPLE 9

a Find the volume of a cube with side length 12 mm.

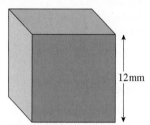

b Find the side length of a cube with volume 2.744 m^2.

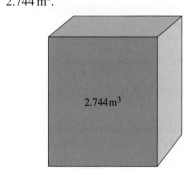

....

a Volume of cube is $12^3 = 1728 \text{ mm}^2$

b Side length of the cube is $\sqrt[3]{2.744} = 1.4 \text{ m}$

From your work on negative numbers you know that $^-2 \times {}^-2 \times {}^-2 = {}^-8$. So if you want to work out the cube root of $^-8$, the answer is $^-2$. Note that you can't find the square root of a negative number on your calculator – you will get an error message if you try.

Exercise 1G

1 Without using a calculator, work out:

    a 3^3 **b** $\sqrt[3]{8}$

    c $(^-1)^3$ **d** $\sqrt[3]{^-125}$

    e $(^-4)^3$ **f** $\sqrt[3]{1}$

2 You may use a calculator to work out:

   a 0.6^3 **b** $\sqrt[3]{1.331}$

   c 4.3^3 **d** $\sqrt[3]{^-21.952}$

3 Rounding the answers to 1 decimal place, work out:

   a 1.43^3 **b** $\sqrt[3]{4.3}$

   c $(^-3.4)^3$ **d** $\sqrt[3]{26.94}$

4 What is the volume of a cube with side length

   a 8.1 mm   **b** 0.7 m   **c** 1.7 cm?

5 What is the side length of a cube with volume

   a $117.649 \, cm^3$

   b $32.768 \, mm^3$

   c $0.512 \, m^3$?

Index notation

You have learned that five cubed or $5 \times 5 \times 5$ can be written as 5^3 for short. This is called **index notation**. 3 is the **power** (also called the **index**). 5 is the **base** number. We can do this with numbers other than squared or cubed numbers.

$2 \times 2 \times 2 \times 2 = 2^4$ *we say 'two to the power four'*

$8 \times 8 \times 8 \times 8 \times 8 \times 8 \times 8 = 8^7$ *we say 'eight to the power seven'*

2^4 can be worked out in your head or with a calculator using the button $\boxed{x^y}$, $\boxed{x^\square}$ or $\boxed{y^x}$ (some calculators require you to press the button $\boxed{\wedge}$ to find the power).

Find out what you need to press on your calculator. Check that your calculator gives $2^4 = 16$.

Exercise 1H

1 Write each of these using index notation:

   a $3 \times 3 \times 3 \times 3 \times 3$

   b $7 \times 7 \times 7 \times 7 \times 7 \times 7 \times 7$

   c $5 \times 5 \times 5 \times 5 \times 5 \times 5 \times 5 \times 5$

   d $^-2 \times {}^-2 \times {}^-2 \times {}^-2$

2 Copy and complete (without a calculator if you can):

   a $2^5 = 2 \times 2 \times \ldots = 32$

   b $2^6 = 2 \times 2 \times \ldots = \square$

   c $4^4 = 4 \times 4 \times \ldots = \square$

   d $5^4 = 5 \times 5 \times \ldots = \square$

   e $1024 = 2 \times 2 \times \ldots = 2^\square$

   f $100\,000 = 10 \times 10 \times \ldots = 10^\square$

3 Using the power key on your calculator, work out:

   a 3.2^4 **b** 1.3^5

   c 2.2^6 **d** 1.1^5

   e $(^-7)^5$

4 Work out:

   a $(^-1)^2$ **b** $(^-1)^3$

   c $(^-1)^4$ **d** $(^-1)^5$

   e When the base is $^-1$, explain how you know whether the answer will be positive or negative for the power you use.

   f Use this rule to predict the answers to $(^-1)^{20}$ and $(^-1)^{37}$. Check these on a calculator. Were you right?

5 On your calculator, try working out 99^{99}. Do you get an error message? Do you get an error message for 99^{33}? (See your teacher if you do not understand the answer your calculator gives.) What is the highest power you can work out on your calculator for the base number 99?

6 What is the highest power you can work out on your calculator for the base number 9999?

1.5 Multiples and factors

The **multiples** of a number are all the numbers in its times table.

The first four multiples of 4 are 4, 8, 12 and 16.

Look at the multiples of 3 and 4:

Multiples of 3 are 3, 6, 9, 12, 15, 18, 21, 24, 27, …

Multiples of 4 are 4, 8, 12, 16, 20, 24, 28, …

Common multiples of 3 and 4 are 12, 24, …

- The **lowest common multiple** of two (or more) numbers is the common multiple that has the lowest value.

The lowest common multiple (LCM) of 3 and 4 is 12.

Factors of a number are the whole numbers that divide into it with no remainder. It is a good idea to look for factors in pairs to make sure that you do not miss any out.

EXAMPLE 10

Find the factors of 24.

....

We find factors by dividing by integers in order starting with 1.

Since $24 \div 1$ is an integer, 1 is a factor of 24 and since $1 \times 24 = 24$ then 24 is also a factor of 24. We look for all other factors of 24 in pairs in the same way.

Try 2: $2 \times 12 = 24$, so 2 and 12 are factors

Then try 3: this gives $3 \times 8 = 24$

Then try 4: this gives $4 \times 6 = 24$

Then try 5: $24 \div 5$ has a remainder so 5 is not a factor.

We already have 6 as a factor, so we know we have found all factors of 24.

1, 2, 3, 4, 6, 8, 12 and 24 are all factors of 24.

Look at the factors of 12 and 8:

> The factors of 8 are 1, 2, 4, 8.

> The factors of 12 are 1, 2, 3, 4, 6, 12.

The **common factors** of 8 and 12 are 1, 2, 4.

• The **highest common factor** of two (or more) numbers is the common factor that has the highest value.

The highest common factor (HCF) of 8 and 12 is 4.

Exercise 1I

1 Write down all the factors of these numbers:
   a i 18, 12
      ii 10, 15
      iii 13, 17
      iv 36, 48
      v 25, 30

   b Find the common factors for each number pair.
   c Find the HCF for each number pair.

2 Write down the first 10 multiples of each number:
   a i 3, 7
      ii 4, 9
      iii 6, 8
      iv 3, 12
      v 8, 16

   b Find the LCM for each number pair.

3 Find the LCM of these pairs of numbers:
   a 24, 68
   b 420, 180

4 Find the HCF of these pairs of numbers:
   a 180, 300
   b 270, 378

5 Find all the pairs of numbers that have a HCF of 30 and an LCM of 3150.

Prime numbers have exactly two different factors, 1 and the number itself. Prime numbers can be found using Eratosthenes' sieve (see Book 1). The prime numbers under 30 are 2, 3, 5, 7, 11, 13, 17, 19, 23 and 29.

A common error is to think that 1 is a prime number. 1 is not a prime number as it only has one factor, not two.

Every number can be written as a product of its prime factors. The two most common ways to do this are by using factor trees and division by primes.

EXAMPLE 11

Write 126 as a product of its prime factors using
a the factor tree method
b the division by primes method.

....

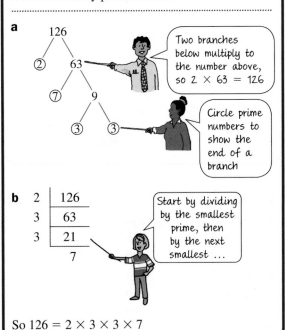

So $126 = 2 \times 3 \times 3 \times 7$

This can also be written in index form. This is useful if there are a lot of repeated prime factors.

$126 = 2 \times 3 \times 3 \times 7$ in index form is $2 \times 3^2 \times 7$

Writing a number as a product of its primes is useful for working out harder HCF and LCM problems.

You may have found Questions **3** and **4** in Exercise 1I hard. Example 12 shows how products of primes can help with questions like these.

EXAMPLE 12

Find **a** the HCF
  b the LCM of 1260 and 600.

....

a Using either factor trees or repeated division write 1260 and 600 as a product of their primes:
$1260 = 2 \times 2 \times 3 \times 3 \times 5 \times 7$
$600 = 2 \times 2 \times 2 \times 3 \times 5 \times 5$

HCF $= 2 \times 2 \times 3 \times 5 = 60$

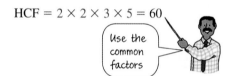

Use the common factors

b Using either factor trees or repeated division write 1260 and 600 as a product of their primes using index notation:

$1260 = 2^2 \times 3^2 \times 5 \times 7$
$600 = 2^3 \times 3 \times 5^2$

LCM $= 2^3 \times 3^2 \times 5^2 \times 7 = 12\,600$

Use the highest power of each common factor.

4 Three strings of different lengths, 240 cm, 318 cm and 426 cm, are to be cut into equal lengths. What is the greatest possible length of each piece?

5 Two lighthouses flash their lights every 20 seconds and 30 seconds respectively. Given that they flashed together at 7 pm, when will they next flash together?

6 A man has a garden measuring 84 m by 56 m. He divides it into the minimum number of square plots. What is the length of each square plot?

7 Neha said she had a different way of working out LCMs. Here is Neha's method for finding the LCM of 90 and 300:

Step 1: Find the HCF first using the product of primes
$90 = 2 \times 3 \times 3 \times 5$
$300 = 2 \times 2 \times 3 \times 5 \times 5$

Common factors are one each of 2, 3 and 5.
So HCF $= 2 \times 3 \times 5 = 30$

Step 2: LCM $=$ HCF $\times$ numbers left over
LCM $= 30 \times 2 \times 3 \times 5 = 900$

a Check Neha's method by using it to work out the answers to Question **2** again.
b Rani said Neha's method wouldn't work if you had to find the LCM of three numbers. Is she right? Try using Neha's method for Question **3**.
c Are there any occasions when Neha's method will work with three numbers?

8 Use the digits 0, 1, 5, 6, 7 and 8 to make two 3-digit numbers with an HCF of 45 and an LCM under 3000.

Exercise 1J

1 Find the HCF of these pairs:
    a 468, 324   **b** 540, 108
    c 450, 990   **d** 330, 910

2 Find the LCM of these pairs:
    a 108, 360   **b** 34, 39
    c 450, 180   **d** 150, 490

3 Find the HCF and LCM of
    a 18, 20 and 30
    b 9, 16 and 12
    c 8, 18 and 50

1.6 Mental strategies
Squares, cubes and roots

You should know all of your square numbers up to 20×20 and all of your cube numbers up to $5 \times 5 \times 5$ without using your calculator. You should also know the corresponding roots and cube roots. You may want to spend some time learning these before attempting the next exercise.

Exercise 1K

Do this exercise without a calculator.
(Where the question asks for a square root you only need to give the positive answer unless the question indicates otherwise.)

1 Write down:
   a 8×8 b $3 \times 3 \times 3$

   c $\sqrt{49}$ d 15^2

   e $\sqrt{324}$

2 $\square \times \square = 144$. Find $\square$.

3 What is the area of a square with side length 11 cm?

4 Write down:

   a 5^3 b $(^-18)^2$

   c $^-6 \times ^-6$ d $\sqrt[3]{64}$

   e $^-1 \times ^-1 \times ^-1$ f both possible values for $\sqrt{100}$

5 $\square \times \square = 361$. Find $\square$.

6 What is the side length of a square with area 169 m²?

7 Write down:
   a $\sqrt{289}$ b $\sqrt[3]{-8}$ c 14×14

8 What is the side length of a cube with volume 27 mm³?

9 Write down both possible values for $\pm \sqrt{36}$.

Rules of arithmetic

You have learned before about **rules of arithmetic** to help make sure that everyone completes calculations the same way. The **order of operations** (BIDMAS) tells us that in calculations we do:

Brackets first	Operations in brackets are completed first
then Indices	Numbers raised to a power (index) are done next
then Division and Multiplication	Divisions and multiplications are completed next – the order you do these doesn't matter
then Addition and Subtraction	Additions and subtractions are completed next – the order you do these doesn't matter

You are now going to extend this to look at more complex calculations.

The best way to set out these calculations is to work down the page, a stage at a time, as shown in Examples 13 and 14.

EXAMPLE 13

Work out:
a $5^2 - 2 \times (5 - ^-1)$

b $(10 + ^-3) + 20 \div 2^2$

....

a $5^2 - 2 \times (5 - ^-1)$ BIDMAS
  $= 5^2 - 2 \times (5 + 1)$ Brackets first
  $= 5^2 - 2 \times 6$ Then Indices
  $= 25 - 2 \times 6$ Then Multiplication
  $= 25 - 12$ Then Subtraction
  $= 13$

b $(10 + ^-3) + 20 \div 2^2$ BIDMAS
  $= (10 - 3) + 20 \div 2^2$ Brackets first
  $= 7 + 20 \div 2^2$ Then Indices
  $= 7 + 20 \div 4$ Then Division
  $= 7 + 5$ Then Addition

When you have a calculation written as a fraction you need to work out the denominator and numerator separately first, before doing any divisions.

EXAMPLE 14

Work out $\dfrac{10 + 18}{9 - 2}$

....

You must work out the numerator and denominator first.

$\dfrac{10 + 18}{9 - 2}$

$= \dfrac{28}{7}$

$= 4$

The long dividing line acts like Brackets.

Then do the Division.

In Example 14, a common error would be to type this into the calculator as $10 + 18 \div 9 - 2$. If you do this the calculator gives the answer as 10, because your calculator follows the rules of arithmetic. Your calculator is doing:

$10 + 18 \div 9 - 2$ Division first
$= 10 + 2 - 2$ Then Addition and Subtraction
$= 10$

That is, if you type in $10 + 18 \div 9 - 2$ the calculator does $10 + \frac{18}{9} - 2$ instead of $\frac{10 + 18}{9 - 2}$.

Note: Some calculators don't follow BIDMAS rules. Check that yours does.

Exercise 1L

Do this exercise without a calculator (apart from Question **5**).

1 Work out:

  a $4 + 2 \times 5$ **b** $10 - 12 \div 3$

  c $4 \times 4 - 20$ **d** $7 + 18 \div 3^2$

  e $5^2 - 3^2 \times 5$ **f** $7^2 \div 7 - {}^-3$

  g $3^3 - ({}^-4 \times 5)$ **h** ${}^-10 - (8 \times {}^-5)$

  i ${}^-5 \times {}^-4 + 100$

2 Work out:

  a $15 - 3 \times 4 + {}^-2$

  b ${}^-3 \times {}^-6 + 2 \times 2^2$

  c $12 \times 2 + ({}^-21 \div 3)$

  d $40 \div {}^-4 + 9 \times 2$

  e $3^3 - 24 \times 3 - 5$

  f $2 - 12 \div {}^-2 + {}^-1$

  g $4^3 - 2 \times {}^-3 \times {}^-2$

  h ${}^-6 + 9 \times 2^2 \div 3$

  i ${}^-8 - 2^2 \times 2^3 + 35$

3 Work out:

  a $({}^-8 - {}^-3) \times ({}^-3 + {}^-2)$

  b $({}^-27 - 33) \div ({}^-25 + 19)$

  c $5 + ({}^-7 - {}^-4) \times 2^2$

4 Work out:

  a $\dfrac{3 + 9}{3}$ **b** $\dfrac{14 + 22}{11 - 2}$

  c $\dfrac{4^2 + 8}{2}$ **d** $\dfrac{2 - {}^-6}{2^2}$

  e $5 + \dfrac{12 - 3}{3}$ **f** $\dfrac{5^3}{5 \times 5}$

5 Repeat Question **4** using a calculator. Try to type in each calculation as one sum (you will probably need brackets to do this, unless you have a calculator that allows calculations to be typed in as fractions).

6 Write brackets to make these correct:

  a $6 + 2^2 \times 10 = 100$

  b $3 + 12 \div 2 - {}^-1 = 7$

  c $10^2 - 10 \times 6 - 4 = 80$

7 When working out the answer to $20 - 2 \times 2^2 + 2$, Odaro has made a mistake:

$$\begin{aligned} & 20 - 2 \times 2^2 + 2 && \text{Indices first} \\ = \; & 20 - 2 \times 4 + 2 && \text{Then Multiplication} \\ = \; & 20 - 8 + 2 && \text{Then Addition} \\ = \; & 20 - 10 && \text{Then Subtraction} \\ = \; & 10 \end{aligned}$$

What mistake has he made?

8 Work out:

  a $7 + (9 - 2^2) \times 2^2 - (3^2 + 5) \div 7$

  b ${}^-5 - (2^2 - 5) - 3 \times 2 - 2^2$

  c $(10 + 2^2 \times {}^-2) +$
       $(({}^-2)^2 + 3) \times (10 + (3^2 \times {}^-2))$

Using known mathematical facts

You should be able to use mathematical facts that you know to derive new facts.

EXAMPLE 15

If $40 \times 37 = 1480$, what is 41×37?

....

$$\begin{aligned} 41 \times 37 &= 40 \times 37 + 1 \times 37 \\ &= 1480 + 37 = 1517 \end{aligned}$$

EXAMPLE 16

Work out 32×26.

....

$$\begin{aligned} 32 \times 26 &= 30 \times 26 + 2 \times 26 \\ &= 780 + 52 = 832 \end{aligned}$$

Exercise 1M

1 If $30 \times 47 = 1410$, what is 31×47?

2 If $20 \times 33 = 660$, what is 23×33?

3 Using the method shown in Example 16, work out:

  a 21×54 **b** 22×31 **c** 31×23

4 If $600 \times 52 = 31\,200$, what is 601×52?

5 If $50 \times 34 = 1700$, what is

  a 5×34 **b** 500×34?

6 If $300 \times 40 = 12\,000$, what is 300×42?

7 If $20 \times 24 = 480$, what is 222×24?

8 Make up some questions of your own like these and give them to your neighbour to solve.

Consolidation

Example 1

Work out:

a $7 + {}^-11$ **b** ${}^-3 + {}^-8$

....

a

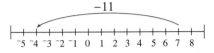

$$7 + {}^-11 = 7 - 11$$
$$= {}^-4$$

b

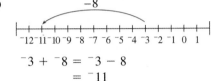

$${}^-3 + {}^-8 = {}^-3 - 8$$
$$= {}^-11$$

Example 2

Work out:

a $7 - {}^-4$ **b** ${}^-7 - {}^-4$

....

a

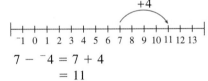

$$7 - {}^-4 = 7 + 4$$
$$= 11$$

b

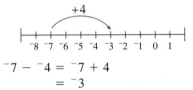

$${}^-7 - {}^-4 = {}^-7 + 4$$
$$= {}^-3$$

Example 3

Work out the HCF and LCM of 300 and 180.

....

Rewrite 300 and 180 as products of primes using the factor tree or repeated division method:

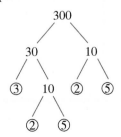

$$300 = 2 \times 2 \times 3 \times 5 \times 5$$

2	180
2	90
3	45
3	15
	5

$$180 = 2 \times 2 \times 3 \times 3 \times 5$$

$$300 = 2 \times 2 \times 3 \times 5 \times 5$$
$$180 = 2 \times 2 \times 3 \times 3 \times 5$$
$$HCF = 2 \times 2 \times 3 \times 5 = 60$$

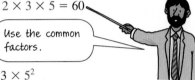
Use the common factors.

$$300 = 2^2 \times 3 \times 5^2$$
$$180 = 2^2 \times 3^2 \times 5$$
$$LCM = 2^2 \times 3^2 \times 5^2 = 900$$

Use the highest power of each prime factor – this is easier to see if the primes are in index form.

Example 4

Work out
$(4 \times 5) - 3 \times 22 + 5$

....

	$(4 \times 5) - 3 \times 22 + 5$	Brackets first
=	$20 - 3 \times 22 + 5$	Then Indices
=	$20 - 3 \times 4 + 5$	Then Multiplication
=	$20 - 12 + 5$	Then Addition and Subtraction
=	13	

Exercise 1

A calculator should **not** be used in this exercise.

1 Work out:
    a ${}^-2 - 4$ **b** ${}^-3 + 7$
    c $8 - 10$ **d** ${}^-12 + 5$
    e ${}^-75 - 0$ **f** $6 - 90$

2 Work out:
    a $3 - {}^-5$ **b** $2 + {}^-6$
    c $4 - {}^-11$ **d** ${}^-9 + {}^-4$
    e ${}^-7 - {}^-5$ **f** $10 - {}^-80$

3 Work out:
  a $2 - {}^-8 + 4$
  b $7 + {}^-5 - {}^-2$
  c $3 - {}^-9 - 4$

4 Work out:
  a 3×5 **b** ${}^-3 \times 4$
  c $3 \times {}^-4$ **d** ${}^-3 \times {}^-4$
  e $5 \times {}^-4$ **f** ${}^-4 \times {}^-6$
  g $7 \times {}^-2$ **h** ${}^-3 \times {}^-7$
  i $11 \times {}^-12$ **j** ${}^-13 \times {}^-14$

5 Work out:
  a $15 \div {}^-3$ **b** ${}^-15 \div 3$
  c ${}^-15 \div {}^-3$ **d** $16 \div {}^-4$
  e ${}^-144 \div {}^-9$ **f** ${}^-182 \div 13$

6 Find the value of:

  a $\dfrac{{}^-7}{7} - \dfrac{{}^-6}{2}$ **b** $\dfrac{{}^-3 \times {}^-9}{{}^-27}$

  c $\dfrac{14}{{}^-7} + \dfrac{18}{{}^-6}$ **d** $\dfrac{{}^-25}{{}^-5} - \dfrac{24}{{}^-4}$

  e ${}^-3 \times {}^-2 \times {}^-5$ **f** $7 \times {}^-4 \times {}^-2$

  g $\dfrac{55}{{}^-11} + \dfrac{{}^-45}{5}$

7 George noticed that the temperature
  was 3 °C before he went to bed. During
  the night the temperature dropped
  by 17 °C.

  What was the temperature in the morning?

8 On Monday I had $23.50 in my bank account.
  On Tuesday I withdrew $39. I paid $22 into
  my account on Wednesday. How much did I
  have in my account on Thursday?

9 The temperature in St Petersburg on 4 December
  was ${}^-2$ °C. What was the temperature in:
  a Bratislava, if it was 9 °C warmer than
  St Petersburg
  b Murmansk, if it was three times colder
  than St Petersburg
  c Volgograd, if it was 3 °C colder than
  St Petersburg?

10 Work out:
  a 3^3 **b** $\sqrt[3]{8}$
  c $({}^-1)^2$ **d** $\pm\sqrt{121}$

11 What is the area of a square of side length 3 m?

12 Work out:
  a 2^7 **b** 4^4 **c** 5^3 **d** 3^4

13 If the volume of a cube is 216 cm^3, what is its
  side length?

14 Jade says that the square root of ${}^-25$ is ${}^-5$.
  Katy says you can't work out the square root
  of a negative number. Who is right?

15 Write these numbers as products of their
  primes:
  a 15 **b** 18 **c** 30
  d 45 **e** 27 **f** 36

Summary

You should know ...

1 a Subtracting a negative number is the
  same as adding a positive number.
  For example: $3 - {}^-4 = 3 + 4 = 7$

Two minuses make a plus.

  b Adding a negative number is the
  same as subtracting a positive number.
  For example: ${}^-2 + {}^-6 = {}^-2 - 6 = {}^-8$

Mix means minus.

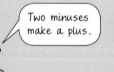

Check out

1 Work out:
  a ${}^-7 - {}^-5$
  b ${}^-3 + {}^-8$
  c ${}^-3 - {}^-10$
  d $15 + {}^-5$

2 a When you multiply two positive numbers together, the answer is positive.
For example: $5 \times 4 = 20$ and $12 \times 8 = 96$

Two minuses make a plus.

b When you multiply two negative numbers together, the answer is positive.
For example: $^-5 \times {}^-3 = 15$ and $^-6 \times {}^-4 = 24$

c When you multiply a positive and a negative number together, the answer is negative.
For example: $^-3 \times 4 = {}^-12$ and $5 \times {}^-6 = {}^-30$

Mix means minus.

2 Calculate:
  a $3 \times {}^-4$
  b $^-3 \times 4$
  c $^-3 \times {}^-4$
  d $^-5 \times 3$
  e $^-4 \times {}^-6$
  f 4×5

3 a When you divide a positive number by a positive number, the answer is positive.
For example: $\dfrac{12}{3} = 4$ and $\dfrac{15}{5} = 3$

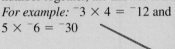
Two minuses make a plus.

b When you divide a negative number by a negative number, the answer is positive.
For example: $\dfrac{^-6}{^-2} = 3$ and $\dfrac{^-30}{^-15} = 2$

c When you divide a positive number by a negative number, the answer is negative.
For example: $\dfrac{15}{^-3} = {}^-5$ and $\dfrac{21}{^-7} = {}^-3$

Mix means minus.

d When you divide a negative number by a positive number, the answer is negative.
For example: $\dfrac{^-14}{7} = {}^-2$ and $\dfrac{^-60}{20} = {}^-3$

Mix means minus.

3 Work out:
  a $15 \div 3$
  b $15 \div {}^-3$
  c $^-15 \div 3$
  d $^-15 \div {}^-3$
  e $^-24 \div 4$
  f $24 \div {}^-4$
  g $^-36 \div {}^-9$
  h $36 \div {}^-9$

4 Any positive number has two square roots; one is positive and one is negative.
For example: The square roots of 16 are 4 and $^-4$.

4 Work out:
  a $(^-3)^2$ **b** 9^2
  c $\pm\sqrt{36}$ **d** $\pm\sqrt{49}$

5 You can use indices or powers to write a sum more simply.
For example: $3 \times 3 \times 3 \times 3 \times 3 = 3^5 = 243$

In 3^5, 3 is the base number, 5 is the power or index.

5 Work out:
  a 4^3 **b** 2^8
  c 5^4 **d** 3^4

6 The cube root of a number means the number which, when multiplied by itself and then multiplied by itself again, makes the original number.

For example: $\sqrt[3]{512} = 8$, since $8 \times 8 \times 8 = 512$

6 Work out:
  a $\sqrt[3]{125}$ **b** $\sqrt[3]{^-8}$
  c $\sqrt[3]{1}$ **d** $\sqrt[3]{27}$

7 Multiples of a number are the numbers in its times table.
For example: Multiples of 5 are 5, 10, 15, 25, …

The factors of a number are the whole numbers that divide into it with no remainder.

For example: The factors of 12 are 1, 2, 3, 4, 6, and 12.

7 Work out:
  a The first five multiples of 8
  b The factors of 30
  c The first five multiples of 12
  d The factors of 32.

8 A number can be written as a product of its prime factors using a factor tree or repeated division by primes.
For example:

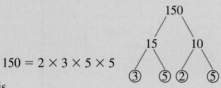

$150 = 2 \times 3 \times 5 \times 5$

In index form this is
$150 = 2 \times 3 \times 5^2$

8 Write these numbers as products of their prime factors:
  a 45   **b** 72
  c 60   **d** 75

9 The highest common factor of two (or more) numbers is the highest factor common to those numbers. Use the common prime factors of each number to calculate it.
For example:
$90 = 2 \times 3 \times 3 \times 5$
$150 = 2 \times 3 \times 5 \times 5$
$HCF = 2 \times 3 \times 5 = 30$

The highest common factor of 90 and 150 is 30.

The lowest common multiple of two (or more) numbers is the lowest number that is a multiple of those numbers. Use the highest power of prime factors to calculate it.
For example:
$90 = 2 \times 3^2 \times 5$
$150 = 2 \times 3 \times 5^2$
$LCM = 2 \times 3^2 \times 5^2 = 450$

The lowest common multiple of 90 and 150 is 450.

For example:
$10 = 2 \times 5$
$15 = 3 \times 5$
$LCM = 2 \times 3 \times 5 = 30$
The lowest common multiple of 10 and 15 is 30.

9 Work out **i** the HCF and **ii** the LCM of
  a 45 and 72
  b 60 and 75

10 The order of operations (BIDMAS).
For example:

$(2 \times 7) - 5 \times 2^2$	Brackets first
$= 14 - 5 \times 2^2$	Then Indices
$= 14 - 5 \times 4$	Then Multiplication
$= 14 - 20$	Then Subtraction
$= {}^-6$	

10 Work out:
  a $4 \times 3^2 - (5 \times 7)$
  b $3 \times 5 - 12 \div (3 + 1)$
  c $(3^3 + 1) - 2^2 \times 3$
  d $\dfrac{3^2 + 5}{7}$

2 Expressions and functions

Objectives

- Simplify or transform linear expressions with integer coefficients; collect like terms; multiply a single term over a bracket.
- Know that algebraic operations, including brackets, follow the same order as arithmetic operations; use index notation for small positive integer powers.
- Know that letters play different roles in equations, formulae and functions; know the meanings of formula and function.
- Construct linear expressions.

What's the point?

Basic formulas are part of your life. When you want to find out how long a journey will take you will be using an algebraic formula:

Journey time = distance ÷ speed

Before you start

You should know ...

1 How to work with negative numbers.
For example:
$7 - {}^-5 = 7 + 5 = 12$ (two minuses make a plus)
${}^-3 + {}^-4 = {}^-3 - 4 = {}^-7$ (mix means minus)

2 How to simplify basic algebra.
For example:
$a + a = 2a$
$3 \times b = 3b$ for short (no need to write the multiplication symbol)
$t \times 5 = 5t$ for short (write the number first, then the letter)

Check in

1 Work out:
a ${}^-2 - {}^-6$
b ${}^-4 + {}^-10$
c ${}^-8 - {}^-3$
d $20 + {}^-10$

2 Write these expressions in a shorter way.
a $m + m + m$
b $6 \times y$
c $r \times 10$
d $c + c + c + c + c$

3 The area of a rectangle, A, is

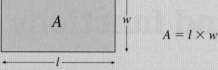

$A = l \times w$

For example:

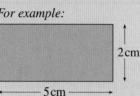

$A = 2\,\text{cm} \times 5\,\text{cm} = 10\,\text{cm}^2$

3 Work out the area of the rectangle with:

a length 2 cm, width 3 cm

b length 6 cm, width 6.5 cm

c length 18 m, width 3.25 m

2.1 Simplifying and expanding

Algebra is really generalised arithmetic. It follows the same rules of arithmetic but uses letters or symbols instead of numbers.

In arithmetic you have:

$$4 + 4 + 4 + 4 + 4 = 5 \times 4$$

while in algebra you have:

$$x + x + x + x + x = 5 \times x = 5x$$

In the same way

$$3 \times 3 \times 3 \times 3 = 3^4$$

while

$$x \times x \times x \times x = x^4$$

You see basic algebraic expressions in many settings. For example:

- Area of a rectangle $= l \times w$
  where l is length and w is width of the rectangle.

To work with algebraic expressions, you need to be able to simplify them.

The basic rule is that you can only add or subtract **like terms**.

In the expression

$$7 + 6x + 4y - 2x$$

$6x$ and $-2x$ are like terms. The number 7 and the term $6x$ are **unlike terms** and cannot be combined.

To simplify expressions you have to combine like terms.

EXAMPLE 1

Simplify:

a $7 + 4x + 6y + 2x + y$

b $xy - y + 2xy + 3y$

. . . .

a $7 + 4x + 6y + 2x + y$
$= 7 + (4x + 2x) + (6y + y)$
$= 7 + 6x + 7y$

b $xy - y + 2xy + 3y$
$= (xy + 2xy) + (3y - y)$
$= 3xy + 2y$

When we simplify the expression $b \times b \times b \times b \times b$ to get b^5, this is using **index notation**. We say "b to the power of 5." b is the **base** and 5 is the **index**. The plural of index is **indices**. Like terms must have exactly the same indices to be combined. For example, x^2y and xy^2 are unlike terms, since x^2y means $x \times x \times y$ and xy^2 means $x \times y \times y$.

Remember the different notation for adding and multiplying letters. For example, $a + a = 2a$ and $a \times a = a^2$. It is quite common to see these written the wrong way around.

The method for simplifying is the same even with more complex expressions.

EXAMPLE 2

Simplify:

a $4x^2y - 3xy^2 + xy + 2x^2y$

b $3abc + abc^2 - 2abc - a^2bc$

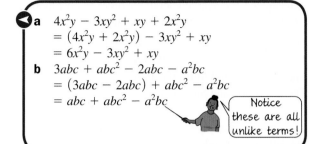

a $4x^2y - 3xy^2 + xy + 2x^2y$
$= (4x^2y + 2x^2y) - 3xy^2 + xy$
$= 6x^2y - 3xy^2 + xy$

b $3abc + abc^2 - 2abc - a^2bc$
$= (3abc - 2abc) + abc^2 - a^2bc$
$= abc + abc^2 - a^2bc$

Notice these are all unlike terms!

Exercise 2A

1 Simplify:
  a $3x + 4x$
  b $5y - 2y$
  c $3x + 4y$
  d $5y - 2x + y$
  e $4x - 3y + 2y$
  f $2 + ab - 2ab + 3ab + 3$
  g $3x - 4y + 2x$
  h $4y - 2x - 2y + x$
  i $1 + 3xy - 2x + 4 + 2xy$
  j $2y - 3x + 3y - 2x$

2 Simplify:
  a $2x^2 - x^2 + 3x^2$
  b $4y - y + 6$
  c $3y - 2y + 4y$
  d $3x^2 - 2x^2 - x^2$
  e $4 + 3y - 2 + 4y$
  f $3 - 3x + 6 - 6x$
  g $3a^2 - 3a - a^2$
  h $4a^2 - 3a^2 + a^2 + a$
  i $3xy - y + 2xy + y$
  j $4a^3 - 2a^2 + 3a^2 + 2a$

3 Sort these terms into pairs of like terms to find the odd one out:
$ab^2 \quad ca^2 \quad bc^2 \quad a^2c \quad a^2b \quad c^2b \quad b^2a$

4 Amy says these are all like terms:
$9n \quad {}^-0.75n \quad 80\,000n \quad 2N \quad \frac{3}{4}n$

Amy is wrong. Which term is the odd one out?

5 Copy the boxes below.
Tick (✓) the pairs which are like terms.

$3x^2pt$	$7x^2p$	$\frac{1}{2}y^2x$	$12m^2n^3$
$0.09tx^2p$	$0.4px^2$	$5xy^2$	$8n^2m^3$

6 Copy and complete this diagram with four more equivalent expressions for $5t - 3m$.

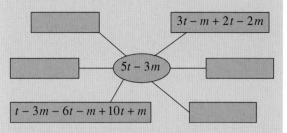

$3t - m + 2t - 2m$

$5t - 3m$

$t - 3m - 6t - m + 10t + m$

7 Copy and complete:
  a $3m + \square - 2m + 4p = m + 7p$
  b $\square - 5t - 6v + 2t = 2v - 3t$
  c $3c - 4r - \square - \square = {}^-5c - 6r$
  d $18p + 15y - 17p - \square p + \square y = 7p + 8y$

8 a To complete the pyramid, the expression in each box is found by adding the two blocks below it. The second row of this pyramid has been completed for you. What goes in the top block? Simplify your answer.

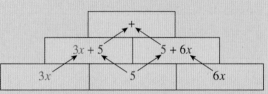

$3x + 5$ $5 + 6x$

$3x$ 5 $6x$

b Fill in the missing blocks in this pyramid.

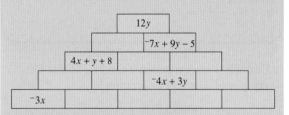

$12y$

${}^-7x + 9y - 5$

$4x + y + 8$

${}^-4x + 3y$

${}^-3x$

💻 TECHNOLOGY

Learn more about simplifying algebraic expressions by visiting the website

www.onlinemathlearning.com

and following the links to 'Algebra, Simplifying Expressions'.
Make sure you watch the videos!

Expanding brackets

You can work out the multiplication

$$6 \times 74$$

using the distributive law:

$$6 \times 74 = 6 \times (70 + 4)$$
$$= 6 \times 70 + 6 \times 4$$
$$= 420 + 24$$
$$= 444$$

In algebraic terms the distributive law is

$$a \times (b + c) = a \times b + a \times c$$

That is, everything inside the brackets is multiplied by what is outside.

This is called multiplying out brackets or **expanding brackets**.

EXAMPLE 3

Expand the brackets.

a $3(x + 2y)$ **b** $x(x + 1)$

....

a $3(x + 2y) = 3 \times x + 3 \times 2y$
$$= 3x + 6y$$

b $x(x + 1) = x \times x + x \times 1$
$$= x^2 + x$$

In Example 3 you expanded $3(x + 2y)$. This is like working out the area of this rectangle:

$x + 2y$

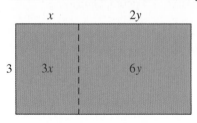

3

which can be divided into two rectangles like this:

	x	$2y$
3	$3x$	$6y$

$$3(x + 2y) = 3x + 6y$$

In Chapter 1 you learned about the order of operations, BIDMAS, and how it applies to numbers:

Brackets first
Then Indices
Then Division and Multiplication
Then Addition and Subtraction

> BIDMAS tells you the order you should do operations in.

The same rules apply to algebra. In the next example, brackets are expanded (or multiplied out) before doing the addition and subtraction, to simplify.

EXAMPLE 4

Simplify:

a $3(a - 2b) + a(4 - 2b)$
b $2a(3 - 2b) - a(4b - 2)$
c $7m + 5(6n + 3m) - 4n$

....

a $3(a - 2b) + a(4 - 2b)$
$$= 3 \times a - 3 \times 2b + a \times 4 - a \times 2b$$
$$= 3a - 6b + 4a - 2ab$$
$$= 3a + 4a - 6b - 2ab$$
$$= 7a - 6b - 2ab$$

b $2a(3 - 2b) - a(4b - 2)$
$$= 2a \times 3 - 2a \times 2b - a \times 4b + a \times 2$$
$$= 6a - 4ab - 4ab + 2a$$
$$= 6a + 2a - 4ab - 4ab$$
$$= 8a - 8ab$$

c $7m + 5(6n + 3m) - 4n$
$$= 7m + 30n + 15m - 4n$$
$$= 22m + 26n$$

> This is positive because multiplying two negatives makes a positive.

Exercise 2B

1 Expand the brackets.
    a $3(x + 2)$
    b $4(2x - 6)$
    c $5(3x - 3)$
    d $6(4 - 3x)$
    e $x(x + 5)$
    f $3x(x + 4)$
    g $2m(3m + 7)$
    h $5p(7 - 2p)$

2 Work out the areas of the rectangles.

a

$x + 7y$

2

b

$x - 3y$

4

c

$2x + 5y$

5

d

$3x + 6y$

7

3 a Draw an area that shows the expression $4(x + 3)$.

  b Write a different expression that gives the same area.

4 a Draw an area that shows the expression $(4p)^2$.

  b Write a different expression that gives the same area.

5 a Draw an area that shows the expression $(t + 4)^2$.

  b Write a different expression that gives the same area.

6 Expand the brackets and simplify.

  a $2(x + 1) + 3(x + 2)$

  b $4(y - 1) + 2(y - 2)$

  c $3(2x + 1) + 4(3x - 4)$

  d $4(1 - 2x) + 3(2 - 3x)$

  e $5(x - 3) - 2(x + 2)$

  f $3(4x - 2y) + 3(2x - 3y)$

  g $2(x - 4y) - 2(x + y)$

  h $3(x - y) - 3(2x + 3y)$

  i $4(2x - 3y) - 2(2x - y)$

  j $5(3y - x) - 4(x - 3y)$

7 Work out the areas of the two shaded regions below. Simplify your expressions.

a

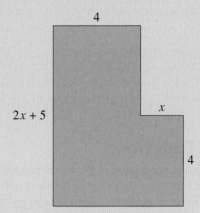
4

$2x + 5$

x

4

b

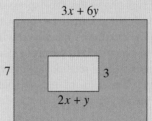
$3x + 6y$

7   3

$2x + y$

8 Pair up equivalent expressions to find the odd one out.

| $^-6x - 5$ | $6x - 11$ |

| $5 - 2(3x + 5)$ | $6 + 15x$ |

$5(3x + 2) - 9x - 7$

$4(2x - 1) - (2x - 7)$

$5 - 3(4 - 5x) + 13$

9 Simplify:

  a $5y^2 - y(1 + 2y)$

  b $x(3 + 2x) + x^2(1 + 2x)$

  c $3x(1 - 2x) + x(x - 1)$

  d $5y(1 - y) - y(y + 3)$

  e $3y(1 + y - y^2) - y^2(2 - 3y)$

  f $7 - (c - 3) - 2c + 3(4 - c)$

  g $3m + 6 \times 2m - 20m$

  h $f - (h - 3f) + 5 \times 4h - 8f$

10 Work out the missing numbers or terms:
  a $2(x - \square) = 2x - 16$
  b $\square(3x + 4) = 18x + 24.$
  c $\square(8g - 12) = 104g - 156$
  d $\square(6 + 13y) = 42 + \square$

11 If the area of a rectangle is $8x + 12$ and the width is 4, what is the length of the rectangle?

12 Remove brackets and simplify:
  a $3(x - 2y) + 2(x + y) - 3(x + 2y)$
  b $4x(1 + y) + 3y(2 - x) - 2(xy + 3y)$
  c $x(x^2 - 3) + x^2(1 - x) - 3x(3 + x)$

2.2 Functions

An **algebraic expression** is one which contains some letters instead of numbers. An **equation** is different from an expression. An equation contains an equals sign. The equals sign shows that the expressions either side of it equal each other. Equations can be solved to find the value of the unknown letters (you will learn more about this in Chapter 8). A **formula** also has an equals sign. It describes the relationship between two (or more) variables. The value of one variable depends on the value (or values) of another.

In Book 1 you learned about function machines and mappings, for example:

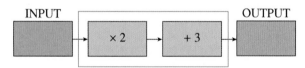

If x is the input, this can also be written as $x \rightarrow 2x + 3$ and if you call the output y, this is written as $y = 2x + 3$.

A **function** is similar to an equation and a formula. Another way of writing the same thing is $f(x) = 2x + 3$. This means the function applied to x is multiply by 2 then add 3. The function tells us about the way in which the output depends on the input.

If you want to find the value of the output when the input is 5, you would do:

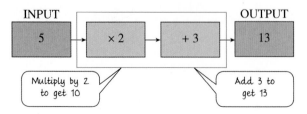

To write this in a shorter way, when $f(x) = 2x + 3$,

$f(5) = 2 \times 5 + 3$
$ = 10 + 3$
$ = 13$

EXAMPLE 5

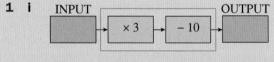

a Write down the function in the form $f(x) = \ldots$ for the function machine.
b Find $f(3)$.

....

a $f(x) = 5x - 1$
b $f(3) = 5 \times 3 - 1$
$ = 15 - 1$
$ = 14$

Exercise 2C

1 **i**
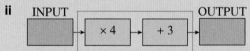

  ii
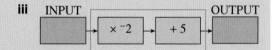

  iii

  iv
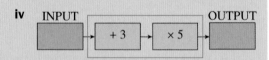

  a Write down each function in the form $f(x) = \ldots$
  b For each of these functions find $f(3)$.
  c For each of these functions find $f(6)$.
  d For each of these functions find $f(0)$.

2 For the function $f(x) = 7x - 3$, find:
  a $f(0)$ **b** $f(1)$
  c $f(^-1)$

3 For the function $f(x) = 2 - 4x$, find:
  a $f(4)$ **b** $f(2)$
  c $f(^-5)$

4 For the function $f(x) = 5(x + 2)$, find:
    a $f(2)$ **b** $f(6)$
    c $f(^-3)$

5 Match one card from the first column with one from the second column. The first is done for you.

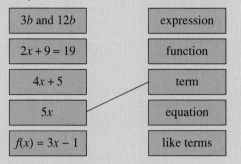

$3b$ and $12b$	expression
$2x + 9 = 19$	function
$4x + 5$	term
$5x$	equation
$f(x) = 3x - 1$	like terms

6 For the function $f(x) = 8x - 3$, the output was 37. What was the input?

7 What do you think $f(x) = 2$ means?

8 If $f(x) = 2(3x - 5)$ when $f(x) = 74$, what is the value of x?

2.3 Constructing expressions

We can write algebraic expressions to help us derive formulae (you will learn more about formulae in Chapter 8). Algebraic expressions are a shorter way of writing something.

In Book 1 you learned how to construct simple expressions like this:

If I have 3 pens costing $4 each the cost of these 3 pens is $3 \times 4 = \$12$

If I have 5 pens costing $4 each the cost of these 5 pens is $5 \times 4 = \$20$

If I have x pens costing $4 each the cost of these x pens is $x \times 4 = \$4x$

$4x$ is an expression that tells us the cost of an unknown number of pens each costing $4.

If you are not sure how to construct the expression, think what sum you would need to do if numbers were involved instead of letters. For instance, in Example 6 you could estimate the number of paper clips in the pot, think what sum to do using that number, then replace that number with a letter.

EXAMPLE 6

Write an expression for the number of paper clips if there are:
a   2 paper clips taken out of the pot
b   5 more paper clips in the pot
c   3 pots of paper clips exactly the same as this one
d   4 people sharing the pot of paper clips equally between them.

....

There is an unknown number of paper clips in this pot, let's say n paper clips.
a $n - 2$
b $n + 5$
c $3n$
d $n \div 4$ or $\dfrac{n}{4}$

> It doesn't matter what letter you use.

EXAMPLE 7

Write an expression for the total time needed to roast a turkey with a mass of k kilograms, if it takes 20 minutes plus 35 minutes for each kilogram.

....

Time, in minutes, is $20 + 35k$

Exercise 2D

1

There are S sugar cubes in a bowl.

Write an expression for the number of sugar cubes if I
a   take out 6 sugar cubes
b   put in 10 sugar cubes
c   have 5 bowls exactly the same
d   take out half of the sugar cubes.

2 Write an expression for the perimeter of these shapes (the distance around the edge). Simplify your expression where possible.

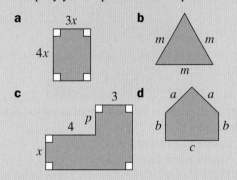

a $3x$, $4x$

b m, m, m

c 3, 4, p, x

d a, a, b, b, c

3 **a** Write an expression for the total cost of k pens at \$3 per pen and p pencils at \$1 per pencil.
b Write an expression for the total cost of k pens at \$$m$ per pen and p pencils at \$$t$ per pencil.

4 Write an expression for the cost of hiring a taxi to travel K kilometres if there is a fixed cost of \$2 plus \$0.50 per kilometre.

5 To change a temperature in degrees Celsius to degrees Fahrenheit, multiply by 1.8 and then add 32. Write an expression to show the temperature in degrees Fahrenheit of something with a temperature of C degrees Celsius.

6 An exchange rate shows that you get 2 New Zealand dollars for every UK pound. Write an expression to show the number of New Zealand dollars you would get for P UK pounds.

7 **a** A book costs V dollars. A CD costs W dollars.
Match each description with the correct expression. The first one is done for you.

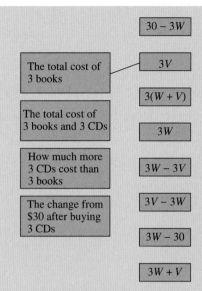

The total cost of 3 books — $3V$
The total cost of 3 books and 3 CDs
How much more 3 CDs cost than 3 books
The change from \$30 after buying 3 CDs

$30 - 3W$
$3V$
$3(W + V)$
$3W$
$3W - 3V$
$3V - 3W$
$3W - 30$
$3W + V$

b Choose any two expressions from **a** that are not paired with a description and write the meaning of those expressions in words.

8 To change kilometres to miles you divide by 8 then multiply by 5. Write an expression for the number of miles in H kilometres.

9 Write using algebra:
a I think of a number w, multiply it by 3 and add 7.
b I think of a number x, add 4 then multiply it by 2.
c I think of a number y, divide it by 5 and then subtract 9.
d I think of a number z, and multiply it by itself.

10 The monthly cost of local calls on a mobile phone is \$8 plus 9 cents per call. Write an expression for the total cost, in dollars, of r local calls in x months.

11 Write an expression for the number of sugar cubes left in the bowl in Question **1** if I take out 10% of the sugar cubes.

⟫ INVESTIGATION

Think of a number, call it A. Think of a different number, call it B. Which of the expressions below are always the same as each other no matter what values you choose for A and B?

AB $A + B$ $\frac{B}{A}$ $2(A + B)$ $\frac{A}{B}$ BA

$A^2 + B^2$ $B - A$ $B + A$ $(A + B)^2$ $A - B$ $2A + 2B$

Consolidation

Example 1

Simplify:

a $2s + 8 + 3t - 5s + 4t - 2$

b $4(3x + 5) - 3(6 - 2x)$

---

a $2s + 8 + 3t - 5s + 4t - 2$
$= 2s - 5s + 3t + 4t + 8 - 2$
$= {}^{-}3s + 7t + 6$

> You get the plus sign here from multiplying together two negatives.

b $4(3x + 5) - 3(6 - 2x)$
$= 12x + 20 - 18 + 6x$
$= 18x + 2$

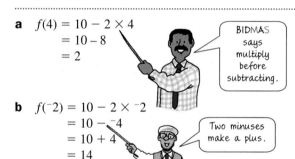

Example 2

For the function $f(x) = 10 - 2x$, find:
a $f(4)$ **b** $f({}^{-}2)$

---

a $f(4) = 10 - 2 \times 4$
$= 10 - 8$
$= 2$

> BIDMAS says multiply before subtracting.

b $f({}^{-}2) = 10 - 2 \times {}^{-}2$
$= 10 - {}^{-}4$
$= 10 + 4$
$= 14$

> Two minuses make a plus.

Example 3

Write an expression for the distance travelled by a man walking at U kilometres per hour for t hours.

Distance = speed × time, so distance is $U \times t$ or Ut.

Exercise 2

1 Simplify:
  a $5x - 4y + 2x$
  b $2y - 4x - 3y - 8x$
  c ${}^{-}4d + 3a + 6d - a$
  d $5R - 2r + 3R + 4 + 8r$

2 Fill in the blanks in the expression
  $3t + \Box h - 5t + \Box t + 4h$
  if it simplifies to give $t + 6h$.

3 Expand the brackets:
  a $5(x - 3)$ **b** $9(2x + 8)$
  c $3(4 + 7y)$ **d** $12(7m - 12p)$

4 Write down four expressions equivalent to
  $5m - 3t + m - 3t$ without using the numbers
  5, 3 or 1. You may use brackets if you want.

5 Expand the brackets and simplify:
  a $3(x + 2) + 4x$
  b $5(2 + 4y) + 3(2y + 6)$
  c $2 + 4(5x - 7)$
  d $3(2t - 1) - 5(4t - 3)$

6 Write down an expression for the areas of
  these rectangles:
  a   **b**
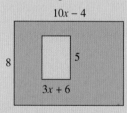

7 Work out the missing numbers or terms:
  a $3(h - \Box) = 3h - 15$
  b $\Box(4x + 3) = 16x + 12$
  c $\Box(4x - 3) = 40x - 30$
  d $\Box(3 + 12m) = 15 + \Box$

8 If the area of a rectangle is $4x + 6$ and the
  width is 2, what is the length?

9 Write down an expression for the shaded area
  in this shape:
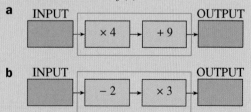

10 Write down the functions for these function
  machines in the form $f(x) = \ldots$.
  a
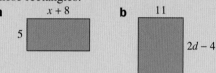
  b

11 Copy and complete this diagram for the
  function $f(x) = \dfrac{2x}{5}$.
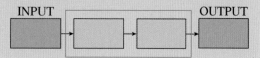

Summary

You should know ...

1 Like terms of an algebraic expression contain the same letters and can be simplified.
For example: $5x + 2y - x - 3y = 4x - y$

2 You can expand brackets and use the order of operations with algebra.
For example: $2 - 2(3x - 4) + 7x$
$$= 2 - 6x + 8 + 7x$$
$$= 10 + x$$

3 The difference between a term, an expression, an equation, a function and a formula.
For example: $3x$ is a term (and also a simple expression)
$f(x) = 4x$ is a function
$5x + 2$ is an expression
$3x + 1 = 13$ is an equation

4 You can construct expressions.
For example:

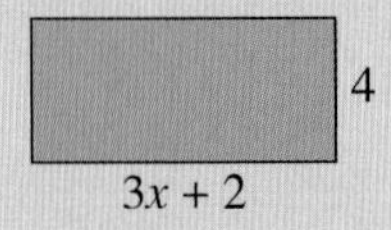

The area is width $\times$ length $= 4 \times (3x + 2)$
$$= 4(3x + 2)$$
$$= 12x + 8$$
The perimeter is the distance around the outside
$$= 4 + 3x + 2 + 4 + 3x + 2$$
$$= 6x + 12$$

Check out

1 Simplify:
a $5m + 3t - 4m - t$
b $3x + 4y + 2x - 9y$
c $3p + 4 - 5p + 7 + 4v$
d $6f - 3g - 10f - 8 + 2f - g$

2 a Expand the brackets:
i $3(5p - 2)$
ii $^-6(4 - 3x)$
iii $x(x + 5)$
iv $3p(5p - 4)$

b Expand the brackets and simplify:
i $7(6p - 8m) + 5(2p + 4m)$
ii $50 + 80T - 10(5T + 2)$

3 Which of the following are expressions?
a $3t - 4 = 11$
b $7x + 4y$
c $2t$
d $f(x) = 3x$

4 Construct expressions for
i the area and
ii the perimeter of these rectangles.

a

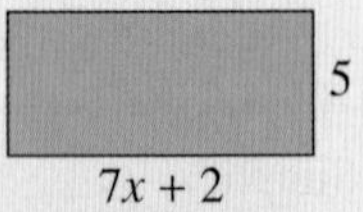

b

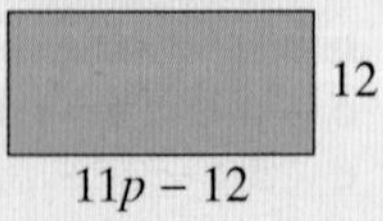

Shapes and mathematical drawings

Objectives

- Use a ruler and compasses to construct
  - circles and arcs
  - a triangle, given three sides (SSS)
  - a triangle, given a right angle, hypotenuse and one side (RHS).
- Know that the longest side of a right-angled triangle is called the hypotenuse.
- Use a straight edge and compasses to construct
  - the midpoint and perpendicular bisector of a line segment
  - the bisector of an angle.

- Know that if two 2D shapes are congruent, corresponding sides and angles are equal.
- Draw simple nets of solids, e.g. cuboid, regular tetrahedron, square-based pyramid, triangular prism.
- Identify all the symmetries of 2D shapes.
- Classify quadrilaterals according to their properties, including diagonal properties.

What's the point?

Architects, draughtsmen, engineers and carpenters are professionals who rely on accurate mathematical drawings for their work. There are many other professions that also use mathematical drawings, without which their work would be extremely difficult.

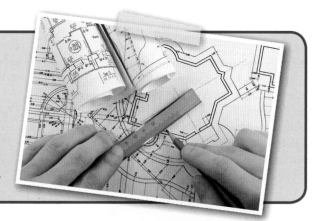

Before you start

You should know ...

1 How to use a protractor to draw and measure angles.

Check in

1 a Use your protractor to draw angles of:
   i   30°   ii 45°
   iii 72°   iv 143°

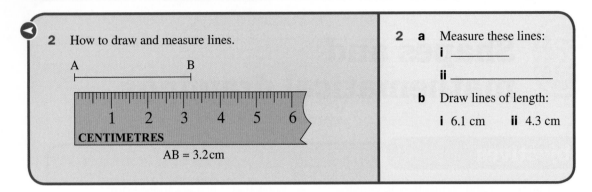

2   How to draw and measure lines.

A   B

CENTIMETRES

AB = 3.2cm

2   **a**   Measure these lines:

   i _____

   ii _____

   b   Draw lines of length:

   i 6.1 cm   **ii** 4.3 cm

3.1 Constructing circles, arcs and triangles

Circles and arcs

When constructing circles using a pair of compasses make sure the compasses are fully tightened and cannot widen or narrow as the circle is drawn. The best compasses are those that don't slip:

A pair of compasses is sometimes wrongly called a 'compass', however a compass is really a navigational tool for indicating the direction of magnetic north. A pair of compasses has two parts, joined by a hinge, which is why it is called a 'pair'. We will use **compasses** for short. If you have compasses that hold a pencil, make sure that when they are closed the pencil point is in line with the metal tip.

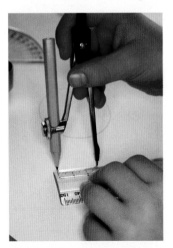

Hold the compasses in one hand and at the very top. Some people find circles are neater when they turn the paper as well as turning the compasses. Practise drawing some neat and accurate circles in the next exercise.

Part of a circle's circumference is called an arc. Arcs are used a lot in constructions – you will be using them later.

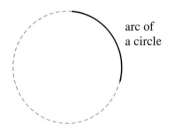

arc of a circle

For all the diagrams you draw in this chapter, make sure that you use a pencil and that your pencil is sharp, so that your work is accurate.

Exercise 3A

1   Construct these circles and semicircles with a pair of compasses. In each case, C is the centre of a circle.

   a

4cm

C

b

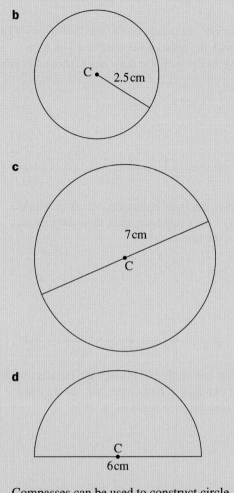

C• 2.5 cm

c

7 cm
C

d

C
6 cm

2 Compasses can be used to construct circle patterns. See if you can draw these patterns.

Constructing triangles

The construction of a triangle given the lengths of its sides is particularly simple.

EXAMPLE 1

Construct a triangle ABC with AB = 8 cm, BC = 6 cm and AC = 4 cm, using a ruler and compasses only.

a Draw AB, making sure it is 8 cm long.

A　　　　　　　B

b With your compasses, draw an arc of radius 6 cm centred at B.

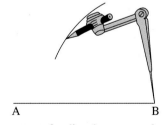
A　　　　　　　B

c Draw an arc of radius 4 cm centred at A.

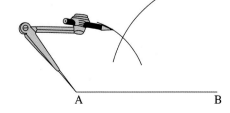
A　　　　　　　B

37

d Label the point where the two arcs intersect as C.
e Join AC and BC.

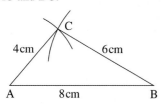

In a right-angled triangle there are two perpendicular sides and a third side. The third side is opposite the right angle and is called the **hypotenuse.** The hypotenuse is the longest side of the right-angled triangle because it is opposite the largest angle in the triangle.

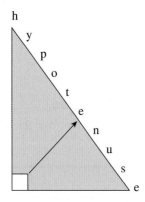

A right-angled triangle can be constructed using a pair of compasses.

EXAMPLE 2

Construct the right-angled triangle PQR where PQ = 6 cm and the hypotenuse QR = 10 cm.
Draw a line longer than 6 cm. Mark P and Q on that line, 6 cm apart. You now need to construct the right angle at P.

To construct the right angle, open the compasses to a short length (e.g. 2 cm), put the point of the compasses on P and draw arcs on the line either side of P.

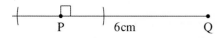

Then open the compasses a little further (say 5 cm) and with the point of the compasses on A (as shown in the diagram) draw an arc above the line.

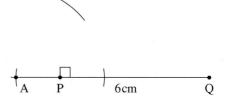

Keeping the same radius, and this time with the point of the compasses on B, draw another arc above the line.

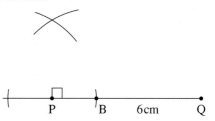

Draw a line up from P through the intersection of these two arcs, J. Remember, you do not know how long this line needs to be yet.

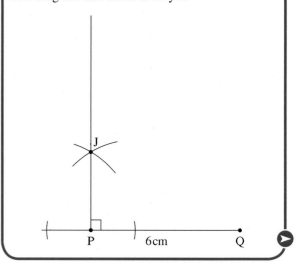

Put the point of the compasses on Q and with the radius set at 10 cm draw an arc on the line from P through J. Where the arc crosses the line is the point R.

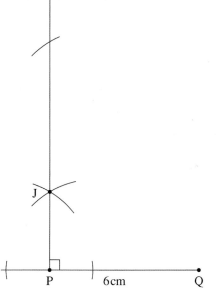

Finally, draw line QR to complete triangle PQR.

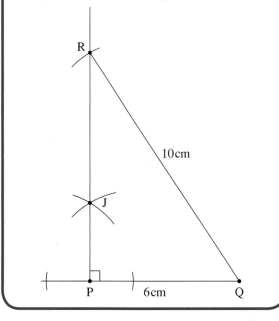

Exercise 3B

1 Construct triangles with sides:
   a 3 cm, 4 cm and 5 cm
   b 6 cm, 8 cm and 10 cm
   c 7 cm, 6 cm and 9 cm
   d 6.3 cm, 4.2 cm and 8.7 cm
   e 11.3 cm, 7.9 cm and 6.3 cm.

2 Construct these right-angled triangles:
   a ABC, where AB = 4 cm and the hypotenuse BC = 5 cm
   b RST, where RS = 5 cm and the hypotenuse ST = 13 cm
   c MNO, where MN = 7 cm and the hypotenuse NO = 8.5 cm

3 Draw the triangle ABC such that:
   a AB = 8 cm, CÂB = 60° and AC = 7 cm
   b AB = 9 cm, CÂB = 45° and AC = 6.8 cm
   c AB = 4.8 cm, CÂB = 30° and AC = 7.3 cm.

You can use a protractor for this question (see Book 1 if you have forgotten how to do this)

4 Construct a right-angled isosceles triangle of your choice.

5

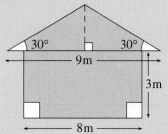

The diagram shows the side view of a house.

Using a scale of 1 cm to represent 1 m, make an accurate drawing of the side view.

6 In the triangle DEF, DF = 10 cm, FX = 6 cm and XE = 4 cm. Construct an accurate copy of triangle DEF.

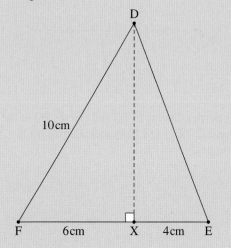

3.2 Bisecting angles and lines

The word **bisect** means to cut exactly in half. If you bisect a line segment you can find the halfway point, called the **midpoint**. If you bisect an angle you will be able to find half the angle. In this section you will learn how to bisect angles and lines.

In your diagrams you will draw lots of arcs to help you. These are called **construction lines** – do not rub them out, as they show the method you have used.

Perpendicular bisector of a line segment

You can bisect the line PQ with a pair of compasses and a ruler.

P   Q

Draw an arc with centre P and radius more than $\frac{1}{2}$ PQ. Draw another arc, centre Q, with the same radius.

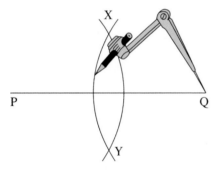

The arcs meet at X and Y. Join XY.

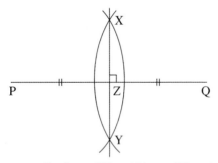

XY is **perpendicular** to PQ and bisects PQ at the midpoint Z. PZ = QZ. Perpendicular means at right angles.

Exercise 3C

1   Draw a line PQ = 8 cm.
    a   Bisect the line.
    b   Each part of your bisected line PQ should be 4 cm long. Check by measuring.

2   Copy the lines and construct their perpendicular bisectors. (When you copy the lines keep them at approximately the angles shown.)

a

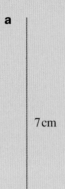

7 cm

b

9 cm

c

6 cm

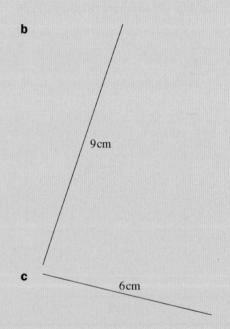

3   **a**   Draw four different triangles.
    b   Construct the perpendicular bisectors of each side of the triangle.
    c   What do you notice?

4 For your triangles in Question **3**, are you able to draw a circle so that all three corners of the triangle lie on the circumference?
(**Hint:** use your answer to part **c**.)

5 Draw an approximate copy of this triangle, about twice as big.

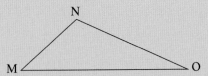

Construct a circle that passes through the three vertices M, N and O. This is called the **circumcircle** of the triangle.

Bisecting an angle

You can bisect the angle ABC with a pair of compasses and a ruler.

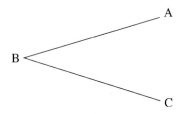

Draw an arc with centre B to cut AB at X and BC at Y.

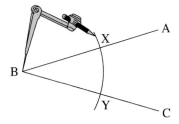

Draw two more arcs with the same radius, centred at X and Y. Label the point where they meet Z.

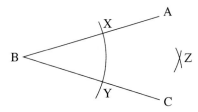

Join BZ. This line bisects AB̂C.

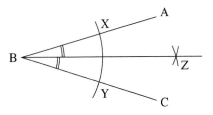

Exercise 3D

1 a Draw an angle of 90° using a protractor.
   b Bisect it to make an angle of 45°.

2 a Draw an angle of 60° using a protractor.
   b Bisect it to make an angle of 30°.

3 Draw an approximate copy of each angle, then bisect it.

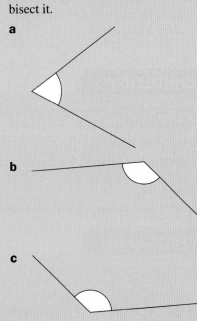

a

b

c

d

4 a Draw a straight line. This represents an angle of 180°.
   b Bisect the angle of 180° to get an angle of 90°.

5 **a** Using only a ruler and pair of compasses, construct these shapes.

i

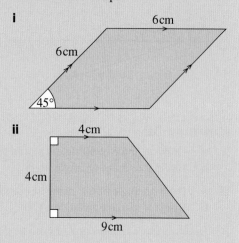

6cm

6cm

45°

ii

4cm

4cm

9cm

b What are the names of the shapes you constructed in part **a**?

TECHNOLOGY

Get a complete review of all these geometric constructions and more by visiting

www.onlinemathlearning.com

Click on the constructions you are interested in and watch the video demonstrations.

It's a great way to learn!

INVESTIGATION

You can construct an angle of 60° at a point X on a line.

Draw a large arc, with centre X, to cut the line at P.

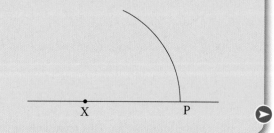

X   P

Using the same radius draw an arc, with centre P, to cut the first arc at Q.
Join XQ.

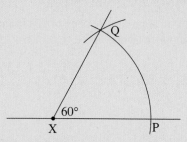

Q

60°

X   P

The angle QXP = 60°.
Investigate how to construct these angles using what you have learned: 45°, 30°, $22\frac{1}{2}°$, 15°, 75° and 135°.

3.3 Congruency

You will need squared paper and tracing paper.

Congruent shapes are the *same shape* **and** the *same size*.

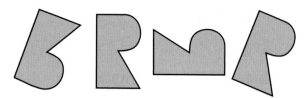

The four shapes above are congruent.

Any shape that is transformed by translation, rotation or reflection will be congruent to its image.

Look at the triangles ABC and PQR.

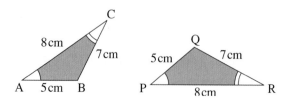

C

8cm   7cm

A   5cm   B

Q

5cm   7cm

P   8cm   R

They are congruent.

You can write

$$\triangle ABC \equiv \triangle PQR$$

Where ≡ means 'is congruent to'.

The order of letters is important.

From the order you can tell

$\hat{A} = \hat{P}, \hat{B} = \hat{Q}, \hat{C} = \hat{R}$

and AB = PQ, AC = PR, BC = QR.

Exercise 3E

1 a Are the two shapes congruent?

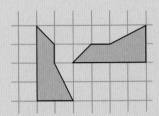

b Make a copy of the shapes on squared paper and draw a third shape that is congruent to them.

2 a Are these two squares congruent?

b Are all 10 cm squares congruent?
c Are all squares congruent?

3 a Are these two shapes congruent?

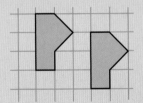

b Are these two shapes congruent?

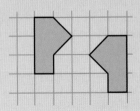

4 a The diagram shows four pairs of congruent triangles. Pick out the pairs.

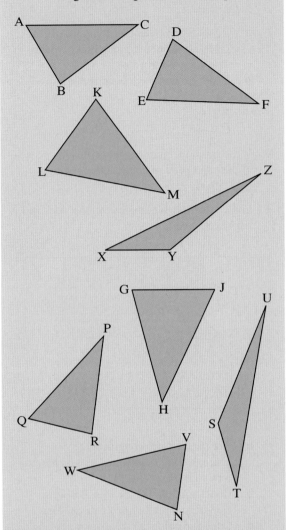

b Use tracing paper to check if you were right.
c For which pairs did you need to turn your tracing over to make the triangles fit exactly?

5 a If you trace triangle ABC in Question **4** and fit it on to triangle PQR, which angle does $\hat{A}$ fit onto?

b You can say $\hat{A}$ **maps** to $\hat{Q}$, or $\hat{A} \rightarrow \hat{Q}$.
Copy and complete the mapping:
$\hat{A} \rightarrow \hat{Q}$
$\hat{B} \rightarrow \square$
$\hat{C} \rightarrow \square$

6 Write down the mappings for the angles in each of the other pairs of congruent triangles in Question **4**.

7 a If you fit the tracing of triangle ABC in Question **4** on to triangle PQR, which side does AB fit onto?

b Copy and complete the mapping:
AB → QR
BC → ☐
CA → ☐

8 Write down the mappings for the sides in each of the other pairs of congruent triangles in Question **4**.

9 Look at these two triangles. All the sides and angles are given.

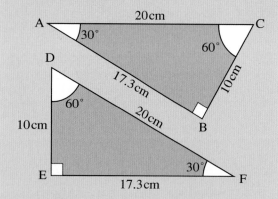

a Are the triangles congruent?
Trace △ABC to find out. Did you have to turn your tracing over?

b Could you tell that the triangles were congruent without tracing them? How?

c Copy and complete the mapping of angles:
Â → ☐, B̂ → ☐, Ĉ → ☐

d Copy and complete the mapping of sides:
AB → ☐, BC → ☐, CA → ☐

e Explain why it is correct to write
△ABC ≡ △FED

10 a Are these two triangles congruent?

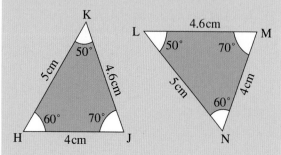

b Copy and complete the mapping:
Ĥ → ☐   HJ → ☐
Ĵ → ☐   JK → ☐
K̂ → ☐   KH → ☐

c Copy and complete the statement:
△JHK ≡

11 Look at these two congruent triangles.

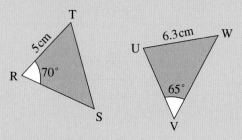

△RST ≡ △UWV
Use the information given in the diagram to answer these questions.

a What are the lengths of RS and UV?

b What are the sizes of **i** RŜT **ii** VÛW?

c Which angle of △UWV is the same size as RŜT? What size is it?

d Find a side which is the same length as WV.

12 Look at these two congruent triangles.

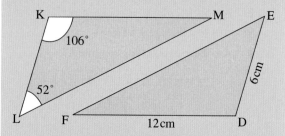

△KLM ≡ △DEF
Using the information given in the diagram, answer these questions.

a What is the size of KM̂L?

b Write down the size of each angle of △DEF.

c Write down the lengths of the sides KL and KM.

13 You are told the following information about two triangles:
△XYZ ≡ △ABC
XY = 7 cm, XZ = 5 cm
YX̂Z = 100°

Draw a sketch of both triangles and write down all you know about the sides and angles of △ABC.

Testing for congruency

Triangles are congruent if any one of these four conditions hold.

1 *Two sides and the included angle of each triangle are equal (SAS).*

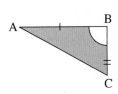 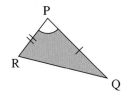

In triangles ABC and PQR:

AB = PQ,
BC = PR

and $A\hat{B}C = R\hat{P}Q$.

So, $\triangle ABC \equiv \triangle QPR$

This condition is called **SAS**.

2 *Two angles and a corresponding side of each triangle are equal (ASA).*

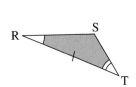 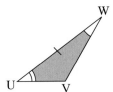

In triangles RST and UVW:

$S\hat{R}T = V\hat{W}U$
$R\hat{T}S = W\hat{U}V$
and RT = UW.

So, $\triangle RST \equiv \triangle WVU$

3 *Three sides of one triangle are equal to three sides of the other triangle (SSS).*

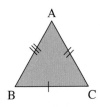 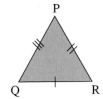

AB = PQ, BC = QR, AC = PR

So, $\triangle ABC \equiv \triangle PQR$

4 *Both triangles have a right angle, and the hypotenuse and another side of each triangle are equal (RHS).*

 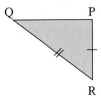

$A\hat{B}C$ and $Q\hat{P}R$ are right angles,

AB = PR and AC = QR

So, $\triangle ABC \equiv \triangle RPQ$

Exercise 3F

1 a Which pairs of the following triangles are congruent? Say which test you have used.
   b For each pair, write down the equal angles and equal sides.

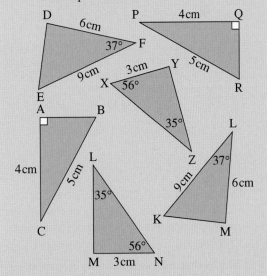

2 In Question **1**, the triangles ABC and QRP are congruent. The **congruency statement** for them is:

$\triangle ABC \equiv \triangle QRP$

Copy and complete the congruency statements for the other triangles:

$\triangle LMN \equiv \ldots$

$\triangle DEF \equiv \ldots$

3 Draw the triangle ABC when:
  a AB = 3 cm, BC = 4 cm, CA = 2 cm
  b AB = 4 cm, BC = 5 cm, B̂ = 30°
  c Â = 50°, B̂ = 70°, BC = 5 cm
  d AB = 5 cm, BC = 4 cm, Ĉ = 90°

4 **a** Try using the measurements in Question **3 a** to draw another triangle with a *different* shape. Can you do it? Will all triangles drawn to these measurements be congruent?
  b Repeat for the other parts of Question **3**.

5 Draw two *different* triangles ABC with AB = 6 cm, B̂ = 30° and AC = 4 cm. Explain why these triangles are not congruent.

6 These three triangles are congruent. Write down the sizes of all the unmarked angles and sides.

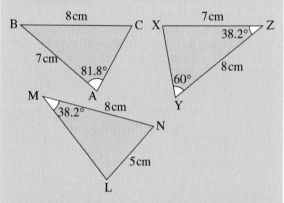

7 Draw a rectangle PQRS. Mark in the diagonal PR. Explain why the two triangles formed are congruent, and write a congruency statement for them.

8 Which of the following triangles are congruent?
  a ΔDEF: ∠D = 54°, ∠E = 77°, DE = 9 cm
  b ΔGHJ: ∠G = 49°, ∠J = 77°, HJ = 9 cm
  c ΔSTV: ∠V = 77°, ∠T = 54°, SV = 9 cm
  d ΔXYZ: ∠X = 54°, ∠Y = 49°, XZ = 9 cm

9 If AC bisects ∠BAD and ∠BCD, show that AB = AD.

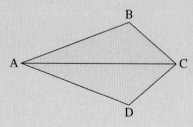

10 If PQ is parallel to TS and PQ = TS, prove that:
  a PR = SR and QR = TR
  b triangles PRT and QRS are congruent.

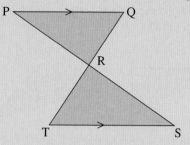

11 If AB = DC and ∠ABC = ∠DCB, prove that AC = BD.

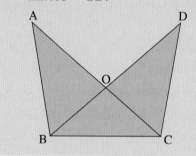

3.4 Drawing nets

Factories use different-shaped boxes for packing goods. They start with a **net**.

- A net is a flat shape that folds to make a solid object.

In a net, sides that join up to each other when folded up must be equal in length.

In the diagrams opposite you can see the net of a cube and a cuboid. Faces which are opposite each other when each net is folded up are the same colour.

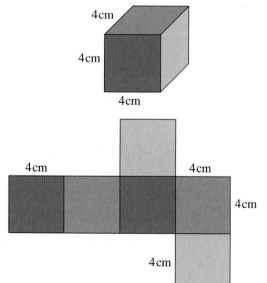

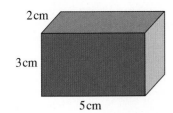

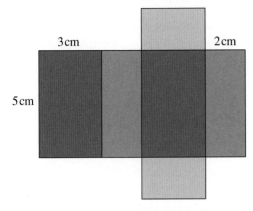

There are many different nets that make the same object. For example, an alternative net for a cube is:

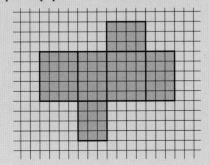

Exercise 3G

You will need squared paper, card, scissors and glue or sticky tape.

1 **a** Make larger copies of these nets on squared paper.

i

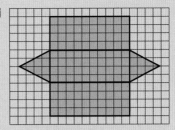

ii

b Cut them out carefully.
c Fold them along the heavy lines.
d Stick the edges together with sticky tape.
e Name the solids you have made.

2 Copy each net onto squared paper. Cut it out and fold it. What shape does it make?

a

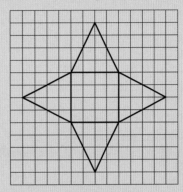

b

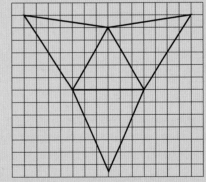

3 **a** Construct this triangle accurately.

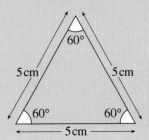

47

b Cut the triangle out. Draw around it on card to make this net:

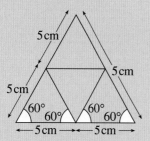

c Draw flaps on every other outside edge:

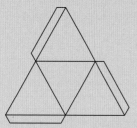

d Fold your net along the lines.
e Glue the flaps carefully.
Your completed solid should look like this.

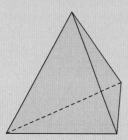

It is called a tetrahedron.
f How many faces, edges and vertices does it have?

4 For each of these objects draw an accurate net to scale. Where there are triangle faces construct these using compasses.
a Cuboid

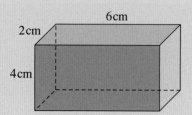

b Regular tetrahedron

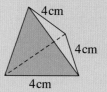

c Square-based pyramid

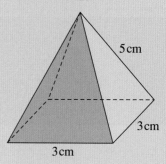

d Triangular prism

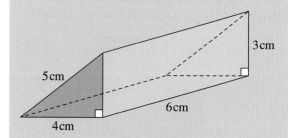

e Triangular prism

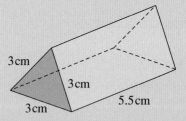

f Hexagonal prism, with regular hexagons for end faces

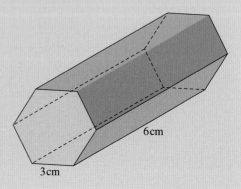

5   James has drawn this:

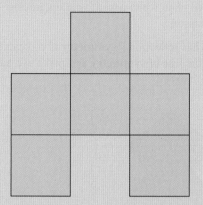

Samirah has drawn this:

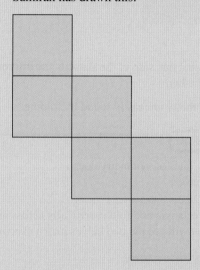

James and Samirah both say they have drawn a net of a cube.

Which of these (if either) are correct nets for a cube?

6   Draw the net of a cuboid measuring 3.5 cm by 4.8 cm by 5.2 cm.

7   Make an approximate copy of this net of a cuboid.

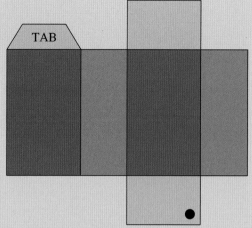

a   On your copy write the letter 'T' on the edge that the tab will stick to.

b   On your copy write a cross (**✗**) on the two corners that will meet the corner marked with the spot (**●**)

8   Sketch the net of a cylinder.

9   Sketch the net of a cone.

▷ **ACTIVITY**

1   Make solids from each of these nets.

a

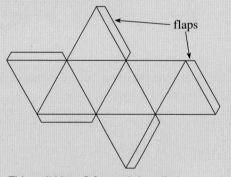

flaps

This solid has 8 faces. It is called an **octahedron**.

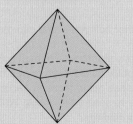

b

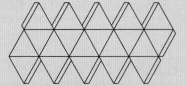

This solid has 20 faces. It is called an **icosahedron**.

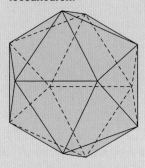

2 Another well-known solid is the **dodecahedron**.

Its net consists of twelve 5-sided shapes.

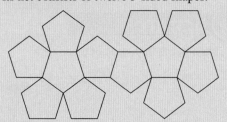

Make the net using these dimensions for each five-sided shape.

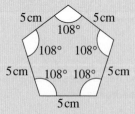

3 Find some other solids that you can make the nets for.

4 Collect as many different cardboard boxes as you can. Carefully unfold them to see how the manufacturer has made them, in particular the tabs they use to secure the boxes. Write a report on your findings.

3.5 Symmetry

You will need plain white paper to cut up, tracing paper and scissors.

• A shape has reflectional **symmetry** if you can fold it along the **line of symmetry** and make two halves that match exactly.

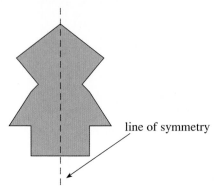

line of symmetry

You can say that one side of the shape is the **mirror image** of the other.

Lines of symmetry are easily found by folding.

EXAMPLE 3

How many lines of symmetry does a rectangle have?

....

Cut out a rectangle and fold it vertically across its middle. You will see the two halves match exactly.

line of symmetry

Repeat, folding horizontally.

line of symmetry

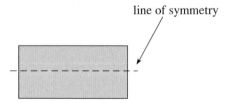

You should see the rectangle has two lines of symmetry.

Some shapes have three lines of symmetry:

Some have four lines of symmetry:

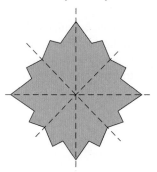

Others have lots of lines of symmetry:

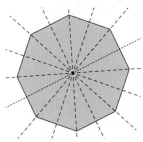

Exercise 3H

1 Make paper cut-outs of these shapes.

a   **b**

c    **d**

By folding, find out how many lines of symmetry each has.

2 Half of each shape is drawn, together with its line of symmetry. Trace and complete each shape.

a 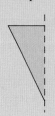   **b**

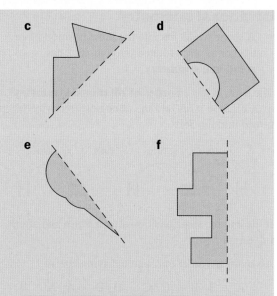

3 **a** Which of these road signs have reflectional symmetry?

i   **ii**

iii   **iv**

b How many lines of symmetry does each sign in part **a** have?

4 Draw each of these shapes and state how many lines of symmetry it has:

a isosceles triangle
b equilateral triangle
c square
d parallelogram
e kite
f regular hexagon
g regular pentagon
h circle

Rotational symmetry

• A shape which fits into the same position more than once when rotated through 360° has **rotational symmetry**.

The letter H has an **order of rotational symmetry** of 2, because it fits into the same position twice when turned through 360°.

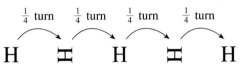

It fits into the same position after a $\frac{1}{2}$ turn and after a full turn.

A square has order of rotational symmetry 4:

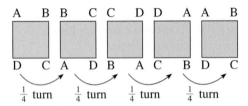

A shape which fits into the same position only once it has turned 360° (that is, back to the start again) is *not* described has having an order of rotational symmetry of 1. We say it has **no rotational symmetry**.

EXAMPLE 4

What is the order of rotational symmetry of these shapes?

a

b

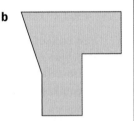

....

a This shape fits into the same position 6 times, so it has order of rotational symmetry 6.

b This shape looks the same only after a full rotation, so it has no rotational symmetry.

Exercise 3I

1 The triangle ABC is equilateral. Its centre is G. Trace the triangle and place your tracing over it. Place the point of your pencil at G, and rotate the tracing in an anticlockwise direction.

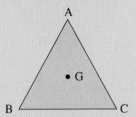

a After what angle of rotation does the triangle fit into the same position?

b How many times does the triangle fit into the same position in one complete turn?

c What is the order of rotational symmetry of the triangle?

2 Write down the order of rotational symmetry of each shape.

a

b

c

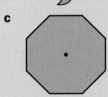

d

3 What is the order of rotational symmetry of:
  a a square
  b a parallelogram
  c an isosceles triangle
  d a rhombus
  e a kite
  f a rectangle
  g an isosceles trapezium
  h a regular pentagon
  i a circle?

(**Hint:** if you are unsure what these shapes look like, refer to the next section on 'Classifying shapes'.)

4 **a** Find some designs with rotational symmetry.
  b Copy them and identify the order of rotational symmetry of each design.

5 Copy and complete these sentences:

The number of lines of symmetry of a regular polygon is equal to …
The order of rotational symmetry of a regular polygon is equal to …

6 A shape has been started below. Copy and complete it so that it has 2 lines of reflectional symmetry and rotational symmetry of order 2.

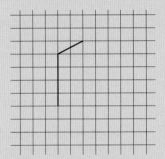

7 Shade in two more squares in this diagram so that it has no rotational symmetry and one line of reflectional symmetry.

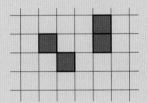

8 Draw several copies of this 2-by-2 grid:

You need to shade 0, 1, 2, 3 or 4 of the squares inside the grid.
  a How many different shapes can you make with
    i 1 line of symmetry
    ii 2 lines of symmetry
    iii 0 lines of symmetry
    iv 4 lines of symmetry
    v 3 lines of symmetry?

b How many different shapes can you make with
    i order of rotational symmetry 2
    ii no rotational symmetry
    iii order of rotational symmetry 4
    iv order of rotational symmetry 3?
  c Can you ever get order of rotational symmetry 3 or 3 lines of symmetry in patterns based **i** inside squares or **ii** using squares?

 TECHNOLOGY

Artists have long used ideas of symmetry to make beautiful paintings and patterns.

The Dutch artist M.C. Escher often used symmetry in his works of art.

Search online for 'Escher patterns' to view some of his work.

Classifying shapes

You can sort shapes in many different ways, for example by:

- the number of corners
- the number of curved edges
- the length of edges
- the size of angles
- the number of lines of symmetry.

Triangles

Triangles are classified by their angles, sides or lines of symmetry.

Triangle		Sides	Angles	Lines of symmetry
Equilateral		All 3 sides equal	All 3 angles equal	3 lines of symmetry
Isosceles		2 sides equal	2 angles equal	1 line of symmetry
Scalene		No equal sides	No equal angles	No lines of symmetry

53

Quadrilaterals

Quadrilaterals can be classified by their sides, angles, or diagonals.

'Bisect' means 'divide into two equal parts'

Quadrilateral		Sides	Angles	Diagonals
Trapezium		1 pair parallel	—	—
Kite		2 pairs of adjacent sides equal	1 pair of opposite angles equal	One diagonal bisects the other at right angles
Parallelogram		Opposite sides parallel and equal	Opposite angles equal	Bisect each other
Rhombus		All sides equal, opposite sides parallel	Opposite angles equal	Bisect each other at right angles. Diagonals bisect angles
Rectangle		Opposite sides parallel and equal	All angles 90°	Bisect each other and equal in length
Square		All sides equal, opposite sides parallel and equal	All angles 90°	Bisect each other at right angles and equal in length. Diagonals bisect angles

Some shapes possess all the properties of another shape. For example, a square is also a rectangle because the square satisfies all the conditions required of a rectangle.

Exercise 3J

1   Copy and complete these sentences about quadrilaterals using words from the cards below. Cards may be used more than once.

| parallelogram | rhombus | trapezium |
| square | kite | rectangle |

   a   A .... has 4 right angles and all the sides are equal in length.
   b   A .... has diagonals equal in length that bisect each other at right angles.
   c   A .... has diagonals that are unequal in length but one diagonal bisects the other at right angles.
   d   A square, a rhombus and a rectangle are also ....s.
   e   A .... has only one pair of parallel sides.
   f   A .... and a .... have diagonals that bisect their angles.

2   Are these statements true or false?
   a   All parallelograms are quadrilaterals.
   b   All rectangles are parallelograms.
   c   All squares are rhombuses.
   d   All rectangles are squares.

3   A pair of identical isosceles triangles are joined along an equal side. Draw diagrams to show that the resulting quadrilateral can be:
   a   a parallelogram
   b   a rhombus
   c   a kite.

4   ABC is a right-angled isosceles triangle with $\hat{B} = 90°$. ACX is an isosceles triangle drawn on the side AC. Describe the triangle ACX if:
   a   ABCX is a kite
   b $A\hat{C}X = 105°$
   c   ABCX is a square

5   AC and BD are diagonals of a quadrilateral intersecting at O. Name the quadrilateral if:
   a   OA = OC, OB > OD and $A\hat{O}B = 90°$
   b   OA = OC, OB = OD and $A\hat{O}B > 90°$
   c   OA = OB = OC = OD

Consolidation

Example 1

Write down four properties of:

a   a parallelogram
b   a square

....

a   Parallelogram

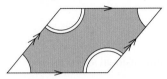

Two pairs of parallel sides
Two pairs of equal sides
Two pairs of equal angles
Diagonals bisect each other

b   Square

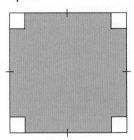

Two pairs of parallel sides
Four sides equal in length
Four right angles
Four lines of symmetry

Example 2

Construct the triangle ABC with BÂC = 90°,
AB = 7 cm and hypotenuse BC = 10 cm.
First draw the line AB.

A   B

Then construct a perpendicular at A.

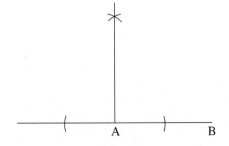

Open the compasses to 10 cm and, with the point of the
compasses on B, draw an arc on the perpendicular line.
Join B and C.

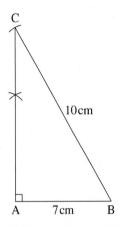

Example 3

Draw the net of this triangular prism.

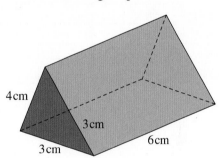

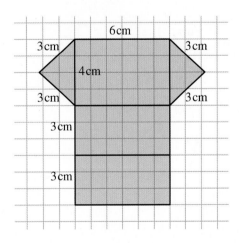

Exercise 3

1 Using ruler and compasses only, construct the triangle XYZ with:

    a   XY = 7 cm, YZ = 8 cm, XZ = 12 cm

    b   XY = 8 cm, XZ = 10 cm, XŶZ = 90°

2

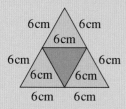

    a   Construct the four equilateral triangles with side 6 cm shown above.

    b   Cut out the large triangle and fold along the inside edges. What three-dimensional shape have you made?

3 Construct a triangle with:

    a   one line of symmetry   **b**   three lines of symmetry

4 Copy and complete this table.

Quadrilateral	Lines of symmetry	Number of right angles	Number of parallel sides	Do diagonals bisect each other?	Are diagonals equal in length?	Do diagonals bisect the angles?
Square						
Parallelogram						
Kite						
Trapezium						

5 Which of these triangles are congruent? Say which test you have used.

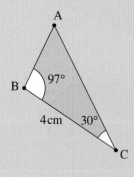

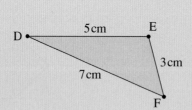

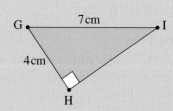

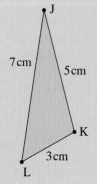

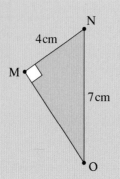

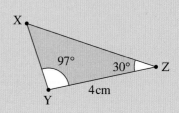

6 Draw a line 7 cm long.
  a Construct the perpendicular bisector of the line.
  b Label the midpoint of the line M.

7 Draw an angle of 58°. Bisect the angle using only a pencil and compasses.

8 **a** Construct the triangle ABC where AB = 4.5 cm, BC = 6 cm and AC = 7.5 cm
  b What sort of triangle have you constructed?
  c What is the special name for side AC in this triangle?

9 Draw the nets of these shapes, constructing triangles with compasses where necessary.
a
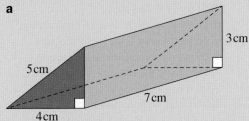
b
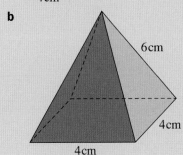
c
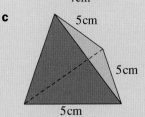
d
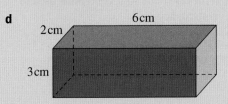

10 Which quadrilateral am I?
  a I have four lines of symmetry.
  b I have one line of symmetry. My diagonals are not equal in length.
  c I have two lines of symmetry. My diagonals are equal in length.
  d I have order of rotational symmetry 2 and four right angles.
  e My diagonals bisect each other at right angles. I contain no right angles.

11 Describe as many properties as you can of a parallelogram. Include rotational symmetry, lines of symmetry and diagonals among the properties you consider.

12 Write down **a** the order of rotational symmetry and **b** the number of lines of symmetry for these shapes:

i
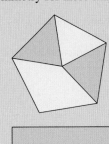

ii
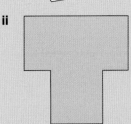

iii
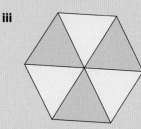

iv
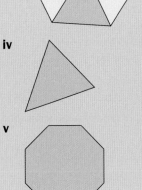

v

vi
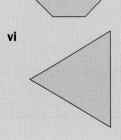

Summary

You should know ...

1 How to classify shapes in terms of:
- number of lines of symmetry
- number of right angles
- number of equal sides
- number of parallel sides
- properties of diagonals
- order of rotational symmetry

2 How to use a ruler and compasses to:
a construct a triangle
b construct the perpendicular bisector of a line
c bisect an angle

3 The longest side of a right-angled triangle is called the hypotenuse.

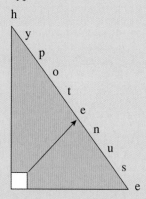

4 Congruent shapes are the same shape and size. *For example:*

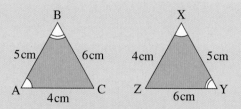

Triangle ABC is congruent to triangle XYZ.
This can be written $\triangle ABC \equiv \triangle XYZ$

Check out

1 Write down three properties of:
a a rhombus
b a rectangle
c a kite
d a parallelogram

2 **a** Construct triangle ABC with AB = 7 cm, AC = 12 cm and $A\hat{B}C = 90°$.
b Construct the triangle with side lengths 5 cm, 7 cm and 8 cm.
c Draw a line 5 cm long. Construct the perpendicular bisector of this line.
d Using a protractor, draw an angle of 40°. Construct the bisector of this angle.

3 Identify the hypotenuse in these triangles.

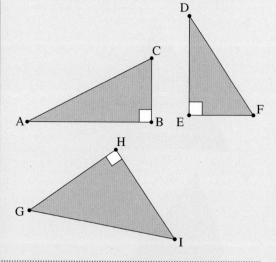

4 Write down the four conditions used to test congruency.

5 How to draw nets of cuboids, regular tetrahedrons, square-based pyramids and triangular prisms.

For example: The net of a regular tetrahedron with side length 3 cm is

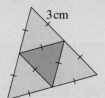

5 a Draw the net of a cuboid measuring 4 cm by 3 cm by 7 cm.

b Draw the net of this triangular prism, using compasses to construct the triangle faces.

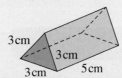

c Draw the net of this square-based pyramid, using compasses to construct the triangle faces.

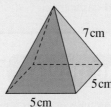

6 How to work out the number of lines of symmetry and the order of rotational symmetry.

For example: The order of rotational symmetry of this shape is 5 because it fits back on itself exactly 5 times when turning through 360°.

The number of lines of symmetry for a rectangle is 2 because it has only 2 fold lines which give the same shape either side.

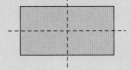

6 Write down **a** the order of rotational symmetry and **b** the number of lines of symmetry for these shapes:

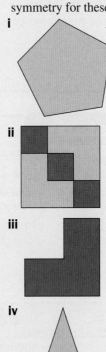

i

ii

iii

iv

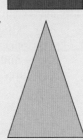

4 Length, mass and capacity

Objectives

- Choose suitable units of measurement to estimate, measure, calculate and solve problems in a range of contexts, including units of mass, length, area, volume and capacity.

- Know that distances in the USA, the UK and some other countries are measured in miles, and that one kilometre is about $\frac{5}{8}$ of a mile.

What's the point?

The ground staff who prepare a cricket pitch and set up the wickets measure lengths and areas regularly. Did you know that the height of the stumps should be 71.12 cm? Or that the distance between the stumps should be 22.86 cm? These measurements must be exact!

Before you start

You should know ...

1 How to multiply an integer by a fraction. *For example:*

> Find $\frac{1}{8}$ of 16 first then multiply by 5

$$16 \times \frac{5}{8} = (16 \div 8) \times 5 = 2 \times 5 = 10$$

2 The following metric abbreviations:

mm – millimetres	kg – kilograms
cm – centimetres	t – tonnes
m – metres	ml – millilitres
km – kilometres	ℓ – litres
g – grams	

Check in

1 Work out:

a $56 \times \frac{5}{8}$

b $25 \times \frac{3}{5}$

c $44 \times \frac{5}{11}$

2 a From the list on the left, which are measurements of **i** length, **ii** mass and **iii** capacity?

b Which measurements are used for the largest objects?

c Which measurements are used for the smallest objects?

4.1 Units of measure

The main metric unit of length is the **metre (m)**.
One metre is about one long stride.

Small lengths are measured in **centimetres (cm)** or **millimetres (mm)**.

1 m = 100 cm
1 cm = 10 mm

A fingernail is about 1 cm or 10 mm long.

Large lengths are measured in **kilometres (km)**.

1000 m = 1 km

You should be able to walk a kilometre in roughly 15 minutes.

For mass, the **kilogram (kg)** is used.

A large pineapple weighs about 1 kilogram

You probably weigh between about 38 and 48 kilograms.

The **gram (g)** is used for small objects.

1 kg = 1000 g

A lemon weighs about 50 grams

A pencil weighs about 5 grams

The **tonne (t)** is used for heavy objects.
1 t = 1000 kg

A large car weighs about 1 tonne

Exercise 4A

1   Write down three things that are:
    a   less than 1 cm long
    b   less than 10 cm long
    c   about 1 m long.

2   On a sheet of paper, mark two dots that you think are 5 cm apart.

    Now measure the distance between them, using a ruler. Were you nearly right? If not, try again.

3   Repeat Question **2** for dots that are:
    a   10 cm apart   b   15 cm apart
    c   3 cm apart   d   1 cm apart

4   Which metric unit would you choose to measure
    a   the length of this book
    b   the height of a lamp post
    c   the length of your little finger
    d   the distance from the Earth to the Moon
    e   the height of the tallest building in your town or village
    f   the length of the hour hand on a watch?

5   Write down three things that have a mass of:
    a   about half a kilogram
    b   about 1 kg
    c   about 2 g

6   Which metric unit would you use to measure the mass of
    a   an elephant   b   this book
    c   your friend   d   a coin
    e   a bus?

61

Converting metric units

Sometimes you need to change metric units from one to another.

EXAMPLE 1

a Alroy is 1 m 87 cm tall. What is his height in centimetres?

b A piece of wood is 263 mm long. What is its length in centimetres?

....

a 1 m 87 cm = 1.87 m = 1.87 × 100 cm
$$= 187 \text{ cm}$$

b 263 mm = 263 ÷ 10 cm = 26.3 cm

You can convert units of mass in the same way.

EXAMPLE 2

An orange weighs 268 g. What is its mass in kilograms?

....

1 kg = 1000 g
so 268 g = 268 ÷ 1000 kg = 0.268 kg

Exercise 4B

1 Copy and complete:
   a ☐ g = 1 kg   **b** ☐ kg = 1 t

2 Copy and complete:
   a ☐ cm = 1 m   **b** ☐ m = 1 km
   c ☐ mm = 1 cm

3 Copy and complete this table.

Metres	Centimetres	Millimetres
6	6 × 100 = ☐	6 × 1000 = ☐
8		
	3400	
	500	
		4000

4 Copy and complete this table.

Kilograms	Grams
9	9 × 1000 = ☐
15	
3.1	
	4000
	800

5 This table shows the length of five objects. Copy and complete the table.

Object	Length (cm and mm)	Length (cm)	Length (mm)
Pencil	18 cm 5 mm	18.5	
Pen		13.3	
Eraser			50
Mat		95	
Fingernail			9

6 **a** Veda is 1.45 m tall. What is her height in centimetres?
   b Veda has a mass of 52.3 kg. How much is this in grams?

7 Copy and complete:
   a 0.24 km = ☐ m   **b** 0.76 kg = ☐ g
   c 5 mm = ☐ cm   **d** 3250 m = ☐ km
   e 2.3 t = ☐ kg

8 A piece of metal is 2.3 m long. A 45 cm length is cut from it. How many metres of metal are left?

9 I walk to the shops, which is a journey of 3.2 km. Then I walk to my friend's house which is 900 m from the shops. How many kilometres have I walked altogether?

10 A sculptor carves a figure from a 4 kg piece of stone. The finished figure has a mass of 2.3 kg. How many grams of stone were removed?

11 The distance around a running track is 420 m. If Merpati runs round the track 6 times how many kilometres has she run in total?

⇒ INVESTIGATION

Write these distances in ascending order (you may need to do some further research):

- The distance you can run in 5 seconds.
- The current men's world record for long jump.
- The distance a car can drive on 5 cm³ of fuel.
- The height of your school.
- The current women's world record for high jump.
- The length of a football field.

4.2 Units of area, volume and capacity

For area, **square centimetres (cm²)** are used.

This square has an area of 1 cm².

For smaller areas, **square millimetres (mm²)** are used.

Since 1 cm = 10 mm,
then 1 cm² = 10 mm × 10 mm = 100 mm²

For larger areas, **square metres (m²)** or **square kilometres (km²)** are used.

A square metre is about the size of half a door.

1 m² = 100 cm × 100 cm = 10 000 cm²

For volume, we use **centimetres cubed (cm³)**.

1 cm³ is a cube with side lengths of 1 cm.

For smaller volumes we use **millimetres cubed (mm³)**.

1 cm³ = 10 mm × 10 mm × 10 mm = 1000 mm³

For larger volumes, we use **metres cubed (m³)** or even **kilometres cubed (km³)**.

1 m³ = 100 cm × 100 cm × 100 cm
= 1 000 000 cm³

For capacity, the **litre (ℓ)** is used.

A measuring jug has a capacity of about 1 litre.

The **millilitre (ml)** is used for smaller capacities.

1000 ml = 1 ℓ

A teaspoon has a capacity of about 5 millilitres.

Note also that 1 ml = 1 cm³.

Exercise 4C

1 Write down three things that have
   a an area of less than 1 m²
   b a capacity of less than 1 ℓ
   c a volume of less than 1 m³.

2 Which unit would you use to measure
   a the area of your town, village or city
   b the capacity of a bottle of water
   c the area of this ink stain:

   d the volume of your classroom
   e the capacity of a bucket

   f the area of the piece of paper you are writing on?

3 Copy and complete:
   a 3200 ml = ☐ ℓ
   b 10 ml = ☐ cm³
   c 23 ℓ = ☐ ml

4 A bottle of medicine contains 250 ml. How many litres of medicine are there in seven of these bottles?

5 A bucket contains 3.4 litres of water. If 2300 ml are poured out, how many litres of water are left in the bucket?

6 Copy and complete:
  a $5\,m^2 = \square\,cm^2$
  b $2.3\,km^2 = \square\,m^2$
  c $0.7\,m^3 = \square\,cm^3$
  d $9.25\,km^3 = \square\,m^3$
  e $420\,000\,cm^2 = \square\,m^2$
  f $7\,000\,000\,mm^2 = \square\,m^2$

4.3   Estimation

The ability to make a good guess or **estimate** for the size of something is very important in mathematics. It helps you check whether an answer is sensible or not.

Exercise 4D

1 **a** Estimate the lengths of these lines:
  i _____
  ii _____
  iii ____
  iv _____
  v _____

  b Check your answers with your ruler. How accurate were you?

2 Estimate in metric units:
  a the width of your classroom
  b the height of your classroom door
  c the length of the blackboard or whiteboard in your classroom
  d the distance from your school to your home
  e the distance from your seat to the principal's office
  f the width of a blade of grass
  g the width of a pencil point
  h the height of your school.

3 Estimate the mass of:
  a a pencil   **b** a large grapefruit
  c yourself   **d** a bucket of water
  e a small car   **f** a mosquito
  g your exercise book

4 Compare your answers to Questions **2** and **3** with a friend.

  Ask your teacher about any you cannot agree on.

5 **a** Which of the following are sensible estimates?
  i The height of a grown man is about 2 m.
  ii The mass of an egg is about 1 kg.
  iii The time to run a kilometre is about 1 hour.
  iv The length of a fly is about 1 cm.
  v Alison walks at about 50 metres per second.
  vi The area of the classroom door is about $2\,m^2$.
  vii A soft drink bottle holds about $\frac{1}{2}\,\ell$.

  b Write a good estimate for each part in **a** which you did not think was sensible.

6 A litre of paint covers about $8\,m^2$ How many litres should be bought to cover a rectangular wall:
  a 12 m by $2\frac{1}{2}$ m
  b 17 m by 3 m?

Approximate and accurate values

Look at the flowers drawn above.
a *Without counting*, say how many flowers you think there are.
b Now count them. How many are there?

In **a**, you **estimated** the number of flowers. You used your eyes and your mind but you did not count.

In **b** you counted and found the exact or **accurate** answer.

Sometimes it is important to find the accurate number.

At other times the **approximate** number is good enough.

Exercise 4E

1

> **New Zealand wins again**
>
> A crowd of 7000 watched for 4 hours
> 22 minutes, in brilliant sunshine, while
> New Zealand scored another victory in
> the Cricket today.

This report is from the Sports page of the
Daily Herald. 7329 people actually watched
the match, so 7329 is the accurate number and
7000 is the approximate number.

a Do you think that the reporter should have
used the accurate number?

b Would the readers be interested in the
exact size of the crowd?

c Would the readers be interested in the
exact length of the game?

2 Which of these statements shows an estimate?
a There are 39 people on this bus.
b I think it is 8 o'clock.
c It is about 5 kilometres to school.
d That dress costs $18.45.
e There are 40 students in this class.
f 75% of students should pass their
maths exam.

3 Say whether you would count the number
accurately or approximately for each of the
following:
a The number of people in your family.
b The number of goals in a football match.
c The number of boxes of oranges harvested
by a farmer.
d The number of oranges in each box that
were harvested.
e The number of marks you scored in your
last history test.
f The population of the world in the year 2050.
g The number of grains in a sack of rice.

4 Look through a newspaper. Try to find:
a three numbers that you think are given
accurately
b three numbers you think are approximate.

Explain your choices.

5 a Ellis wanted to measure the mass of a
coconut. So he put it on a machine for
weighing people.

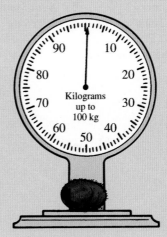

i What mass does each division on the
dial represent?
ii What is the mass of the coconut?

b Ellis took the coconut home. He weighed
it on the kitchen scales.

i What does each small division on the
dial represent?
ii What is the mass of the coconut?
c Which scale gives a more accurate reading?

6 Which of these would you use to time
someone running the 100 m sprint? Why?
a The Town Hall clock.
b A wristwatch.
c A stopwatch.
d An alarm clock.

7 Look at the centimetres and millimetres on your ruler. Which of these measurements is it possible to get, using your ruler?

a 4 cm	**b** 4 cm 5 mm
c 32 mm	**d** 32.5 mm
e 32.54 mm	

Explain your answers.

8

Which of these would you *not* measure using your ruler? Why?

a The thickness of a hair
b The length of this book
c The length of a football field
d The width of your desk
e The length of a diamond

Some people, like engineers and scientists, need to measure as accurately as possible in their work. For other people, less accurate measurements are often good enough.

• The sign for *approximately equals* is ≈.
  For example, the distance from earth to the moon ≈ 400 000 km.

Exercise 4F

1 For each pair of people, say which one needs to measure more accurately, and why:

a A nurse in a hospital giving an injection; a cook giving out soup in a restaurant.
b A farmer measuring the length of his field; a carpenter measuring wood to make a chair.
c A chemist weighing an important new medicine he has made; a woman weighing out rice in the market.
d A builder shovelling sand to make cement; a tailor measuring a man for a new suit.

2 If you asked for a metre of cloth in a shop, would you expect to get exactly 100 cm 0 mm? Why?

3 Say whether you would measure roughly or accurately, and explain your answer:

a The mass of a diamond
b The volume of water in a lake
c The winning time for a 400 m race
d The distance from New York to London
e The surface area of Nigeria

4 Which of these measurements are silly? Why?

a Jo is 13 years, 7 months, 2 weeks, 3 days, 4 hours, 5 minutes and 10 seconds old.
b That truck weighs 3 t 507 kg 200 g.
c The Essequibo river is 1090 km long.
d It is 3 km 430 m 31 cm from my house to the park.
e Eva is 1 m 57 cm 4 mm tall.
f Emil's personal best time for the 100 m sprint is 10.48 s.
g It took me $4\frac{1}{2}$ hours to walk from one village to the next.
h The Earth is 400 381 km 30 m from the Moon.

5 Rewrite all the measurements in Question **4** using the ≈ sign.

4.4 Imperial units

Some countries, including the USA and the UK, use a different system for measuring, called **imperial** measurements.

One imperial measurement of length is the mile. You can convert between miles and kilometres using the following approximation:

$$1 \text{ km} \approx \frac{5}{8} \text{ mile}$$

The word 'mile' comes from the Latin word *mille* (thousand), used over two thousand years ago by Romans to describe a distance of one thousand paces (*mille passuum*).

EXAMPLE 3

Convert **a** 48 km to miles **b** 45 miles to kilometres.

....

a $1\,\text{km} \approx \dfrac{5}{8}$ mile

$48 \div 8 = 6$

$48\,\text{km} \approx 48 \times \dfrac{5}{8} \text{ miles} = 6 \times 5 \text{ miles}$
$\qquad\qquad\qquad\qquad\quad = 30 \text{ miles}$

$48\,\text{km} \approx 30 \text{ miles}$

b $1 \text{ mile} \approx \dfrac{8}{5}\text{km}$

$45 \div 5 = 9$

$45 \text{ miles} \approx 45 \times \dfrac{8}{5}\text{km} = 9 \times 8 \text{ km}$
$\qquad\qquad\qquad\qquad\qquad = 72 \text{ km}$

$45 \text{ miles} \approx 72 \text{ km}$

Some people forget whether they should be multiplying by $\frac{8}{5}$ or $\frac{5}{8}$. The most important thing to remember is that a mile is longer than a kilometre. So the number of kilometres will be larger than the number of miles if you have converted correctly.

Exercise 4G

1   Which of these conversions are clearly wrong?
   a   80 miles ≈ 50 km   **b**   160 km ≈ 100 miles
   c   30 km ≈ 50 miles   **d**   40 miles ≈ 25 km
   e   20 miles ≈ 32 km

2   Copy and complete:
   a   35 miles ≈ ☐ km   **b**   32 km ≈ ☐ miles
   c   40 km ≈ ☐ miles   **d**   65 miles ≈ ☐ km
   e   120 miles ≈ ☐ km

3   Sujatmi said that multiplying by $\frac{8}{5}$ was the same as multiplying by 1.6, so her method for changing 30 miles to kilometres was:

$1.6 \times 30 = 1.6 \times 10 \times 3 = 16 \times 3$

$\begin{array}{r} 16 \\ \times\ \ 3 \\ \hline 48 \end{array}$

so 30 miles ≈ 48 km

Use Sujatmi's method to convert **a** 20 miles, **b** 40 miles and **c** 90 miles to kilometres.

4   The distance from Wellington in New Zealand to Jakarta in Indonesia is roughly 4800 miles. Approximately how many kilometres is this?

5   The distance from Abu Dhabi in UAE to Cairo in Egypt is roughly 2368 kilometres. Approximately how many miles is this?

6   Copy and complete:
   a   24 miles ≈ ☐ km   **b**   50 km ≈ ☐ miles
   c   32 miles ≈ ☐ km   **d**   35 km ≈ ☐ miles

7   You may use a calculator for this question.

So far we have used the conversion $1\,\text{km} \approx \frac{5}{8}$ mile. As a decimal, $\frac{5}{8} = 0.625$. The decimal conversion is more accurately $1\,\text{km} \approx 0.621371$ miles. Using $1\,\text{km} \approx \frac{5}{8}$ mile makes calculations quicker and easier when you don't have a calculator. If you want more accurate answers then you need to use a more accurate decimal.

The distance from the earth to the sun is about 149.6 million kilometres.
   a   Convert this distance to miles using $1\,\text{km} \approx \frac{5}{8}$ mile
   b   Convert this distance to miles using $1\,\text{km} \approx 0.621371$ miles
   c   Subtract to find the difference between the two converted measurements in **a** and **b**. This is the error.
   d   Repeat parts **a** to **c** for a distance of 50 km
   e   When does the greater error happen?
   f   Can you find a more accurate conversion than $1\,\text{km} \approx 0.621371$ miles?

8   The area of Cambridge, UK is about 45 square miles. Approximately how many square kilometres is this?

9   Jamil said that he was going to use a ratio method to convert between miles and kilometres. He said 8 km ≈ 5 miles, so the ratio of kilometres to miles is 8:5. So if you have the distance in miles and it is a multiple of 5 then the distance in kilometres will be the same multiple of 8. For example, 15 miles is the third multiple of 5 so the answer in kilometres is the third multiple of 8, which is 24 km.

Use Jamil's method to repeat Question **2**.

Consolidation

Example 1

What is:
a 6.3 m in centimetres
b 253 g in kilograms?

....

a 100 cm = 1 m
So 6.3 m = 6.3 × 100 cm
= 630 cm
b 1000 g = 1 kg
So 253 g = 253 ÷ 1000 kg
= 0.253 kg

Example 2

Which is the most sensible unit to use out of mm, cm, m and km for
a The length of an ant
b The width of your classroom
c The length of your hand
d The distance between England and France?

....

a The length of an ant: mm
b The width of your classroom: m
c The length of your hand: cm
d The distance between England and France: km

Exercise 4

1 a Write in centimetres:
  i 5 m   **ii** 28 m   **iii** 7.2 m
  iv 3 km   **v** 13.45 km
b Write in metres:
  i 25 cm   **ii** 293 cm
  iii 8.1 km   **iv** 0.32 km
  v 2615 mm

c Write in grams:
  i 4 kg   **ii** 0.4 kg
  iii 3.2 kg   **iv** 0.49 kg
  v 15.2 kg
d Write in kilograms:
  i 500 g   **ii** 75 g
  iii 5000 g   **iv** 3168 g
  v 13 459 g

2 Write in ml:
a 5.3 ℓ   **b** 230 ℓ   **c** 52 cm^3

3 Write in ℓ:
a 280 ml   **b** 40 000 ml   **c** 3000 cm^3

4 A piece of wood is 3.1 m long. A 38 cm length is cut from it. How many centimetres are left?

5 Copy and complete:
a 7 m^2 = □ cm^2 **b** 8.1 km^3 = □ m^3
c 0.4 km^2 = □ m^2 **d** 0.08 m^3 = □ cm^3

6 A bottle of water contains 300 ml. How many litres of water are there in six of these bottles?

7 Estimate:
a the height of your classroom
b the length of your pencil
c the capacity of the drink you had for breakfast
d the mass of your pen.

8 Copy and complete:
a 32 km ≈ □ miles
b 10 miles ≈ □ km
c 120 km ≈ □ miles
d 120 miles ≈ □ km

Summary

You should know ...

1 The metric conversions for length, mass and capacity:
10 mm = 1 cm   1000 g = 1 kg
100 cm = 1 m   1000 kg = 1 t
1000 m = 1 km   1000 ml = 1 ℓ

For example: How many kilograms are there in 0.4 tonnes?
1000 kg = 1 t
0.4 × 1000 = 400 kg

Check out

1 Copy and complete:
a 4200 ml = □ ℓ
b 1.7 m = □ cm
c 3000 mm = □ cm
d 19 300 m = □ km
e 2.5 t = □ kg
f 9.9 kg = □ g

2 The metric conversions for area and volume:
$1\,cm^2 = 100\,mm^2$
$1\,m^2 = 10\,000\,cm^2$
$1\,cm^3 = 1000\,mm^3$
$1\,m^3 = 1\,000\,000\,cm^3$

2 Copy and complete:
a $7\,cm^2 = \square\,mm^2$
b $24\,000\,cm^2 = \square\,m^2$
c $5\,cm^3 = \square\,mm^3$
d $4\,000\,000\,cm^3 = \square\,m^3$
e $3\,m^2 = \square\,mm^2$

3 The most appropriate unit of measurement.
For example: It would not be sensible to measure the mass of an elephant in grams. Tonnes is a more sensible unit.

3 Match the measurement to the item. One has been done for you.

Item	Measurement
Mass of a rhino	200ml
Capacity of a cup	70kg
Length of a lion	9cm
Capacity of a car fuel tank	2.3m
Length of a phone	1.5t
Mass of a man	60ℓ

4 The symbol $\approx$ means *approximately equals*. There is a difference between accurate and approximate amounts.
For example: "The bus is 10.3 metres long" is an approximation as it is sensible not to measure a bus to the nearest mm.

4 Which of these statements gives an approximate value and which gives an accurate value?
a Jane scored 63 marks in her last test.
b Jane has a mass of 44 kg.

5 That distances in the USA, the UK and some other countries are measured in miles, and that one kilometre is about $\frac{5}{8}$ of a mile.
For example: Convert 56 km into miles.

$1\,km \approx \frac{5}{8}$ mile

$56\,km \approx 56 \times \frac{5}{8}$ miles $= 7 \times 5$ miles
$= 35$ miles

$56\,km \approx 35$ miles

5 Copy and complete:
a $24\,km \approx \square$ miles
b 30 miles $\approx \square\,km$
c $88\,km \approx \square$ miles
d 25 miles $\approx \square\,km$

5 Number and calculation 2

Objectives

- Round whole numbers to a positive integer power of 10, e.g. 10, 100 or 1000, and decimals to the nearest whole number or to one or two decimal places.

- Order decimals, including measurements, making use of the $=$, $\neq$, $>$ and $<$ signs.

- Read and write positive integer powers of 10; multiply and divide integers and decimals by 0.1 and 0.01.

- Use known facts and place value to multiply and divide simple decimals, e.g. 0.07×9, $2.4 \div 3$.

- Divide integers and decimals by a single-digit number, continuing the division to a specified number of decimal places, e.g. $68 \div 7$.

What's the point?

Decimals allow you to perform calculations efficiently and easily, especially if you use a calculator or a computer. Decimals are also very useful for comparing and ordering numbers which otherwise are harder to compare, for example fractions such as $\frac{7}{15}$, $\frac{15}{32}$ and $\frac{8}{17}$.

Before you start

You should know ...

1 The place value of a number.
For example:
the value of the 6 in 361 243 is 60 000.

2 How to read and write numbers.
For example:
31 058 is thirty-one thousand and fifty-eight.

Check in

1 What is the value of the 4 in:
  a 24   b 403
  c 2145   d 6430
  e 34 218   f 4 013 625?

2 a Write in words:
    i 16 037   ii 2 150 263
  b Write in figures:
    i one hundred and fifteen thousand, two hundred and six
    ii one million and five.

3 The metric conversions:

10 mm = 1 cm
100 cm = 1 m
1000 m = 1 km
1000 g = 1 kg
1000 kg = 1 t
1000 ml = 1 ℓ

For example: How many grams are there in 0.27 kg?

1000 g = 1 kg
0.27 × 1000 = 270 g

3 Copy and complete:
   a 0.3 m = ☐ cm
   b 800 ml = ☐ ℓ
   c 200 mm = ☐ cm
   d 0.2 t = ☐ kg
   e 40 000 m = ☐ km
   f 3.81 kg = ☐ g

5.1 Rounding numbers

It is not always necessary to give an exact number. For example, you would not be expected to give the *exact* number of people living in a city or attending a sporting event.

Instead you could give an approximate number, rounded to the nearest thousand or ten thousand.

Rounding a whole number means writing it correct to the nearest ten, hundred, thousand, etc. For example, 124 is nearer to 120 than to 130.

124 is 120, correct to the nearest ten.

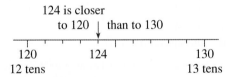

To round a whole number without drawing a number line:

1 Look at the number and find the place you want to round it to – tens, hundreds, etc.
2 Now look at the digit to the right of this place.
3 If this digit is 0, 1, 2, 3 or 4 the number must be rounded *down*. If it is 5, 6, 7, 8 or 9 the number must be rounded *up*.

EXAMPLE 1

a Write 372 and 125 correct to the nearest ten.
b Write 1378 and 26 425 correct to the nearest hundred.

....

a The tens place is underlined:

37② $\xrightarrow{\text{round down}}$ 370 to the nearest ten.

12⑤ $\xrightarrow{\text{round up}}$ 130 to the nearest ten.

---

b The hundreds place is underlined:

13⑦8 $\xrightarrow{\text{round up}}$ 1400

264②5 $\xrightarrow{\text{round down}}$ 26 400

Exercise 5A

1 Use the number line to write these numbers correct to the nearest ten:
   a 231 **b** 238 **c** 258
   d 243 **e** 250 **f** 251
   g 236 **h** 245 **i** 255

2 Write correct to the nearest ten:
   a 56 **b** 45 **c** 71
   d 68 **e** 65 **f** 144
   g 292 **h** 1411 **i** 1045
   j 2542 **k** 13 785 **l** 14 299

3 Write correct to the nearest hundred:
   a 173 **b** 809 **c** 1486
   d 2545 **e** 4555 **f** 4499
   g 11 659 **h** 12 111 **i** 13 011
   j 20 068 **k** 130 506 **l** 220 087

4 Write correct to the nearest thousand:
   a 1054 **b** 1154 **c** 1621
   d 1995 **e** 4350 **f** 4512
   g 4482 **h** 4921 **i** 13 316
   j 12 051 **k** 437 012 **l** 718 954

5 13 742 people watched a cricket match. How many is this to the nearest
   a ten **b** hundred **c** thousand?

71

6

PICK OF THE WEEK
Tel: 724761

5 doors, 5 gears, 90 mph

291 to get TTEC wards

Crisis costing $91000 a day

500 on strike

2500 people gather

Look at the newspaper headlines. Which numbers have been reported exactly? Which have been rounded? Give reasons for your answers.

7 249 is the largest whole number which, when rounded to the nearest hundred, gives 200. What is the smallest such number?

8 What are the largest and smallest whole numbers that when rounded to the
a nearest ten give 30
b nearest hundred give 1500
c nearest thousand give 8000?

Decimal places

You can round a decimal number to the nearest tenth or hundredth. Look at the number line:

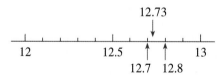

12.73 is 12.7 when rounded to the nearest tenth or correct to 1 decimal place (1 d.p.).

EXAMPLE 2

a Write 5.64 and 3.1824 correct to 1 d.p.
b Write 2.144 and 3.2556 correct to 2 d.p.
....
a The first decimal place is underlined:

$5.6\underline{6}④$ $\xrightarrow{\text{round down}}$ 5.6 correct to 1 d.p.

$3.1\underline{8}24$ $\xrightarrow{\text{round up}}$ 3.2 correct to 1 d.p.

b The second decimal place is underlined:

$2.14\underline{4}$ $\xrightarrow{\text{round down}}$ 2.14 correct to 2 d.p.

$3.25\underline{5}6$ $\xrightarrow{\text{round up}}$ 3.26 correct to 2 d.p.

Exercise 5B

21.7	21.8	21.9	22

1 Use the number line to write these numbers correct to the nearest tenth (correct to 1 d.p.).
a 21.73 **b** 21.86 **c** 21.78
d 21.94 **e** 21.99 **f** 21.77
g 21.84 **h** 21.80 **i** 21.95

2 Write correct to one decimal place:
a 1.31 **b** 1.35 **c** 1.42
d 3.71 **e** 5.461 **f** 2.202
g 12.394 **h** 13.011 **i** 0.057

3 Write correct to two decimal places:
a 1.443 **b** 2.504 **c** 10.4001
d 0.3031 **e** 10.089 **f** 0.008
g 0.1084 **h** 13.327 **i** 15.401

4 A rectangular postcard has length 18.4 cm and width 11.6 cm. What is its area correct to one decimal place?

5 Repeat Question **3** but this time round to the nearest whole number.

6 Work these out on your calculator, giving all answers to 2 d.p.
a 1 ÷ 6
b 3 ÷ 11
c 5 ÷ 13
d the square root of 47

7 Two of the calculations below give the same answer when rounded to 1 d.p. Which two are they?

7.3 ÷ 2.4	15 ÷ 4.8

2.32 × 1.4	24.5 ÷ 8.19

0.91 × 3.2

8 What are the smallest numbers that when written to:
a one decimal place give 3.6
b two decimal places give 3.62

9 Six friends share a meal costing $173. They decide to share the bill equally between them.
a What do they each owe, to the nearest cent?
b If they each pay this amount they will not cover the bill. Why is this?
c What could they do instead?

10 Round:
a 124.98 to 1 d.p.
b 95.895 to 2 d.p.
c 199.999 to 1 d.p.
d 99.99 to the nearest whole number
e 43 995 to the nearest 1000
f 7.9949 to 2 d.p.

11 In Book 1 you learned that sometimes you need to use common sense with rounding. Round these figures as stated then say what is wrong with them:

a A bridge can safely carry a weight of 12.85 tonnes. In a safety leaflet someone rounded this figure to the nearest whole number.

b A stadium has 68 850 seats. A ticket agent rounded this figure to the nearest thousand and printed this many tickets.

c A company made $5349 profit in a month. They decided to round this figure to the nearest hundred. Profits are shared equally among the 100 employees in the company.

5.2 Ordering decimals

In Book 1 you learned how to write decimals in order of size by considering the value of the numbers in each of the place value columns. In particular, you learned that a good way of comparing decimals is to write all the numbers with the same number of decimal places.

EXAMPLE 3

Write these numbers in order of size, smallest first:
2.4   2.405   2.44   2.442   2.04   2.044

2.4   →   2.400
2.405 →   2.405
2.44   →   2.440   *Line up the decimal points.*
2.442 →   2.442   *Write zeros where there*
2.04   →   2.040   *are no hundredths and no*
2.044 →   2.044   *thousandths.*

In order, smallest first, this is:
2.04, 2.044, 2.4, 2.405, 2.44, 2.442

Inequalities

'5 is a larger number than 3', '5 is greater than 3', '3 is smaller than 5', '3 is not equal to 5', are all sentences describing 3 and 5. They can all be written in a shorter way using the symbols $<$, $>$ and $\neq$.

$<$ means *is less than*
$>$ means *is greater than*
$\neq$ means *is not equal to*

$5 > 3$ means: 5 is greater than (or bigger than) 3
$3 < 5$ means: 3 is less than (or smaller than) 5
$3 \neq 5$ means: 3 is not equal to 5

From Example 3, you can see that $2.04 < 2.044$ and $2.44 > 2.405$.

Notice that the symbols $<$ and $>$ both have a narrow, pointed end and a wide, open end. The wide end is always next to the bigger number of the two being compared.

$<$ and $>$ are **inequality** symbols. They show that two numbers are not equal to each other and which of the two is larger. Inequality symbols can be used to compare more than two numbers. The numbers in Example 3 can be written in ascending (increasing) order separated by *is less than* symbols, like this:

$2.04 < 2.044 < 2.4 < 2.405 < 2.44 < 2.442$

or in descending (decreasing) order using the *is greater than* symbol, like this:

$2.442 > 2.44 > 2.405 > 2.4 > 2.044 > 2.04$

Exercise 5C

1 Copy and complete, using $<$ or $>$ in place of the boxes.
a 7.06 $\square$ 7.6   b 0.31 $\square$ 0.131
c 20.02 $\square$ 20.002   d 1.333 $\square$ 1.4

2 Are these statements true or false?
a $2.65 > 2.7$ b $0.028 < 0.27$
c $3.04 = 3.040$ d $2.08 \neq 2.80$

3 Using the numbers below, copy and complete the inequality statement.

3.404, 3.5, 3.501, 3.4

$\square < \square < 3.41 < \square < \square$

4 Using the numbers below, copy and complete the inequality statement

7.06, 7.62, 7.7, 7.6

$\square > \square > 7.612 > \square > \square$

5 □ < 0.041

Which numbers could you place inside the box?

6 Copy and complete, using <, > or =.
a 0.4 km □ 40 m
b 250 g □ 0.25 kg
c 7200 ml □ 72 ℓ
d 24 mm □ 0.24 cm

7 Copy and complete these inequality statements using the numbers below.

87 ml, 870 g, 870 ℓ, 8 t, 0.87 ℓ, 80 700 g, 8.07 g

a □ > 8700 ml > □ > □ > 0.87 ml
b □ < 80.7 g < □ < 8.07 kg < □ < □

8 Write these measurements in descending order of size using > symbols.

350 cm, 300 mm, 220 cm, 2.4 m, 0.25 m, 0.2 km

9 Write these measurements in ascending order of size using < symbols.

2400 mm, 140 cm, 1.2 cm, 0.14 m, 2.24 km, 24.20 m

10 0.13 < □ < 0.131

Which numbers could you place inside the box?

11 Copy and complete the following using <, > or =.
a 6 × 8 □ 4 × 12
b 7 + 4 □ 15 − 3
c 52 □ 5 × 2
d 7 × 0.1 □ 7 ÷ 10
e 3 − 10 □ 7 − 8
f 8 ÷ 100 □ 8 × 0.01

5.3 Multiplying decimals

In Book 1 you learned how to multiply and divide whole numbers and decimals by 10, 100 and 1000. 10, 100 and 1000 are examples of powers of 10 where the power is a positive integer. For example, $100 = 10^2 = 10 \times 10$ and $1000 = 10^3 = 10 \times 10 \times 10$.

When the power gets higher the numbers are harder to say. In the US and UK, commas are often used to help write large numbers (some calculators have commas on their display). We can also use spaces. To say large numbers, think what word goes in place of each comma or space.

$10^7 = 10\,000\,000$, or 10 million, is sometimes written 10,000,000.

EXAMPLE 4

Say the number 4560802.36

This number is more easily read when written with spaces or commas. We group the figures in threes from the decimal point, working to the left:
4 560 802.36 or 4,560,802.36

4 560 802.36

This space separates the millions

The space closest to the decimal point separates the thousands

We say *four million, five hundred and sixty thousand, eight hundred and two point three six*. Notice we do not say *point thirty-six* at the end!

You can multiply and divide by these powers of 10.

EXAMPLE 5

Work out:
a 12.4 × 100
b 21 ÷ 1000

a 12.4 × 100 = 1240

Move all digits two places to the left to make them worth 100 times their original value.

b 21 ÷ 1000 = 0.021

Move all digits three places to the right to make them worth a 1000th of their original value.

Exercise 5D

1 Write down in figures and words the powers of 10. The first is done for you.
a $10^3 = 1000 = $ one thousand
b $10^5 = $
c $10^2 = $
d $10^6 = $
e $10^4 = $
f $10^8 = $

2 Write these numbers in words:
a 2 700 340
b 23 819.45
c 410 007
d 21 345.98

3 Write these numbers in figures:
   a Three million, two hundred and forty thousand and seventy-one
   b Five million, sixty-two thousand and seven point four three
   c Two million and eighteen
   d Ninety-one million, three hundred thousand and seventeen point one

4 Work out:
   a 4.2×100 **b** 3.7×10
   c 0.2×1000 **d** $53 \div 100$
   e $2.8 \div 10$ **f** $350 \div 1000$
   g 0.021×100 **h** 0.4×10
   i 2.65×1000 **j** $42.4 \div 100$
   k $0.34 \div 10$ **l** $92.1 \div 1000$

5 From your work on fractions and decimals in Book 1 you learned that
$$3 \times \frac{1}{10} = \frac{3}{10} = 0.3$$
Write the answers to these calculations as decimals:
   a $7 \times \frac{1}{10}$ **b** $9 \times \frac{1}{10}$
   c $4 \times \frac{1}{100}$ **d** $17 \times \frac{1}{100}$
   e $18 \times \frac{1}{10}$ **f** $172 \times \frac{1}{100}$
   g $63 \times \frac{1}{10}$ **h** $256 \times \frac{1}{100}$

6 Look at this pattern:
$$25 \times 1000 = 25\,000$$
$$25 \times 100 = 2500$$
$$25 \times 10 = 250$$
$$25 \times 1 = 25$$
$$25 \times \frac{1}{10} = 2.5$$
$$25 \times \frac{1}{100} = 0.25$$
Copy and complete this pattern:
$$175 \times 1000 =$$
$$175 \times 100 =$$
$$175 \times 10 =$$
$$175 \times 1 =$$
$$175 \times \frac{1}{10} =$$
$$175 \times \frac{1}{100} =$$

7 Work out:
   a $7 \times \frac{1}{10}$ **b** $7 \div 10$
   c $21 \times \frac{1}{100}$ **d** $21 \div 100$
   e $24 \div 10$ **f** $24 \times \frac{1}{10}$
   g $315 \div 100$ **h** $315 \times \frac{1}{100}$

Comment on your answers.

8 Copy and complete these sentences:
Multiplying by $\frac{1}{10}$ is the same as dividing by... .
Multiplying by $\frac{1}{100}$ is the same as dividing by... .

Multiplying integers and decimals by 0.1 and 0.01

You know from Book 1 that $0.1 = \frac{1}{10}$ and $0.01 = \frac{1}{100}$.

- Multiplying by 0.1 is the same as multiplying by $\frac{1}{10}$ or dividing by 10.
- Multiplying by 0.01 is the same as multiplying by $\frac{1}{100}$ or dividing by 100.

EXAMPLE 6

Work out:
a 24×0.1
b 173.4×0.01
c 0.08×0.1

....

a $24 \times 0.1 = 24 \div 10 = 2.4$
b $173.4 \times 0.01 = 173.4 \div 100 = 1.734$
c $0.08 \times 0.1 = 0.08 \div 10 = 0.008$

We can use known facts and place value to help us multiply simple decimals in our heads.

EXAMPLE 7

Work out:
a 3×0.05 **b** 12×0.4

....

a For 3×0.05, we know that $0.05 = 5 \times 0.01$
So $3 \times 0.05 = 3 \times 5 \times 0.01 = 15 \times 0.01 = 15 \div 100 = 0.15$
b For 12×0.4, we know that $0.4 = 4 \times 0.1$
So $12 \times 0.4 = 12 \times 4 \times 0.1 = 48 \times 0.1 = 48 \div 10 = 4.8$

Exercise 5E

1 Work out:

   a 32×0.1 **b** 256.1×0.01

   c 0.1×0.04 **d** 356×0.1

   e 28×0.01 **f** 5×0.1

   g 0.01×4.1 **h** 7×0.01

   i 0.2×0.1 **j** 2300×0.01

   k 0.1×4560 **l** $30\,900 \times 0.01$

2 Work out:

   a 8×0.6 **b** 3×0.07

   c 0.5×9 **d** 7×0.8

   e 4×0.03 **f** 5×0.4

   g 0.02×11 **h** 7×0.09

   i 12×0.3 **j** 0.06×9

   k 7×0.4 **l** 12×0.05

3 Jane had a different method for multiplying 3 by 0.05. Here is Jane's method:

   3×0.05

   • Do the calculation ignoring the decimal places: $3 \times 5 = 15$
   • Count the number of decimal places in the question: $3 \times 0.\underline{05}$. The two underlined digits show two decimal places in the question
   • Put the same number of decimal places in the answer: $15 \rightarrow 0.15$

   So $3 \times 0.05 = 0.15$

   Use Jane's method to work out:

   a 8×0.2 **b** 7×0.07

   c 0.3×12 **d** 5×0.06

4 Govinda had a different method. Here is Govinda's method for multiplying 12 by 0.4:

   $12 \times 0.4 = 12 \times \dfrac{4}{10} = \dfrac{48}{10} = 4.8$

   Use Govinda's method to work out:

   a 6×0.4 **b** 12×0.05

   c 8×0.3 **d** 0.07×9

5 Ife wrote the answers below in her homework for these questions:

   a 0.08×5 **b** 6×0.05

   a $8 \times 5 = 40$
      so $0.08 \times 5 = 0.04$

   b $6 \times 0.05 = 6 \times \dfrac{5}{10} = \dfrac{30}{10} = 3$

   Ife has made a mistake in both questions. What mistakes has she made?

6 A glass holds 0.3 litres of water. How many litres would there be in nine glasses of water?

7 A sheet of paper is 0.1 mm thick. How thick would 360 sheets of paper be

   a in mm   **b** in cm?

8 A box has a mass of 0.4 kg. What is the mass of seven of these boxes

   a in kg   **b** in g?

9 Work out:

   a 15×0.7 **b** 14×0.05

   c 0.3×23 **d** 32×0.04

   e 12×0.9 **f** 0.07×25

10 Look at this pattern:

   $3200 \div 1000 = 3.2$
   $3200 \div 100 = 32$
   $3200 \div 10 = 320$
   $3200 \div 1 = 3200$

   $3200 \div \dfrac{1}{10} = 32\,000$

   $3200 \div \dfrac{1}{100} = 320\,000$

   Copy and complete this pattern:
   $410 \div 1000 =$
   $410 \div 100 =$
   $410 \div 10 =$
   $410 \div 1 =$

   $410 \div \dfrac{1}{10} =$

   $410 \div \dfrac{1}{100} =$

11 Copy and complete these sentences.

   Dividing by $\frac{1}{10}$ is the same as multiplying by ... .

   Dividing by $\frac{1}{100}$ is the same as multiplying by ... .

5.4   Dividing decimals

Dividing integers and decimals by 0.1 and 0.01

Since $0.1 = \frac{1}{10}$ and $0.01 = \frac{1}{100}$,

• dividing by 0.1 is the same as dividing by $\frac{1}{10}$ or multiplying by 10
• dividing by 0.01 is the same as dividing by $\frac{1}{100}$ or multiplying by 100.

EXAMPLE 8

Work out:
a $12 \div 0.1$
b $0.17 \div 0.1$
c $0.43 \div 0.01$

....

a $12 \div 0.1 = 12 \times 10 = 120$
b $0.17 \div 0.1 = 0.17 \times 10 = 1.7$
c $0.43 \div 0.01 = 0.43 \times 100 = 43$

Dividing decimals by whole numbers

Dividing decimals by whole numbers is similar to dividing whole numbers by whole numbers.

EXAMPLE 9

Work out $0.175 \div 5$

....

$$
\begin{array}{r}
0.035 \\
5\,\overline{)\,0.1\overset{1}{7}\overset{2}{5}}
\end{array}
$$

↑
line up the
decimal points

EXAMPLE 10

Work out $3.8 \div 4$

....

$$
\begin{array}{r}
0.95 \\
4\,\overline{)\,3.\overset{3}{8}\overset{2}{0}}
\end{array}
$$

line up the
decimal points

add an extra
at the end

Exercise 5F

1   Work out:

a $59 \div 0.1$ b $35.21 \div 0.01$
c $0.07 \div 0.1$ d $249 \div 0.1$
e $76 \div 0.01$ f $8 \div 0.1$
g $3.7 \div 0.01$ h $15 \div 0.01$
i $0.5 \div 0.1$ j $4100 \div 0.01$
k $7290 \div 0.1$ l $19\,300 \div 0.01$

2   Work out:

a $5.76 \div 4$ b $6.25 \div 5$
c $9.66 \div 7$ d $2.75 \div 5$
e $1.38 \div 6$ f $2.712 \div 8$
g $0.64 \div 4$ h $0.23 \div 5$
i $0.0784 \div 7$

3   Work out:

a $15.1 \div 4$ b $19.32 \div 6$
c $15.664 \div 11$ d $21.69 \div 9$
e $28.5 \div 12$ f $31.36 \div 8$

4   Four boys shared \$63 among them equally. How much did each receive?

5   A 3 kg piece of cheese sells for \$14.52. What is the price of cheese per kilogram?

6   A book made from 400 sheets of paper is 2.5 cm thick. What is the thickness of each sheet of paper?

7   The area of a rectangular field is $2136\,\text{m}^2$. What is the width of the field if its length is 60 m?

8   Copy and complete, filling in the blanks with 0.01, 0.1, 10 or 100:

a $74 \div \square = 740$ b $6.43 \times \square = 0.643$
c $0.8 \div \square = 0.08$ d $329 \div \square = 3.29$
e $85 \times \square = 8.5$ f $2.1 \div \square = 210$
g $1.23 \times \square = 12.3$ h $41 \div \square = 0.41$

9   Jasmine used this method to work out $0.6 \div 5$:

$$
0.6 \div 5 = \frac{0.6}{5} \overset{\times 2}{\underset{\times 2}{=}} \frac{1.2}{10} = 0.12
$$

Use Jasmine's method to work out:

a $0.7 \div 5$ b $0.8 \div 5$
c $0.03 \div 5$ d $0.9 \div 50$
e $0.4 \div 2$

10   Nasim decided to use this method to work out $0.9 \div 4$:

$$
\begin{aligned}
0.9 \div 4 &= 0.9 \times \frac{1}{4} \\
&= 0.9 \times 0.25 \\
&= 9 \div 10 \times 25 \div 100 \\
&= 9 \times 25 \div 10 \div 100 \\
&= 9 \times 25 \div 1000 \\
&= 225 \div 1000 \\
&= 0.225
\end{aligned}
$$

Use Nasim's method to work out:

a $0.7 \div 4$ b $0.2 \div 5$
c $0.03 \div 4$ d $0.5 \div 5$

When you divide some decimals there is a remainder. For example, $68 \div 7$ is 9 remainder 5. You learned in Book 1 that this can also be written as a fraction: $68 \div 7 = 9\frac{5}{7}$. If you want the answer as a decimal you need to add zeros after the decimal point in the division sum in order to continue the division:

$$\frac{9.7\ 1\ 4\ 2\ 8\ 5\ 7\ ...}{7)6^6 8^5.0^1 0^3 0^2 0^6 0^4 0^5 0\ ...}$$

The dots show that this division continues forever. You can see that the last digit carried over is 5, so this decimal will be the recurring (or repeating) digits 714285. The question will say how many decimal places you need to write down so that you do not need to carry on dividing forever. If you need to write the answer to 2 decimal places (2 d.p.), you must work out the next decimal place along – the third decimal place – so that you know whether you need to round up or down.

EXAMPLE 11

Calculate

a $26 \div 7$ to 1 d.p. **b** $7.6 \div 6$ to 2 d.p.

a
$$\frac{3.7\ 1}{7)2^2 6.^5 0^1 0}$$

Since the digit at the second decimal place is 1 round down to 3.7

$26 \div 7 = 3.7$ to 1 d.p.

b
$$\frac{1.2\ 6\ 6}{6)7.^1 6^4 0^4 0}$$

Since the digit at the third decimal place is 6 round up to 1.27
$7.6 \div 6 = 1.27$ to 2 d.p.

Exercise 5G

1 Calculate to 1 d.p.:
   a $35 \div 3$ **b** $50 \div 7$ **c** $85 \div 9$
   d $249 \div 8$ **e** $125 \div 6$ **f** $8.2 \div 5$
   g $7.63 \div 2$

2 Calculate to 2 d.p.:
   a $2.457 \div 2$ **b** $136 \div 3$
   c $9.12 \div 5$ **d** $9.81 \div 7$
   e $0.5 \div 6$ **f** $1.4 \div 9$
   g $17 \div 8$

3 Calculate to 2 d.p.:
   a $247 shared equally between 7 people.
   b What is the missing side length of a rectangle with area $47\,cm^2$ and width $3\,cm$?

c Neya's netball team has 7 players. Their total mass is 507 kg. What is the mean mass of the players?

d A 14.5 m length of metal is cut into 8 equal pieces. What is the length of each piece of metal?

e 6 identical glasses have a total capacity of 2 litres. What is the capacity of one glass, in millilitres?

f The total mass of six bread rolls is 400 g and the total cost for six bread rolls is $0.85.

  i What is the mass of one bread roll?
  ii What is the cost of one bread roll?

g A pile of 8 exercise books is 37 mm high. What is the height of one book?

⇒ INVESTIGATION

In France in the eighteenth century, the decimalised metric system for measurements was invented, to make calculations with length, mass and capacity easier. At the same time the French experimented with decimal time.

The clocks on the left show our time and the clocks on the right show the equivalent French decimal time.

Our time   French decimal time

 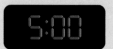

Our time   French decimal time

Explain how French decimal time works.

Write these times in French decimal time: 08:47, 14:15 and 23:59.

🖥 TECHNOLOGY

Visit the website

www.mathsisfun.com

Follow the links to 'Numbers', 'Decimals Menu' for a full review of decimal numbers.

Consolidation

Example 1

Round
a   23456 to the nearest 1000
b   4.7158 to 2 d.p.

....

a   23$\underline{4}$56
     The next number after the thousands column is
     a 4, so round down to 23000.
b   4.71$\underline{5}$8 to 2 d.p.
     The next number after the second decimal place is
     a 5, so round up to 4.72.

Example 2

Are these statements true or false?
a $7.3 > 3.7$ b $0.012\,kg < 120\,g$
c $0.4\,\ell = 40\,ml$ d $70\,m \neq 700\,cm$

....

a $7.3 > 3.7$ is true because 7.3 is greater than 3.7.
b $0.012\,kg < 120\,g$ is true because 0.012 kg is 12 g,
     which is less than 120 g.
c $0.4\,\ell = 40\,ml$ is false because 0.4 litres is 400 ml,
     which is not equal to 40 ml.
d $70\,m \neq 700\,cm$ is true because 70 m is 7000 cm,
     which is not equal to 700 cm.

Example 3

Work out:
a 17×0.01 b $53 \div 0.1$
c 9×0.04 d $5.6 \div 7$
e $7.2 \div 7$ to 2 d.p.

....

a $17 \times 0.01 = 17 \div 100 = 0.17$
b $53 \div 0.1 = 53 \times 10 = 530$
c $9 \times 0.04 = 9 \times 4 \times 0.01 = 36 \times 0.01$
     $= 36 \div 100 = 0.36$
d $5.6 \div 7 = \begin{array}{r} 0.\,8 \\ \hline 7)\overline{5.^56} \end{array}$
e $7.2 \div 7$ to 2 d.p. $= \begin{array}{r} 1.0\;2\;8\;... \\ \hline 7)\overline{7.2^20^60}\,... \end{array}$

     The number in the third decimal place is 8,
     so 1.02$\underline{8}$... rounds up to 1.03.
     $7.2 \div 7 = 1.03$ to 2 d.p.

Exercise 5

1   Round:
     a   365 to the nearest 10
     b   173.56 to the nearest whole number
     c   490 832 to the nearest 1000
     d   0.6678 to 2 d.p.
     e   0.412 to 1 d.p.
     f   1689 to the nearest 100
     g   0.21653 to 2 d.p.
     h   197.483 to 1 d.p.
     i   12.467 to the nearest whole number

2   Copy and complete using $<$, $>$ or $=$ for the
     boxes
     a $8\,m \;\square\; 80\,cm$ b $0.42 \;\square\; 0.242$
     c $0.2\,t \;\square\; 2000\,kg$ d $500\,mm \;\square\; 0.5\,m$

3   Using the numbers below, copy and complete
     the inequality statement.
     7.7, 7.6, 7.635, 7.061
     $\square < \square < 7.63 < \square < \square$

4   Write these measurements in descending order
     of size, using $>$ symbols.
     550 cm, 500 mm, 520 cm, 5.4 m, 5.25 m, 5.2 km

5   Work out:
     a 29×0.01 b 8×0.1
     c $97 \div 0.1$ d $3 \div 0.01$
     e 3200×0.01 f 900×0.1
     g $0.4 \div 0.1$ h $0.035 \div 0.01$

6   Work out:
     a 6×0.04 b 8×0.6
     c 3×0.05 d 6×0.7
     e 7×0.02 f 8×0.8
     g 9×0.3

7   Work out:
     a $6.5 \div 5$ b $85.2 \div 4$
     c $111.6 \div 9$ d $17.12 \div 8$
     e $20.52 \div 6$ f $0.42 \div 3$
     g $0.255 \div 5$ h $394.1 \div 7$

8   Work out:
     a $28 \div 6$ to 1 d.p.
     b $1.4 \div 8$ to 1 d.p.
     c $93.2 \div 7$ to 2 d.p.
     d $0.41 \div 3$ to 2 d.p.
     e $1 \div 7$ to 2 d.p.
     f $1.7 \div 9$ to 2 d.p.

9 How much does each person get if $746.10 is shared equally between 6 people?

10 What is the missing side length of a rectangle with area $99.32\,cm^2$ and width $8\,cm$? Give your answer to 1 d.p.

11 Copy and complete:
  a $61 \div \square = 6100$
  b $\square \times 0.1 = 0.23$
  c $\square \div 10 = 7.1$
  d $\square \times 100 = 408$
  e $234 \times \square = 2.34$
  f $\square \div 0.01 = 780$

12 A $22.5\,m$ length of material is cut into 7 equal pieces. What is the length of each piece of material, to 2 d.p.?

13 The total mass of eight cakes is $625\,g$. What is the mass of one cake, to the nearest gram?

14 These were the numbers of students in school on one particular day:

Year 7	128
Year 8	125
Year 9	117
Year 10	123
Year 11	114

The school magazine reported attendance for that day was approximately 120 for one particular year group. Which year groups could they have been writing about? Explain your answer.

Summary

You should know ...

1 How to round numbers.
For example:
To round 5798 to the nearest 100: the next number after the hundreds column is a 9, so round up to 5800.
To round 12.3217 to 1 d.p.: the next number after the tenths column (the first decimal place) is a 2, so round down to 12.3.

2 How to order decimals, including measurements, using the signs $=$, $\neq$, $>$ and $<$.
For example:
$17.4 > 2.3$ reads *17.4 is greater than 2.3*
$42\,cm < 4.2\,m$ reads *42 cm is less than 4.2 m* (because $4.2\,m = 420\,cm$)
$0.7\,kg \neq 70\,g$ reads *0.7 kg is not equal to 70 g* (because $0.7\,kg = 700\,g$)

3 How to multiply and divide decimals.
For example:
  a $92 \times 0.1 = 92 \div 10 = 9.2$
  b $0.31 \div 0.01 = 0.31 \times 100 = 31$
  c $7 \times 0.4 = 7 \times 4 \times 0.1 = 28 \times 0.1 = 28 \div 10 = 2.8$
  d $0.45 \div 9 = \dfrac{0.0\,5}{9\overline{)0.4^45}}$
  e $8.9 \div 7$ to 2 d.p. $= \dfrac{1.\,2\,7\,1\,...}{7\overline{)8.^19^50^10\,...}}$

Since the number in the third decimal place is 1, round down to 1.27
$8.9 \div 7 = 1.27$ to 2 d.p.

Check out

1 Round:
  a 4164 to the nearest 10
  b 23 675 to the nearest 1000
  c 98.34523 to 2 d.p.
  d 0.2213 to 1 d.p.
  e 749.99 to the nearest 100

2 Copy and complete, using $<$, $>$ or $=$.
  a $6\,kg \;\square\; 6000\,g$
  b $0.28 \;\square\; 0.185$
  c 7.3 litres $\square\; 730\,ml$
  d $22\,cm \;\square\; 2.2\,m$

3 Work out:
  a 18×0.01
  b $0.27 \div 0.1$
  c 6×0.8
  d $7.2 \div 6$
  e $1.05 \div 8$ to 2 d.p.

Planning, collecting and processing data

What's the point?

How can you tell who is the best sportsman or sportswoman? You would need to collect data – the results of previous sporting events. Then you could look at the data and use it to make a decision about which sportsperson you would pick for your team or squad.

Before you start

You should know ...

1 That a tally is a quick way to count frequencies.
   For example:
   ℍℍ ‖
   represents 7

2 How to use the order of operations.
   For example:
   $1 + 3 + 4 + 8 \div 4 = 10$ BIDMAS tells us to do the division first then the addition
   $(1 + 3 + 4 + 8) \div 4 = 4$ BIDMAS tells us to do the brackets first then the division

3 The difference between data collection sheets and questionnaires (you may want to look back at Book 1, Chapter 6).

Check in

1 What do these tallies represent?
   a ℍℍ ℍℍ ‖
   b ℍℍ ‖‖‖‖
   c ℍℍ ℍℍ ℍℍ ℍℍ ‖

2 Work out:
   a $2 + 2 + 5 + 6 + 10 \div 5$
   b $(2 + 2 + 5 + 6 + 10) \div 5$

3 Give two advantages of a questionnaire over a data collection sheet.

6.1 Discrete and continuous data

Quantitative data is numerical data. 'Data' simply means pieces of information. There are two different types of quantitative data, **discrete** and **continuous**.

- Discrete data can take only certain values (within a given range).

For example:
Number of children in a family: 1, 2, 3, etc. – you can't have 3.4 children.

Shoe sizes in a shop in the UK: 3, $3\frac{1}{2}$, 4, $4\frac{1}{2}$, etc. – you can't have a shoe size of 4.23.

Whenever you count something, it is discrete data, as the numbers are integers only.

- Continuous data can take any value (within a given range).

For example:
Your height may be 1.52 metres. If you had a more accurate way of measuring, you might find you are 1.523 metres, 1.5234 metres, etc.

The time taken to run a race may be 11 seconds, but with a more accurate stopwatch you could find it was 11.4 seconds, 11.42 seconds, etc.

Whenever you measure something, it is continuous data. You are limited only by the accuracy of the instrument you are using to measure with. Continuous data is rounded when you write it down.

Qualitative data is non-numerical data.

For example:
Hair colour: brown, black, etc.

Gender: male, female.

Exercise 6A

1 Is the following data quantitative or qualitative?
   a The age of children
   b Eye colour
   c The wingspan of a bird

   d The mass of a person
   e The country you live in

2 Is the following data discrete or continuous?
   a The speed of a train
   b The number of suitcases lost by an airline
   c A score in a maths test
   d The mass of a baby
   e The time it takes to walk to school
   f The length of a leaf
   g The volume of water in a bath
   h The number of books on a bookshelf
   i Clothes size

3 Luke thinks age is discrete. Edward thinks age is continuous.
   a Why might Luke think age is discrete?
   b Why might Edward think age is continuous?
   c Who is right?

6.2 Collecting data

The purpose of collecting data is usually to answer a question.

For example:
Who is better at maths, boys or girls?

Which topics in maths are the hardest?

Which cricketer should open the batting?

What is the most common height of students (to help with school uniform orders)?

In order to answer these questions you need to collect data. You have a few things to decide.

First, the method of data collection. Some possible methods are:
- Questionnaires or data collection sheets (you learned about these in Book 1)
- Interviews
- Getting the data from a book or the internet.

Second, the sample size. To find out who is better at maths between boys or girls you could test every single person in your town, school or class.

Finally, what degree of accuracy you need for the measurements. To find out what the most common height of students is to help with school uniform orders, you probably need measurements only to the nearest centimetre.

You are going to look at each of these areas separately by looking at case studies. A case study is a real situation for you to learn from. When answering the questions in each case study, justify your answers. This means write the reasons why you have given your answers.

Method of data collection

Exercise 6B – Case study 1

1 Mr Appleton wanted to find out what hobbies students liked in order to set up a new lunchtime club. Here is part of the interview he carried out with one of his students:

Mr Appleton: I am conducting a survey to find out which hobbies students like most, as I want to set up a lunchtime club which I hope will be popular. What hobbies do you like and what sort of lunchtime club do you think I should run?

Student: I really like horse riding but I think that may be a bit expensive as the school has no stables or horses! I also like swimming but the school doesn't have a pool so again that is no good. I also like reading and I usually go to the book club in the school library on a Friday but I would enjoy another book club.

This is only part of the interview – the whole conversation lasted 6 minutes.

a What was wrong with Mr Appleton's question?

b What do you need to think about before conducting an interview?

c What would you ask instead?

d Mr Appleton asked all of his Year 10 class their opinions. There are 30 students in his class. Estimate how long you think it may have taken him to talk to everyone.

e 23 of his class said they really enjoyed chess. Mr Appleton can't play chess. Does this matter? What if they had suggested something else Mr Appleton knew nothing about?

f What are the advantages of conducting an interview?

g What are the disadvantages of conducting an interview?

2 Mrs Appleton suggested a questionnaire might be better than an interview. She wrote this questionnaire for her husband's students:

- How old are you?
  - ☐ 11 or younger ☐ 12 ☐ 13
  - ☐ 14 ☐ 15 ☐ 16 or older
- What gender are you?
  - ☐ Male ☐ Female

- Which lunchtime club(s) would you be interested in joining? (Please tick all that apply)
  - ☐ Book club
  - ☐ Football training
  - ☐ Drama club
  - ☐ Homework help club
  - ☐ Puzzle club
  - ☐ Film club
  - ☐ Art club
  - ☐ Environmental club
- What day would you prefer the club to be on? (Please tick only one)
  - ☐ Monday ☐ Wednesday
  - ☐ Tuesday ☐ Thursday

a Mr Appleton has a staff meeting on a Friday and can't run clubs that day. How does the questionnaire take account of this?

b What other questions would you add to this questionnaire?

c Mr Appleton gives the questionnaire to his Year 10 class, which has 30 students. Estimate how long the survey would take this time. How much time do you think he would save doing a questionnaire instead of interviews?

d What disadvantages are there in doing a questionnaire instead of an interview?

e What advantages are there in doing a questionnaire instead of an interview?

f Do you think Mr Appleton chose a good sample? Justify your answer.

g Mrs Appleton originally had this question in her questionnaire:
Who is your class tutor?
Why do you think she removed this question?

3 Imagine you are trying to find out how students feel about the cost of school dinners and the quality of food in your school cafeteria.

a Design a questionnaire to find out about this. You may include up to 10 questions in your questionnaire.

b Often people ask things in a questionnaire that are not actually relevant to the original question. Swap with a friend and compare your questionnaires. Did either of you include any questions that could be taken out? (To help with this, imagine you were limited to only 5 questions. Which would you miss out?)

4 Write down the following four headings:
Advantages of an interview
Disadvantages of an interview
Advantages of a questionnaire
Disadvantages of a questionnaire
  a Write these statements under the appropriate heading:
  • Questions can be made clearer if not understood
  • Time consuming and more expensive to do
  • People have more time to think about their answers carefully
  • Possible to get biased results if not carried out carefully
  • Easier to analyse responses
  • Can get a poor response rate
  • Questions may not be understood
  • Can be used for sensitive or private topics
  • Likely to get a better response rate
  • Answers can be given in more detail
  • People may miss questions out
  b Can you think of any other advantages or disadvantages of an interview or questionnaire? Add them to your lists.

Sample size

A **population** is any complete collection of people, animals, plants or things from which we may collect data. It is the entire group we are interested in, which we wish to describe or draw conclusions about. It should not be confused with the population of a country. If you were interested in finding out about the reading habits of Year 11 in your school, the population is every person in Year 11 in your school.

A **sample** is a group selected from a larger group (the population). By studying the sample it is hoped to draw valid conclusions about the population. So, for the example given, the sample might be everyone in Year 11 whose surname begins with A, B, C or D.

Exercise 6C – Case study 2

1 Mrs Shah thinks that boys spend more money on sweets than girls do. She decides to test the theory by asking her Year 9 class to keep a record of every time they buy sweets during a week. She gives them all a data collection sheet to fill in the cost every time they buy sweets. There are 30 students in her class, 17 boys and 13 girls. She finds that the boys spend an average of $2.25 a week on sweets whereas the girls spend an average of $2.32 a week on sweets.
  a When Mrs Shah says she thinks boys spend more than girls, do you think she means the Year 9 boys in her class, all boys in the school or something else?
  b Why do you think she sampled only her Year 9 class?
  c Would the result be fairer if she had asked all the students in the school?
  d Mrs Shah didn't ask all the students in the school. Why do you think she didn't?
  e How do you think she could find out the spending habits of boys and girls in another school? Does she need to do this to answer her question?
  f Mrs Shah looked at students only in her own school. Why do you think that was?

2 Mrs Shah wasn't convinced that picking all the people in one Year 9 class was a fair sample, one that represented the entire school.
  a Suggest three other possible samples she could have taken.
  b Compare your answers to part **a** with a friend's answers. Can you think of any reasons why the samples you have suggested might be biased? (**Biased** means unfair and not likely to be a true representation of the entire school.)

3 The most frequently asked question concerning sampling is *What size sample do I need?* The answer depends on many things, including the purpose of the study, the population size, the risk of selecting a 'bad' sample, and the allowable sampling error.
  a Try to find out what some of the terms in the paragraph above mean (you can use books or the internet for your research).
  b Copy and complete these sentences:
    i The .... the sample size, the more accurate your results will be.

ii Taking a sample is often done because it is .... than taking a census. (A **census** is when you ask everyone in the population.)

c One teacher said '*Never have a sample size of less than 25*', another said '*Have a sample size of 10% to 20% of the population*'. If the survey is finding out about how people in New Zealand feel about the education system in their country, the size of the population is 4.4 million (the total population of New Zealand). Discuss the teachers' suggestions for this case.

d Why do you think there is no easy rule about how big a sample should be?

Degree of accuracy of measurement

Exercise 6D – Case study 3

1 Annabel is a fashion designer. She wants to design a new school uniform for her local school. She needs to know some of the measurements of the students at the school in order to make sure that her designs suit the students' different body shapes. She asks the students to complete the following data collection sheet:

Gender	Male / Female (please circle)
Waist measurement	
Hip measurement	
Height	

a Annabel forgot to include the units she wanted students to use to measure themselves. Here are two of the data collection sheets she got back:

Gender	Male / ⟨Female⟩ (please circle)
Waist measurement	68.5 cm
Hip measurement	95 cm
Height	151.5 cm

Gender	⟨Male⟩ / Female (please circle)
Waist measurement	0.7 m
Hip measurement	0.9 m
Height	1.6 m

Which of these returned sheets is the most helpful to Annabel? Why?

b How accurate do you think the measurements need to be? Think carefully about what you think she plans to do with the data she collects.

c Do you think people would be happy giving this information? Do you think anyone would lie? Is there a better way to ask for it?

d Do you think she needed to ask for any information that is not on her data collection sheet?

2 Annabel's friend Kate was helping her with costing the uniform and wanted to ask parents what they would be willing to spend on it. This was one of the questions from Kate's survey:

- How much are you willing to spend on a school skirt or school trousers? (Please tick one box.)
  - ☐ Maximum $10
  - ☐ Maximum $15
  - ☐ Maximum $20
  - ☐ Maximum $25
  - ☐ No maximum

a Do you think the units of measurement are accurate enough?

b Can you see a problem with this question?

c Do you think it would be better to leave the question open and simply put: *How much are you willing to spend on a school skirt or trousers?* Explain your answer.

3 a Write down three surveys for which you might be interested in measurements.

b How accurate should your measurements be?

c Compare your answers with a friend.

6.3 Two-way tables and frequency tables

Two-way tables

Nnamdi conducted a survey to try to find out which gender was more likely to wear glasses. These were his results from 8 people:

I am a girl who wears glasses

I am a boy who doesn't wear glasses

I am a boy who wears glasses

I am a girl who wears glasses

I am a girl who doesn't wear glasses

I am a boy who doesn't wear glasses

I am a girl who wears glasses

I am a girl who doesn't wear glasses

The data is very difficult to analyse when it is written in this way. It is better to gather the data into a two-way table. A two-way table can be used whenever data is split into two main categories. Here is a two-way table showing Nnamdi's data sorted out more clearly:

	Wears glasses	Does not wear glasses
Female	I am a girl who wears glasses I am a girl who wears glasses I am a girl who wears glasses	I am a girl who doesn't wear glasses I am a girl who doesn't wear glasses
Male	I am a boy who wears glasses	I am a boy who doesn't wear glasses I am a boy who doesn't wear glasses

The table is a 'two-way' table because there are two categories: 'gender' and 'do they wear glasses?' Normally we show only the figures in a two-way table. Below is the two-way table for Nnamdi's data with total columns also included:

		Do they wear glasses?		Total
		Wears glasses	Does not wear glasses	
Gender	**Female**	3	2	5
	Male	1	2	3
Total		4	4	8

Exercise 6E

1   The children in a school are to have extra swimming lessons if they cannot swim. The table below gives information about the children in Years 7, 8 and 9.

	Can swim	Can't swim	Total
Year 7	115	45	160
Year 8	138	24	162
Year 9	141	18	159
Total	394	87	481

   a   How many children are there in Year 7?
   b   How many children in Year 8 can swim?
   c   How many children are there in total in Years 7, 8 and 9?

   d   How many children in Year 9 can't swim?
   e   How many children need the extra swimming lessons?

2   The table below shows which science subjects students in Year 10 like the most. Copy and complete this table.

	Biology	Physics	Chemistry	Total
Boys	32		34	90
Girls				
Total	61	49		175

3   During August, exactly half of the 210 babies born in a hospital were boys, and 55 of the babies had a mass of 4 kg or more. There were 20 baby girls who had a mass of 4 kg or more. Copy the table below and use the information to complete it.

	Less than 4 kg	4 kg or more	Total
Boys			
Girls			
Total			

4   In a class of 33 pupils, there were 3 girls who were in the school quiz team and 14 boys who were not in the school quiz team.
   a   How many boys were there in the school quiz team if there were 17 girls in the class?
   b   Draw a two-way table to represent the data above.
   c   How many students were not in the school quiz team?

Frequency tables

In Book 1 you learned how to use frequency tables to gather discrete data into groups. Now you are going to learn how to do the same thing with continuous data.

Often tables are labelled in the same way for discrete data as they are for continuous data, but there is an important difference. With continuous data you can have any value in a given range, so values such as 12.23, 7.4, 80.015 can now be included in the table.

There are many ways in which the classes for data can be written. Both of these tables mean the same thing:

Mass (kg)	Frequency
0-10	3
10-20	8
20-30	2

Mass (kg)	Frequency
0-	3
10-	8
20-	2

The difficulty is in deciding which class numbers like 10 and 20, in this example, belong in. If 10 appears to be in two classes (as it does in the middle table), place it in the class where it is the lower limit. So 10 would go in 10−20 or 10−

Sometimes the same table may be written as:

Mass (kg)	Frequency
0-9	3
10-19	8
20-29	2

This is not such a good table for continuous data. The difficulty here is deciding where numbers such as 9.6 and 19.2 should go. The best way to decide is to round them to the nearest whole number and put them in the appropriate group. 9.6 rounds up to 10 so should go in the group 10–19. 19.2 rounds down to 19 so should also go in the group 10–19.

Exercise 6F

1 a The heights in centimetres of 40 plants in are:

4.2	8.4	11	12.7	9.8
14.9	15.2	4.9	13.5	13.1
20	16.4	10	18.8	6.2
14	5	8.3	11.4	3.8
15.1	23.2	5.1	14.5	8
6.4	11.9	14.3	16.2	21.4
17.4	7	4.1	7.2	19.8
12.1	10	18.1	13	6.2

Copy and complete the grouped frequency table below to gather this continuous data.

Height (cm)	Tally	Frequency
0-5		
5-10		
10-15		
15-20		
20-25		

b Which class had the highest frequency?

2 Kamil was gathering data into a grouped frequency table. He was looking at the masses of items, in kilograms. Which of these four items would go in the group 80–90?

a 79.2 **b** 80
c 83.4 **d** 90

3 Jamil drew this frequency table with equal class widths:

Group	Mass (kg)	Frequency
A	30−	
B	40−	
C	50−	
D	60−	

Which group should these masses go in?
a 40 **b** 69 **c** 73

4 The heights, in centimetres, of 30 students in a class are given below. Draw your own frequency table to gather this data into suitable class widths.

148	172.5	163	149	176
159	157.5	154	161	158.5
151	154.5	162	155.5	150.5
153.5	167.5	158	172.5	166
153	149.5	150	161	180
162	171	177	145	157.5

 **INVESTIGATION**

You will need a stopwatch and something heavy to hold.

Ask a member of your class to hold the heavy object in one hand, with their arm held out straight in front of them at shoulder height. Time how long they can hold the heavy object for. Stop the stopwatch when their arm starts to lower. Ask each student in your class to hold the object in turn. Write down the time for every student. Gather the results into a grouped frequency table, deciding on suitable class widths yourself. Make sure that the class widths are equal. Present your findings to your teacher.

6.4 Averages and range

In Book 1 you met three different types of average: the **mode**, the **median** and the **mean**.

The mode

The shoe sizes of 30 children are listed:

3, 5, 2, 4, 7, 6, 6, 5, 3, 2, 6, 5, 2, 8, 3,
4, 4, 5, 4, 3, 4, 5, 5, 4, 5, 6, 3, 8, 5, 3.

Shoe size	Frequency
2	3
3	6
4	6
5	8
6	4
7	1
8	2

From the table you can see that size 2 appears three times. Three is the frequency of size 2.
- The size with the greatest frequency is the **mode**.

Size 5 is the mode because it occurs eight times.

When data is grouped you find the **modal class**. This is the *class* with the highest frequency.

The heights of 20 children are shown in the table.

Height (cm)	Frequency
130-135	1
135-140	1
140-145	5
145-150	6
150-155	4
155-160	2
160-165	1

The modal class is 145–150 because it has the highest frequency, 6.

The median

- When a set of values is arranged in order, the middle term is called the **median**.

Bessie carried out a survey to find the number of people living in the 19 houses in her street. The results are shown in the table below.

Number of occupants	Tally	Total
less than 5	$\cancel{||||}$ ||	7
5	$\cancel{||||}$	5
6					3
7				2
more than 7	||	2
	Total	19

To find the median, Bessie arranged the numbers in order:
* * * * * * * 5 5 5 5 5 6 6 6 7 7 * *
$\uparrow$

The stars represent *less than 5* and *more than 7*. So you can see that seven of the houses have less than 5 people living in them.

The arrow points to the middle term.
The median number of occupants is 5.

When there is an even number of values, the median is half-way between the middle two. For example, the median for the data 7, 11, 16, 18, 20, 25 is

$$17 \left(\text{do } \frac{16 + 18}{2} = 17 \right).$$

The mean

- To find the **mean** you add up all the values and divide by the number of values.

$$\text{Mean} = \frac{\text{sum of values}}{\text{number of values}}$$

The masses of five babies, in kilograms, are: 3.6, 2.9, 3.4, 3.2, 3.7

The mean mass of these five babies is:

$$\frac{3.6 + 2.9 + 3.4 + 3.2 + 3.7}{5} = \frac{16.8}{5} = 3.36 \,\text{kg}$$

The mean can be calculated directly from a frequency table. The following frequency table shows how to find the mean number of peanuts in 27 pods.

Peanuts per pod	Frequency	Totals
2	5	$5 \times 2 = 10$
3	15	$15 \times 3 = 45$
4	7	$7 \times 4 = 28$
Total	27	83

$$\text{Mean} = \frac{83}{27} = 3.1 \text{ to 1 d.p.}$$

The range

The **range** is a measure of how spread out data is. It is the difference between the highest and lowest values. For the frequency table above, the highest number of peanuts per pod is 4 and the lowest is 2. Subtract to find the difference between these values: $4 - 2 = 2$. So the range for the number of peanuts per pod is 2. When the range is low it shows there is not much variation in the data. You would not expect much variation in the number of peanuts per pod.

Exercise 6G

1 The frequency table shows the number of pairs of shoes owned by a group of 30 children.

Number of pairs	Frequency
1	1
2	5
3	4
4	9
5	3
6	4
7	2
8	2

What is the mode?

2 Find **i** the median and **ii** the range of each set of numbers.

a 7, 19, 15, 18, 24, 17, 7, 21, 17, 14, 21, 16, 12

b 6.2, 6.8, 6.7, 6.02, 6.28, 6.82, 6.08, 6.72, 6.27

c 5, 3, 2, 5, 4, 5, 2, 3, 5, 3, 4

3 On six working days, a garage mended 6, 5, 2, 0, 3 and 2 punctures. Calculate the mean number of punctures mended per day by the garage.

4 Here is a frequency table showing the heights of a sample of people:

Height (cm)	150	151	152	153	154	155
Frequency	1	5	10	16	6	2

a What is the modal height?

b What is the range of heights?

5

14 women were asked how many pairs of shoes they each had.

The bar graph shows the results:

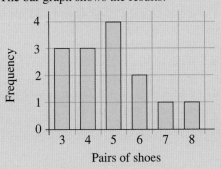

Pairs of shoes

a Write down, in order, the numbers of pairs each woman had.

b What is the median?

c What is the modal number of pairs of shoes?

d What is the range?

6 In a test marked out of 40, the marks of 10 students were 32, 20, 7, 30, 25, 22, 18, 35, 10, 11.

a Calculate the mean mark for the test.

b Find the range.

7 36 women in a women's club in Wellington were asked about the number of children they had. The results are shown in this bar graph.

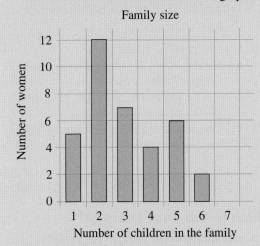

Family size

a Write down the numbers of children the women have, in order.

b What is the median?

c Is the median the same as the mode, for the number of children per family?

8 The number of people in 20 families is shown in the table.

Number of people in family	Frequency
3	3
4	9
5	5
6	2
7	1

Calculate the mean number of people in a family.

9 Here are the heights, in centimetres, of a group of 10 children:
119, 120, 121, 121, 121, 123, 124, 124, 125, 128
  a What is the mode?
  b What is the median?

10 The masses of 20 children are shown in the table.

Mass (kg)	Frequency
30–40	1
40–50	8
50–60	9
60–70	2

Copy and complete this sentence:
The modal class for the masses of
children is ………. .

11 Find the missing number.
  a For the scores 25, 23, 15, 18, 17 and 22, the mean score is ☐.
  b For the scores 15, 19, 18, ☐, 13, 21 and 14, the mean score is 17.
  c For the scores 26, 31, 54, ☐ and 49, the mean score is 38.
  d For the scores 18, 54, 37, 21, ☐, 41, 14 and 15, the mean score is 33.

Which is the best average?

There are three different averages for a reason. Which average we decide to use depends on what we want to find out. They each have advantages and disadvantages. (Note that there are other measures of spread as well, but for now you have to know only about the range.)

Calculation	Advantage	Disadvantage
Mean	• Uses all data values in the calculation. • Can be calculated exactly. • Most people understand this average and it is widely used.	• Hardest average to calculate. • Can be distorted by extreme values. • Can give a decimal answer when the data should be a whole number. • For grouped data the mean is an estimate (you will learn about this in Book 3).
Median	• Relatively easy to find, particularly if data is already in order. • Not affected by extreme values.	• Does not use all data values in the calculation. • Can give a decimal answer when the data should be a whole number. • For grouped data the median is an estimate (you will learn about this when you are older).
Mode	• Easy to find. • Not affected by extreme values. • Will always be a possible value for the data. • Useful for manufacturers, e.g. makers of shoes, clothes, etc.	• Not all data values are used in calculating it. • Often data will not have a mode or there may be several modes. • In small groups or in groups which have an unusual pattern of distribution it may not be representative of the group. • For grouped data only a modal group can be given.
Range	• Easy to work out. • Useful for comparing two sets of data.	• Can be distorted by extreme values. • Not all data values are used in calculating it.

Exercise 6H will help you decide which is the best average to use and why. There are also some harder questions to do with averages in the exercise.

Exercise 6H

1 The scores on a test taken by five students were: 95, 25, 30, 20, 30.
  a What is the mean score?
  b What is the median score?
  c What is the mode?
  d Which average is the most appropriate representation of the data? What difficulties do you see?

2 Look at this advertisement:

> **Wanted**
>
> Young person to join small team
>
> Average salary $40 000

You know that the firm has 6 employees whose salaries are $12 000, $14 000, $16 000, $18 000, $20 000 and $160 000.

a Find the median salary.

b Is there a modal salary.

c Find the mean salary.

d Which average has the company used in the advertisement?

e If you knew that your starting salary would be $12 000, would you say that the advertisement is misleading?
Give reasons for your answer.

3 Mrs Cheyne has four pet cats. One of them had an accident and had to have a leg removed by the vet. The number of legs on each cat is now 4, 4, 4 and 3.

a Find the modal number of legs per cat.

b Find the median number of legs per cat.

c Find the mean number of legs per cat.

d Which average would you *not* use, and why?

4 Aaron works in a shoe shop. He has sold 25 pairs of shoes in one hour this afternoon. He needs to order some more shoes to fill his shelves again. He decides to work out the average shoe size and order more of these.

The sizes of the shoes he sold in this hour are shown in the table.

Shoe size	Frequency
3	6
4	6
5	4
6	3
7	3
8	2
9	1

a Find the modal shoe size.

b Find the median shoe size.

c Find the mean shoe size.

d Which average should Aaron use and why?

5 a Work out the means to find the best batsman:

   i Jimmy who scored 8, 25, 82, 10, 0, 32, 9, 41, 0 in 9 attempts, or

   ii Mahendra who scored 75, 83, 104, 0, 2, 68, 0, 0, 1, 7 in 10 attempts, or

   iii Rikki who scored 14, 56, 102, 37, 76 in 5 attempts.

b Is the mean the best average to use? Why?

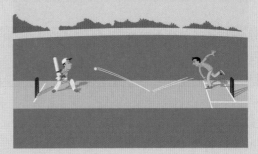

6 For each of the following sets of data write down which average (mean, median or mode) is the best one to use. Give a reason for your answer.

a 78, 30, 74, 76, 83, 73

b 33, 21, 25, 30, 37, 23

c 32, 38, 32, 37, 32, 38, 39, 98, 32

7 The mean of 5, 8 and y is the same as the mean of 2, 4, 7, 8 and 9. Find the value of *y*.

8 The average age of a mother and her three children is 10 years. If the ages of the children are 1, 4 and 7 years, how old is the mother?

9 This frequency table shows the heights of a sample of people.

Height (cm)	Frequency
149	1
150	4
151	9
152	10
153	15
154	7
155	1
156	2
157	1

a Calculate the mean height.

b Determine the median height.

c State the modal height.

d What is the probability that a person has a height of

   i 153 cm   **ii** more than 155 cm?

Consolidation

Example 1

The masses of 20 children in kilograms are:
38, 41, 37, 42, 45, 53, 39, 54, 55, 60
45, 48, 52, 54, 39, 54, 47, 58, 61, 59

a Construct a frequency table using intervals 36–40, etc. to show the data.

b Find the modal class.

....

a

Weight (kg)	36-40	41-45	46-50	51-55	56-60	61-65
Tally	IIII	IIII	II	HII	III	I
Frequency	4	4	2	6	3	1

b The model class is 51–55 kg.

Example 2

The ages of 20 children are shown in the table:

Age (years)	12	13	14	15	16
Frequency	2	4	7	4	3

Find the **a** mean age **b** median age **c** modal age **d** range.

....

a Mean age $= \dfrac{\text{sum of ages}}{\text{number of children}}$

$$= \frac{\begin{array}{c}12 + 12 + 13 + 13 + 13 + 13 + 14 + 14 + 14 + 14 \\ + 14 + 14 + 14 + 15 + 15 + 15 + 15 + 16 + 16 + 16\end{array}}{20}$$

$= 14.1$ years

b To find the median age write the ages in order of size:
12, 12, 13, 13, 13, 13, 14, 14, 14, 14, 14, 14, 14,
15, 15, 15, 15, 16, 16, 16
The two middle numbers are both 14 so the median age is 14 years.

c Mode = most frequently occurring age = 14 years.

d Range = highest age − lowest age
$= 16 - 12$
$= 4$ years

Example 3

The table below shows which flavour ice cream students in Years 8 and 9 like the best.

	Vanilla	Chocolate	Strawberry	Total
Year 8	65	45	20	130
Year 9	70	50	25	145
Total	135	95	45	275

a How many students are there in Years 8 and 9, altogether?

b How many Year 9s like strawberry ice cream the best?

c How many students are there in Year 8?

....

a There are 275 students in Years 8 and 9 altogether.

b 25 Year 9s like strawberry ice cream the best.

c There are 130 students in Year 8.

Exercise 6

1 The lengths of leaves from a tree, measured in millimetres, are:
43, 47, 63, 49, 52, 58, 47, 61, 60, 57
39, 42, 57, 56, 54, 63, 62, 58, 55, 37

a Construct a frequency table to show the data using intervals 35–40, 40–45, etc.

b Find the modal class.

2 The times in seconds the 24 children in Form 3 took to run 200 m are:
29, 33, 38, 32, 34, 36, 27, 29, 30, 32, 40, 41
33, 28, 31, 33, 35, 29, 34, 34, 31, 35, 30, 31
Find the **a** mean **b** mode **c** median times.

3 In 10 innings, Michael's batting scores were
18, 6, 89, 4, 42, 105, 0, 37, 4, 15.

a What was his total score?

b What was his mean score?

c Write down the modal and median scores.

d Which average gives the best idea of how well he batted? Explain why.

e In his eleventh innings Michael scored 43. What does that make his mean score?

4 Is the following data discrete or continuous?
  a The mass of your teacher
  b The number of people in your class
  c The time it takes to run 100 m
  d The length of this book
  e The volume of water in a swimming pool
  f The number of books in your classroom

5 The frequency table shows the results of a mathematics test for 40 students.

Marks	Number of students
0	2
1	1
2	4
3	3
4	5
5	
6	7
7	4
8	2
9	2
10	1

  a How many students scored 5 marks?
  b Calculate:
    i the mean mark
    ii the median mark
    iii the modal mark.
  c Calculate the fraction of students who scored less than 4.

6 Students in Year 7 use a mixture of ink pens and ballpoint pens. Some use black pens and some use blue. Complete the table below to show which pens students are using.

	Ink	Ballpoint	Total
Blue	100		
Black		38	64
Total			210

7 a The ages of the members of a family are 30, 35, 14, 12, 18 and 5 years.
Find the mean age.
  b The heights of 6 people are 157, 149, 158, 160, 152 and 154 cm. Find the median height.
  c Write down the mode of this set of numbers: 10, 7, 8, 9, 9, 6, 7, 8, 9, 9

Summary

You should know ...

1 How to use frequency tables with intervals.
For example:
The fifteen test scores
9, 6, 3, 13, 12, 18, 14, 12, 8, 11, 19, 12, 4, 9, 18
can be shown on a table as:

Score	Tally	Frequency
1–5	||	2
6–10	||||	4
11–15	ⅢⅠ |	6
16–20	|||	3

Check out

1 a The heights in centimetres of 20 plants in a garden are:

4.1	8.3	11.8	13.6
13.9	15.3	11.6	4.8
9.7	22.1	15.2	5.0
14.2	6.5	11	17.1
14.2	8.3	18.1	22.7

Copy and complete the grouped frequency table below to gather this continuous data.

Height (cm)	Tally	Frequency
0–5		
5–10		
10–15		
15–20		
20–25		

2 How to work out the mean, median and mode for a set of data.
For example:
72, 68, 65, 70, 75, 79, 73, 70, 85
In order, the data is: 65, 68, 70, 70, 72, 73, 75, 79, 85
The median is the middle number, 72.
The mode is the number occurring most often, 70.
The mean is the sum of all the values divided by the number of values:
$(65 + 68 + 70 + 70 + 72 + 73 + 75 + 79 + 85) \div 9 = 657 \div 9 = 73$

2 Find the mean, median and mode of this data:
7, 4, 1, 8, 4, 5, 10, 2, 13

3 The range of a set of results is the difference between the highest and lowest numbers.
For example:
The heights of 5 children are
130 cm, 142 cm, 126 cm, 148 cm, 135 cm
The smallest height = 126 cm
The tallest height = 148 cm
Range = 148 − 126 = 22 cm

3 The marks gained by a group of students in a geography test were:
4, 6, 3, 6, 7, 9, 7, 8, 4, 5, 8, 7, 6, 5
What is the range of marks?

4 How to calculate the mean from a frequency table.
For example:

Number	Frequency	No. × Freq.
4	4	16
5	4	20
6	7	42
Total	15	78

The mean for the numbers in this frequency table is $\frac{78}{15} = 5.2$

4 The savings over three months for a group of children is given in the table.

Savings ($)	Frequency
35.00	3
39.50	4
42.00	5
45.50	3
47.00	3
52.00	2

Calculate the mean savings.

5 Discrete data can take only certain values.
For example: The number of children in a class can be 31, 32, 30, etc. You can't have 31.42 children.

Continuous data can take any value.
For example: The mass of a person can be 80 kg, or 80.3 kg or 80.32 kg, etc., depending how accurately you measure it.

5 Is the following data discrete or continuous?
a The speed of a car
b The number of people in a theatre
c The score out of 10 in a spelling test
d The mass of a packet of biscuits
e The time it takes to walk to the shop
f The length of a classroom

6 The purpose of collecting data is usually to answer a question. You have to decide on:
- the method of data collection
- the sample size needed
- the degree of accuracy needed in the measurements.

6 Imagine you are trying to find out whether boys or girls are taller, on average, in your year group at school. Discuss with a friend: the method of data collection, the sample size you need and the degree of accuracy you need for the measurements.

7 A two-way table can be used whenever data is split into two main categories.
For example:
This table shows the number of students who chose maths or science as their favourite subject.

	Maths	Science	Total
Boys	18	13	31
Girls	16	15	31
Total	34	28	62

The table shows that there were 15 girls who chose science as their favourite subject and 18 boys who chose maths as their favourite subject. There were 62 students altogether.

7 The table shows which musical instruments students in Year 8 play. Copy and complete the table.

	Piano	Flute	Total
Boys			20
Girls		23	
Total	25		55

8 Which is the best average to use.
For example:
Some averages are affected by extreme values, e.g. the mean, and some averages are better for manufacturers, e.g. the mode.

8 Discuss with a friend for what types of data sets it is best to use
a the mean
b the median
c the mode.

Review A

1 This number line shows $^-2 + 5 = 3$

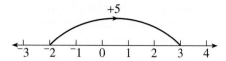

Draw number lines to show:
- **a** $^-5 + 2$
- **b** $^-3 + 4$
- **c** $^-3 + 1$
- **d** $^-4 + 4$
- **e** $^-16 + 9$
- **f** $^-60 + 27$

2 Simplify:
- **a** $3x + 2y + 5x - 4y + x$
- **b** $7a - b + 3c - 2a - 3b + 5c$
- **c** $5(2x + y) - 4(x - 5y)$
- **d** $3x(y + z) - y(z + x) + 2z(x + y)$

3 Draw a quadrilateral with:
- **a** One line of symmetry
- **b** Two lines of symmetry
- **c** Four lines of symmetry

4 Make a good estimate in metric units of:
- **a** The mass of a newborn baby
- **b** The mass of a small car
- **c** The height of your classroom
- **d** How far you can walk in 1 hour
- **e** The area of your desktop.

5 This number line shows $2 - 5 = ^-3$

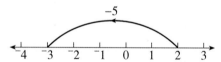

Draw number lines to show:
- **a** $5 - 7$
- **b** $^-1 - 3$
- **c** $0 - 2$
- **d** $3 - 6$
- **e** $7 - 16$
- **f** $34 - 50$

6 What is the area of a square with side length $0.4\,\text{m}$?

7 Draw a line 8 cm long. Construct the perpendicular bisector of that line.

8 Rewrite using index notation:
- **a** $3 \times 3 \times 3$
- **b** $7 \times 7 \times 7 \times 7 \times 7$
- **c** $5 \times 5 \times 5 \times 5 \times 5 \times 5$

9 Draw the net of a cube with side length 3 cm

10 A bottle of lemonade contains 330 ml. How many litres of water are there in 5 of these bottles?

11 Write an expression for the total time taken to roast meat with a mass of k kilograms if it takes 15 minutes plus 40 minutes for each kilogram.

12 Is the following data discrete or continuous?
- **a** The length of your desk
- **b** The number of people in your school
- **c** The time it takes to run 400 m
- **d** The mass of this book
- **e** The volume of water in a pond
- **f** The number of pencils in your pencil case

13 Without a calculator, work out:
- **a** 2^3
- **b** $\sqrt[3]{27}$
- **c** $(-1)^3$

14 Copy and complete:
- **a** $0.13\,\text{km} = \square\,\text{m}$
- **b** $0.7\,\text{kg} = \square\,\text{g}$
- **c** $940\,\text{mm} = \square\,\text{cm}$
- **d** $79\,500\,\text{m} = \square\,\text{km}$
- **e** $1.27\,\text{t} = \square\,\text{kg}$
- **f** $12\,\text{cm}\ 3\,\text{mm} = \square\,\text{mm}$

15 What is the side length of a cube with volume $54.872\,\text{cm}^3$?

16 Simplify:
- **a** $8 - (x - 5) - 5(3 - x)$
- **b** $4t + 6 \times 3t - 50t$
- **c** $m - (4 - 2m) + 5 \times 3m$

17 Write down **a** the order of rotational symmetry and **b** the number of lines of symmetry for these shapes.

i

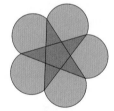

ii

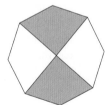

iii

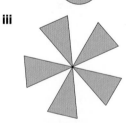

iv

v

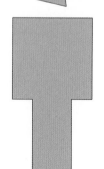

vi

18 Round:
  a 475 to the nearest 10
  b 5.476 to the nearest whole number
  c 32 859 to the nearest 1000
  d 0.24574 to 2 d.p.
  e 0.245 to 1 d.p.
  f 21 350 to the nearest 100
  g 0.2623 to 2 d.p.
  h 7.792 to 1 d.p.
  i 132.49 to the nearest whole number

19 a Copy and complete this two-way table about the hair colour of a group of 32 students.

	Light hair	Dark hair	Total
Female	7		
Male		11	
Total	13		32

  b How many boys are there?
  c How many girls are there?
  d How many girls have dark hair?
  e How many boys have light hair?

20 Work out:
  a $2 - {}^-3$ **b** ${}^-7 - {}^-5$
  c $0 - {}^-3$ **d** ${}^-2 - {}^-3 + 8$

21 i 16, 12   **ii** 20, 15   **iii** 19, 17
  a Write down all the factors of the number pairs above.
  b Find the common factors for each number pair.
  c Find the highest common factor for each number pair.

22 Work out the areas of the rectangles.

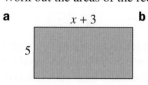
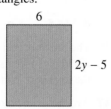

23 Using your protractor, draw an angle of 80°. Construct the perpendicular bisector of this angle.

24 Copy and complete:
  a 75 miles ≈ ☐ km   **b** 24 km ≈ ☐ miles
  c 80 km ≈ ☐ miles   **d** 85 miles ≈ ☐ km

25 Write these numbers in descending size order using > symbols:
8.05, 8.57, 8.8, 7.5, 8.785

26 Workout:
  a $3 + {}^-7$ **b** $7 + {}^-3$ **c** ${}^-3 + {}^-7$
  d ${}^-2 + 7$ **e** $17 + {}^-8$ **f** $7 + {}^-18$

27 A cube has the numerals 1, 2, 3, 4, 5, 6 painted on its faces. Here are two views of the cube:

  a Draw this net of the cube and put in the missing numbers.

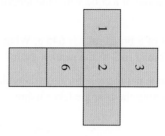

  b What are the numbers on the faces that touch both 5 and 6?

28 What is the side length of a square with area 5.76 cm²?

29 Work out the area of this shape.

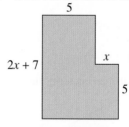

30 Construct triangles with sides
  a 4 cm, 6 cm and 5 cm
  b 5.5 cm, 4 cm and 6 cm

31 Copy and complete:
  a $2.7\,\ell = \square\,\text{ml}$ **b** $435\,\ell = \square\,\text{ml}$
  c $2300\,\text{cm}^3 = \square\,\ell$ **d** $450\,\text{ml} = \square\,\ell$
  e $78\,\text{cm}^3 = \square\,\text{ml}$ **f** $70\,000\,\text{ml} = \square\,\ell$

32 I walk to my friend's house, which is a journey of 2.4 km. Then we walk to school, which is 800 m from my friend's house. How many kilometres have I walked altogether?

33 Copy and complete the table below with as many advantages and disadvantages as you can.

Calculation	Advantages	Disadvantages
Mean		
Median		
Mode		
Range		

34 Copy and complete using < or >
  a 9.03 ☐ 9.3   **b** 0.45 ☐ 0.543
  c 70.07 ☐ 70.007   **d** 7.399 ☐ 7.4

35 What is the volume of a cube with side length 2.5 mm?

36 The total mass of 8 bread rolls is 645 g. What is the mass of 1 bread roll, to the nearest gram?

37 Draw an accurate net to scale for this triangular prism, constructing the triangle faces using compasses.

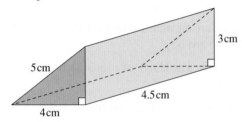

38 Work out, to 2 decimal places:
  a 4.2^4 **b** 2.3^5
  c 1.7^6 **d** 1.9^5

39 Write an expression for the perimeter of these shapes.
  a   **b**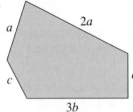

40 Simplify:
  a $2 + {}^-1 + 2$ **b** ${}^-3 - 3 + 2$
  c $4 - 5 + {}^-2$ **d** ${}^-1 + 5 + {}^-4$
  e ${}^-4 + 5 - 1$ **f** ${}^-5 + 4 + {}^-1$

41 Jane goes to the city every 36 days. Indra goes to the same city every 24 days. They met each other one day. How many days later will they get the chance to meet each other again?

42 i

INPUT   × 5   − 3   OUTPUT

  ii

INPUT   × 2   + 6   OUTPUT

  a Write down the functions in the form $f(x) = \ldots$ for these function machines.
  b Find $f(2)$.
  c Find $f(5)$.

43 A tin of peas has a mass of 350 g. What is the mass, in kilograms, of 4 tins of peas?

44 The heights in centimetres of 20 flowers are:

4.2	7.3	17.6	14.7	8.6	23.2	17.4	15	18.3	7.7
4.9	16.8	10.6	4.8	13.6	8.9	12	16.4	19.8	23.8

  a Copy and complete the grouped frequency table below to gather this continuous data.

Height (cm)	Tally	Frequency
0–5		
5–10		
10–15		
15–20		
20–25		

  b What is the modal class?

45 Round the answers to these to 1 decimal place:
  a 0.27^2 **b** $\sqrt{7.2}$ **c** 4.21^2
  d $\sqrt{4.5}$ **e** 3.23^3 **f** $\sqrt[3]{7.3}$

46 The monthly cost of local calls on a mobile phone is $7.50 plus 8 cents per call. Write an expression for the total cost, in dollars, of p local calls in n months.

47 Are these statements true or false?
  a $400.95 > 500.94$ **b** $0.032 < 0.31$
  c $3^2 = 6$ **d** $4.07 \neq 4.70$

48 **a** Make an accurate drawing of this triangle:

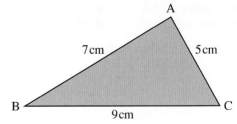

  b Bisect CÂB.

49 Work out:
   a $4 + 3 \times 2$ **b** $10 - 25 \div 5$
   c $7 + 16 \div 2^2$ **d** $2^3 - (^-3 \times 5)$

50 A piece of metal is 2.3 m long. A 45 cm length is cut from it. How many centimetres of the piece of metal are left?

51 Work out:
   a $\dfrac{15}{^-5} + \dfrac{^-12}{2}$ **b** $\dfrac{^-64}{^-8} + \dfrac{^-10}{5}$
   c $\dfrac{^-20}{4} - \dfrac{^-50}{10}$

52 Draw the net of a tetrahedron with side length 3.5 cm.

53 Without a calculator, work out:
   a 350×0.01 **b** 72×0.1
   c $68 \div 0.1$ **d** $9 \div 0.01$
   e 4100×0.01 **f** 300×0.1
   g $0.6 \div 0.1$ **h** $0.25 \div 0.01$

54 Copy and complete this table.

Subtract	2	$^-3$	$^-1$	4
3	1		4	
$^-1$				
4				
$^-2$				

Second number (column header)   *First number* (row header)

55 Work out the following without using a calculator (write both possible answers where there are two):
   a 8^2 **b** $\pm\sqrt{169}$
   c 12^2 **d** $\pm\sqrt{10000}$
   e 11^2 **f** $\pm\sqrt{81}$

56 Work out the shaded area in this shape:

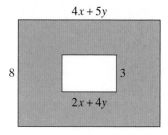

$4x + 5y$ (top), 8 (left), 3 (right), $2x + 4y$ (bottom)

57 Draw triangle ABC such that
   a $AB = 8$ cm, $C\hat{A}B = 60°$ and $AC = 6$ cm
   b $AB = 7.5$ cm, $C\hat{A}B = 50°$ and $AC = 6.3$ cm

58 Without using a calculator, work out:
   a 3×0.08 **b** 7×0.5
   c 4×0.06 **d** 0.9×5
   e 8×0.02 **f** 0.7×9
   g 8×0.4

59 The distance from Wellington in New Zealand to Jakarta in Indonesia is roughly 7680 kilometres. Approximately how many miles is this?

60 For which of the data sets below is it best to calculate **a** the mean, **b** the median or **c** the mode?
   A 37, 109, 33, 35, 42, 32
   B 33, 22, 26, 30, 38, 24
   C 61, 68, 61, 67, 61, 75, 73, 2, 61, 138

Give a reason for each of your answers.

61 Work out:
   a $18 - 2 \times 6 + ^-1$
   b $-2 \times ^-9 + 2 \times 3^2$
   c $8 \times 3 + (^-18 \div 9)$

62 Draw an accurate net to scale for this cuboid:

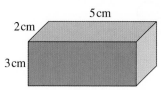

5 cm, 2 cm, 3 cm

63 Without using a calculator, work out:
   a $7.5 \div 5$ **b** $53.6 \div 4$
   c $191.7 \div 9$ **d** $25.84 \div 8$
   e $0.72 \div 3$

64 **i** 5, 7   **ii** 4, 12   **iii** 6, 15
   a Write down the first ten multiples of each number.
   b Find the lowest common multiple for each number pair.

65 Construct the right-angled triangle ABC, where $AB = 4$ cm and the hypotenuse $BC = 5$ cm.

66 Without using a calculator, work out:
   a $22 \div 6$ to 1 d.p.   **b** $1.2 \div 8$ to 1 d.p.
   c $0.44 \div 3$ to 2 d.p. **d** $2 \div 7$ to 2 d.p.
   e $4.4 \div 9$ to 2 d.p.

67 Describe as many properties as you can of a rhombus (include among the properties you consider rotational symmetry, lines of symmetry and diagonals).

68 Work out:

a $\dfrac{4 + 8}{3}$ **b** $\dfrac{10 + 26}{11 - 2}$ **c** $\dfrac{3^2 + 11}{2}$

69 a In this diagram, the number in the square on each side of the triangle is the sum of the numbers in the circles.

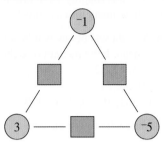

Find the numbers in the squares.

b This time, find the numbers in the circles so that the sum on each side of the triangle is the number in the square:

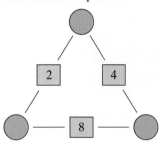

70 Four girls shared $73 equally between them. How much did they each get?

71 Draw an accurate net to scale for this square-based pyramid, constructing the triangle faces using compasses:

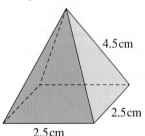

4.5 cm

2.5 cm

2.5 cm

72 Copy and complete:

a $73 \div \square = 7300$
b $\square \times 0.1 = 0.48$
c $\square \div 10 = 8.9$
d $154 \times \square = 1.54$
e $\square \div 0.01 = 230$

73 Which of these statements are true?

a A rectangle has 3 lines of symmetry.
b A parallelogram has rotational symmetry of order 4.
c A kite has diagonals equal in length.
d The diagonals of a rectangle bisect each other.

74 Write as a product of prime factors:

a 360 **b** 630 **c** 500

75 Work out:

a $4 \times {}^-3$ **b** ${}^-10 \times {}^-5$
c $2 \times {}^-8$ **d** ${}^-7 \times {}^-5$
e $12 \div {}^-3$ **f** ${}^-40 \div {}^-5$
g $16 \div {}^-8$ **h** ${}^-63 \div 7$

76 Write these measurements in ascending size order using < symbols:
64 ml, 640 ℓ, 0.64 ℓ, 6400 ml, 0.64 ml

77 How much does each person get if $338.24 is shared equally between 7 people?

78 Copy and complete using $<$, $>$ or $=$

a $0.8\,\text{km} \square 80\,\text{m}$ **b** $760\,\text{g} \square 0.76\,\text{kg}$
c $3400\,\text{ml} \square 34\,\ell$ **d** $92\,\text{mm} \square 0.92\,\text{cm}$

7 Fractions

Objectives

- Add and subtract fractions and mixed numbers; calculate fractions of quantities (fraction answers); multiply and divide an integer by a fraction.
- Recall simple equivalent fractions.

- Use known facts and place value to multiply and divide simple fractions.
- Use the laws of arithmetic and inverse operations to simplify calculations with integers and fractions.

What's the point?

Carpenters and masons use whole numbers and fractions on a daily basis in their work.

Before you start

You should know ...

1 Equivalent fractions show the same fraction using different numbers.

For example: $\dfrac{3}{4} = \dfrac{6}{8} = \dfrac{9}{12} = \dfrac{12}{16}$

2 Fractions can be simplified by dividing.
For example:

$$\dfrac{15}{27} = \dfrac{5}{9}$$

$\div 3$ — numerator

$\div 3$ — denominator

Check in

1 Copy and complete:

a $\dfrac{1}{4} = \dfrac{\square}{8}$ b $\dfrac{3}{8} = \dfrac{\square}{16}$

c $\dfrac{2}{3} = \dfrac{\square}{9}$ d $\dfrac{5}{8} = \dfrac{\square}{24}$

2 Simplify:

a $\dfrac{10}{15}$ b $\dfrac{12}{18}$ c $\dfrac{14}{21}$

d $\dfrac{32}{48}$ e $\dfrac{49}{77}$ f $\dfrac{72}{108}$

3 A mixed number is made up of a whole number and a fraction.

For example:

$$2\frac{2}{5} = \frac{2 \times 5 + 2}{5} = \frac{12}{5}$$

mixed number   improper fraction

4 Improper fractions can be changed to mixed numbers by dividing.

For example:

$\frac{14}{5}$ means $14 \div 5$

$= 2$ remainder $4 = 2\frac{4}{5}$

---

3 Write as improper fractions:

a $2\frac{2}{3}$ **b** $3\frac{1}{4}$ **c** $4\frac{3}{4}$

d $1\frac{1}{6}$ **e** $4\frac{5}{6}$ **f** $6\frac{5}{8}$

4 Write as mixed numbers:

a $\frac{9}{4}$ **b** $\frac{23}{10}$ **c** $\frac{21}{5}$

d $\frac{18}{7}$ **e** $\frac{58}{8}$ **f** $\frac{70}{13}$

7.1 Addition and subtraction of fractions

You can always add or subtract amounts of the same object.

For example, 2 tables + 3 tables = 5 tables

In the same way:

2 eighths + 3 eighths = 5 eighths

or $\frac{2}{8}$ + $\frac{3}{8}$ = $\frac{5}{8}$

If you are adding or subtracting mixed numbers you can add or subtract the whole numbers first.

EXAMPLE 1

Work out:

a $4\frac{3}{5} + 1\frac{4}{5}$ **b** $3\frac{3}{4} - 2\frac{1}{4}$

....

a $4\frac{3}{5} + 1\frac{4}{5} = 4 + 1 + \frac{3}{5} + \frac{4}{5}$

$= 5 + \frac{7}{5}$ Change the improper fraction to a mixed number.

$= 5 + 1\frac{2}{5}$ Subtract the whole numbers (3 − 2)

$= 6\frac{2}{5}$

b $3\frac{3}{4} - 2\frac{1}{4} = 1\frac{2}{4}$

$= 1\frac{1}{2}$ Subtract the fractions $\left(\frac{3}{4} - \frac{1}{4}\right)$

Always simplify fractions.

Sometimes, when you subtract fractions it is easier to turn them into improper fractions first (you will see this later).

Exercise 7A

1 Work out:

a $\frac{1}{8} + \frac{3}{8}$ **b** $\frac{2}{6} + \frac{3}{6}$

c $\frac{2}{7} + \frac{4}{7}$ **d** $\frac{2}{3} + \frac{2}{3}$

e $\frac{5}{6} + \frac{3}{6}$ **f** $\frac{7}{8} + \frac{5}{8}$

g $2\frac{6}{9} + 1\frac{7}{9}$ **h** $3\frac{10}{12} + 7\frac{11}{12}$

i $2\frac{3}{4} + 1\frac{3}{4}$ **j** $4\frac{5}{8} + 3\frac{7}{8}$

2 Work out these subtractions:

a $\frac{3}{4} - \frac{1}{4}$ **b** $\frac{6}{8} - \frac{5}{8}$

c $\frac{4}{5} - \frac{2}{5}$ **d** $\frac{8}{12} - \frac{5}{12}$

e $1\frac{3}{4} - \frac{1}{4}$ **f** $3\frac{2}{3} - 1\frac{1}{3}$

g $2\frac{2}{5} - 1\frac{1}{5}$ **h** $1\frac{3}{5} - \frac{1}{5}$

i $4\frac{3}{8} - 2\frac{2}{8}$ **j** $3\frac{7}{12} - 1\frac{5}{12}$

3 In Mrs Bruno's class $\frac{5}{8}$ of the children walk to school, $\frac{2}{8}$ take a bus and the rest cycle to school.

   a What fraction of the class walk or take a bus?

   b What fraction cycle?

4 Two pieces of carpet $3\frac{1}{4}$ metres and $2\frac{3}{4}$ metres long are joined together. How long is the joined carpet?

5 Al saves $\frac{2}{7}$ of what he earns.
What fraction of his earnings does Al spend?

6 Write down two fractions that add up to make $2\frac{5}{8}$.

▶ ACTIVITY

Equivalent fraction puzzles

1 I am equivalent to $\frac{1}{2}$. The sum of my numerator and denominator is 27. Which fraction am I?

2 I am equivalent to $\frac{52}{182}$. My denominator is a prime number. Which fraction am I?

3 I am equivalent to $\frac{3}{5}$. The product of my numerator and denominator is 135. Which fraction am I?

4 I am equivalent to $\frac{588}{798}$. My denominator is less than 20. Which fraction am I?

5 I am equivalent to $\frac{3}{4}$. My denominator is 5 more than my numerator. Which fraction am I?

6 I am equivalent to $\frac{39}{65}$. My numerator is less than 10. Which fraction am I?

7 I am equivalent to $\frac{4}{9}$. My numerator is 20 less than my denominator. Which fraction am I?

🖥 TECHNOLOGY

Review your knowledge of equivalent fractions by visiting the site

www.learningplanet.com

and following the links to Students and 7th Grade. Play the game 'Fraction Frenzy'.

Adding fractions with different denominators

You **cannot** add or subtract numbers of **different** objects.

For example, you cannot work out 2 tables + 3 chairs.

In the same way, 2 sevenths + 3 fifths or $\frac{2}{7} + \frac{3}{5}$

cannot be added directly.

You have to use equivalent fractions to find a common denominator before you can add different kinds of fractions.

EXAMPLE 2

Work out $\dfrac{2}{7} + \dfrac{3}{5}$

$$\frac{2}{7} + \frac{3}{5} \qquad \text{LCM of 7 and 5 is 35}$$

$$= \frac{10}{35} + \frac{21}{35} \qquad \frac{2}{7} = \frac{10}{35}, \quad \frac{3}{5} = \frac{21}{35}$$

$$= \frac{31}{35}$$

Exercise 7B

1 Copy and complete:

a $\dfrac{3}{4} + \dfrac{3}{16} = \dfrac{\square}{16} + \dfrac{\square}{16} = \dfrac{\square}{16}$

b $\dfrac{3}{8} + \dfrac{1}{4} = \dfrac{\square}{8} + \dfrac{\square}{8} = \dfrac{\square}{8}$

c $\dfrac{1}{6} + \dfrac{2}{3} = \dfrac{\square}{6} + \dfrac{\square}{6} = \dfrac{\square}{6}$

d $\dfrac{2}{3} + \dfrac{1}{4} = \dfrac{\square}{12} + \dfrac{\square}{12} = \dfrac{\square}{12}$

e $\dfrac{2}{5} + \dfrac{2}{7} = \dfrac{\square}{35} + \dfrac{\square}{35} = \dfrac{\square}{35}$

2 Work out:

a $\dfrac{2}{7} + \dfrac{1}{3}$ **b** $\dfrac{3}{8} + \dfrac{1}{2}$ **c** $\dfrac{2}{5} + \dfrac{1}{4}$

d $\dfrac{3}{10} + \dfrac{2}{5}$ **e** $\dfrac{4}{9} + \dfrac{1}{3}$ **f** $\dfrac{3}{7} + \dfrac{1}{2}$

g $\dfrac{3}{4} + \dfrac{1}{8}$ **h** $\dfrac{2}{8} + \dfrac{2}{5}$ **i** $\dfrac{5}{7} + \dfrac{2}{9}$

j $\dfrac{4}{6} + \dfrac{2}{9}$

3 Katy eats a quarter of an orange and her friend eats another $\frac{3}{8}$ of it. What fraction of the orange is eaten?

4 In a class, one third of the boys prefer football and another two fifths prefer cricket.
What fraction like either game?

INVESTIGATION

Choose two fractions, for example $\frac{2}{5}$ and $\frac{3}{11}$. Add these fractions: $\frac{2}{5} + \frac{3}{11} = \frac{37}{55}$

Swap the numerators and add the new fractions: $\frac{3}{5} + \frac{2}{11} = \frac{43}{55}$.

Try some more fractions. Investigate what happens when you swap numerators around. Does the sum of the fractions increase, decrease or stay the same? Write down what you notice.

Adding mixed numbers

Mixed numbers can be added in a similar way.

EXAMPLE 3

Work out $3\frac{2}{3} + 1\frac{3}{4}$

$$3\frac{2}{3} + 1\frac{3}{4}$$

Add the whole numbers first, $3 + 1 = 4$

$$= 4 + \frac{2}{3} + \frac{3}{4}$$
$$= 4 + \frac{8}{12} + \frac{9}{12}$$
$$= 4 + \frac{17}{12}$$
$$= 4 + 1\frac{5}{12}$$
$$= 5\frac{5}{12}$$

$$\frac{2}{3} = \frac{8}{12}, \quad \frac{3}{4} = \frac{9}{12}$$
($\times 4$) ($\times 3$)

Exercise 7C

1 Use the method from Example 3 to work out:

a $3\frac{1}{2} + 2\frac{1}{4}$ b $3\frac{2}{3} + 2\frac{1}{4}$ c $4\frac{1}{5} + 2\frac{3}{4}$

d $2\frac{3}{8} + 3\frac{1}{3}$ e $4\frac{3}{4} + 3\frac{2}{5}$ f $5\frac{2}{3} + 2\frac{5}{8}$

g $4\frac{4}{5} + 3\frac{3}{7}$ h $6\frac{1}{2} + 7\frac{3}{4}$ i $4\frac{3}{7} + 3\frac{3}{5}$

j $2\frac{6}{7} + 3\frac{5}{6}$

2 Work out:

a $2\frac{1}{2} + 3\frac{1}{4} + 2\frac{1}{8}$ b $3\frac{2}{3} + 4\frac{5}{6} + 3\frac{7}{9}$

c $4\frac{7}{20} + 5\frac{2}{5} + 4\frac{9}{10}$ d $2\frac{1}{2} + 3\frac{1}{3} + 4\frac{1}{4}$

e $4\frac{3}{7} + 2\frac{2}{5} + 3\frac{2}{3}$

3 Mr Durant's car has $4\frac{1}{3}$ litres of petrol in its tank. He puts a further $2\frac{3}{4}$ litres in it. How much petrol is in the tank now?

4 Adio wishes to post three parcels with masses $1\frac{1}{2}$ kg, $2\frac{1}{3}$ kg and $4\frac{3}{4}$ kg. What is the total mass of his parcels?

5 Ambrose has $2\frac{1}{4}$ hectares of land. His brother Anselm has $1\frac{2}{3}$ hectares more than Ambrose. How much land do they have altogether?

Subtracting fractions with different denominators

You need to make sure that the denominators of the fractions are the same before subtracting by using equivalent fractions.

EXAMPLE 4

Work out $\frac{8}{9} - \frac{5}{6}$

$$\frac{8}{9} - \frac{5}{6} \qquad \text{LCM of 9 and 6 is 18}$$

$$= \frac{16}{18} - \frac{15}{18} \qquad \frac{8}{9} = \frac{16}{18}, \quad \frac{5}{6} = \frac{15}{18}$$

$$= \frac{1}{18}$$

Exercise 7D

1 Copy and complete:

a $\dfrac{3}{8} - \dfrac{1}{4} = \dfrac{\square}{8} - \dfrac{\square}{8} = \dfrac{\square}{8}$

b $\dfrac{7}{9} - \dfrac{2}{3} = \dfrac{\square}{9} - \dfrac{\square}{9} = \dfrac{\square}{9}$

c $\dfrac{3}{4} - \dfrac{2}{3} = \dfrac{\square}{12} - \dfrac{\square}{12} = \dfrac{\square}{12}$

d $\dfrac{7}{8} - \dfrac{3}{5} = \dfrac{\square}{40} - \dfrac{\square}{40} = \dfrac{\square}{40}$

e $\dfrac{6}{7} - \dfrac{3}{4} = \dfrac{\square}{28} - \dfrac{\square}{28} = \dfrac{\square}{28}$

2 Work out:

a $\dfrac{4}{5} - \dfrac{2}{3}$ b $\dfrac{6}{7} - \dfrac{5}{8}$ c $\dfrac{2}{5} - \dfrac{1}{4}$

d $\dfrac{9}{10} - \dfrac{4}{5}$ e $\dfrac{8}{11} - \dfrac{3}{5}$ f $\dfrac{7}{10} - \dfrac{2}{15}$

g $\dfrac{4}{7} - \dfrac{2}{9}$ h $\dfrac{9}{14} - \dfrac{4}{21}$ i $\dfrac{8}{13} - \dfrac{5}{11}$

j $\dfrac{8}{12} - \dfrac{3}{8}$

3 Work out:

a $\dfrac{5}{8} - \dfrac{2}{9}$ b $\dfrac{11}{12} - \dfrac{3}{7}$ c $\dfrac{7}{9} - \dfrac{5}{12}$

d $\dfrac{11}{16} - \dfrac{4}{9}$ e $\dfrac{11}{12} - \dfrac{9}{10}$ f $\dfrac{7}{10} - \dfrac{4}{15}$

g $1 - \dfrac{1}{6}$ h $1 - \dfrac{3}{4}$ i $1 - \dfrac{10}{17}$

j $1 - \dfrac{23}{24}$

Subtracting mixed numbers

To subtract mixed numbers it is a good idea to turn them both into improper fractions.

EXAMPLE 5

Work out:

a $2\frac{1}{5} - 1\frac{4}{5}$ b $3\frac{2}{3} - 1\frac{5}{9}$ c $2\frac{5}{6} - 1\frac{7}{8}$

....

a $2\frac{1}{5} - 1\frac{4}{5} = \dfrac{11}{5} - \dfrac{9}{5} = \dfrac{2}{5}$

b $3\frac{2}{3} - 1\frac{5}{9} = \dfrac{11}{3} - \dfrac{14}{9}$ $\left(\dfrac{11}{3} = \dfrac{33}{9}\right)$ Improper fractions

$= \dfrac{33}{9} - \dfrac{14}{9}$

$= \dfrac{19}{9}$

$= 2\frac{1}{9}$

c $2\frac{5}{6} - 1\frac{7}{8} = \dfrac{17}{6} - \dfrac{15}{8}$

$= \dfrac{68}{24} - \dfrac{45}{24}$

$= \dfrac{23}{24}$

Exercise 7E

1 Work out:

a $1\frac{1}{3} - \frac{2}{3}$ b $2\frac{3}{4} - 1\frac{1}{4}$ c $4\frac{5}{8} - 1\frac{7}{8}$

d $6\frac{1}{12} - 4\frac{7}{12}$ e $2\frac{1}{3} - 1\frac{5}{9}$ f $4\frac{11}{12} - 1\frac{5}{6}$

2 Work out:

a $6\frac{1}{3} - 2\frac{1}{6}$ b $2\frac{1}{6} - \frac{3}{4}$ c $3\frac{1}{2} - 1\frac{3}{5}$

d $4\frac{1}{7} - 1\frac{3}{4}$ e $6\frac{1}{9} - 5\frac{3}{4}$ f $5\frac{2}{5} - 3\frac{5}{8}$

g $3 - 2\frac{2}{3}$ h $4 - 2\frac{3}{5}$ i $3 - 2\frac{11}{12}$

j $5 - 4\frac{11}{17}$

3 A water tank holds 100 litres. Mrs Shaw uses $6\frac{3}{4}$ litres. How much water is left in the tank?

4 Kimani's home is $6\frac{1}{2}$ km from school. She walks $\frac{5}{8}$ km to the bus stop and takes a bus for the rest of the journey. How far is her school from the bus stop?

5 Olive has 3 parcels to post. The first parcel has a mass of $2\frac{1}{3}$ kg. The second parcel has a mass of $\frac{3}{4}$ kg. The parcels have a mass of 5 kg in total. What is the mass of the third parcel?

6 Write down two fractions with different denominators that add up to make $1\frac{3}{8}$.

7 Work out:

a $4\frac{1}{4} + 1\frac{3}{5} - 2\frac{1}{2}$

b $3\frac{2}{3} + 2\frac{1}{4} - \frac{1}{2}$

c $11\frac{5}{8} - 8\frac{2}{5} + 1\frac{1}{20}$

8 Fill in the missing numbers:

a $\square + 4\frac{2}{3} = 6\frac{7}{15}$

b $5\frac{1}{2} - \square = 2\frac{9}{10}$

9 Jane has made a mistake in her homework. She has written:

$1\frac{1}{4} + 2\frac{3}{8} = \frac{5}{4} + \frac{19}{8} = \frac{24}{12} = 2$

a Jane's friend Ben said it was obvious that this answer was wrong, without adding the fractions. What does he mean?

b What mistake has Jane made?

c Jane's teacher told her there was a step in her working she didn't need to do. What do you think that was?

10 Copy and complete the diagram, so that the sum of the numbers in any two circles equals the number in the square between them.

11 Find the value of the letters in the following:

a $1\frac{a}{5} + \frac{2}{5} = 2\frac{1}{5}$ b $2\frac{1}{6} + \frac{b}{3} = 2\frac{5}{6}$

c $2\frac{c}{8} + 1\frac{7}{8} = 4\frac{1}{2}$ d $4\frac{7}{9} + \frac{d}{18} = 5\frac{1}{6}$

12 Put these numbers into groups of three so that each group has a total of 2.

$\frac{2}{3}$ $\frac{3}{5}$ $\frac{4}{5}$ $\frac{3}{4}$ $\frac{5}{12}$ $\frac{7}{10}$ 1 $\frac{13}{20}$ $\frac{7}{12}$ $\frac{11}{24}$ $\frac{1}{2}$ $\frac{7}{8}$

7.2 Multiplication of fractions

What is $4 \times \frac{2}{3}$?

$$4 \times \frac{2}{3} = \frac{2}{3} + \frac{2}{3} + \frac{2}{3} + \frac{2}{3} = \frac{8}{3} = 2\frac{2}{3}$$

You should see that

$$4 \times \frac{2}{3} = \frac{4}{1} \times \frac{2}{3} = \frac{8}{3} = 2\frac{2}{3}$$

EXAMPLE 6

Work out:

a $\frac{3}{8} \times 5$ b $3\frac{2}{5} \times 4$ c $\frac{3}{4}$ of 30 kg

d $\frac{3}{5} \times \frac{2}{7}$

....

a $\frac{3}{8} \times 5 = \frac{3}{8} \times \frac{5}{1} = \frac{15}{8} = 1\frac{7}{8}$

Write the whole number with a denominator of 1

Multiply numerators and denominators.

b $3\frac{2}{5} \times 4 = \frac{17}{5} \times \frac{4}{1} = \frac{68}{5} = 13\frac{3}{5}$

Use improper fractions.

c $\frac{3}{4}$ of 30 kg $= \frac{3}{4} \times \frac{30}{1} = \frac{90}{4} = \frac{45}{2} = 22\frac{1}{2}$ kg

Replace 'of' with 'x'

Simplify fractions

Convert to mixed number answer

d $\frac{3}{5} \times \frac{2}{7} = \frac{6}{35}$

Exercise 7F

1 Work out:

a $4 \times \frac{1}{4}$ b $5 \times \frac{1}{2}$ c $8 \times \frac{3}{4}$

d $7 \times \frac{2}{5}$ e $6 \times \frac{3}{8}$ f $12 \times \frac{7}{8}$

g $\frac{4}{5} \times \frac{3}{11}$ h $\frac{7}{10} \times \frac{1}{5}$ i $\frac{6}{7} \times \frac{2}{9}$

j $\frac{4}{9} \times \frac{3}{8}$

2 Work out:

a $2\frac{1}{2} \times 7$ b $2\frac{2}{3} \times 5$ c $4\frac{1}{4} \times 3$

d $2 \times 1\frac{2}{3}$ e $6 \times 3\frac{3}{4}$ f $4\frac{2}{3} \times 7$

3 Find:

a $\frac{3}{10}$ of 8 m   b $\frac{3}{4}$ of 42 km

c $\frac{2}{3}$ of 200 ml   d $\frac{4}{5}$ of 3 t

4 Anisha has 4 litres of paint. She uses $\frac{3}{8}$ of it to paint a room.

a How much paint did she use?

b How much paint does she have left?

5 It takes a mechanic 2 hours to service a car. If the mechanic takes one third of this time to change the oil, how long does he take to perform the other tasks?

6 Tarek's school is 6 km from his home.
Every day he walks $\frac{2}{5}$ of the way and takes a bus for the remaining distance.
He does the same on the return journey.

a How far does he walk altogether in one week?

b How far does he travel by bus during the week?

7 A rectangle measures 4 cm by $2\frac{3}{5}$ cm.
What is the area of the rectangle?
(**Remember:** area = length × width)

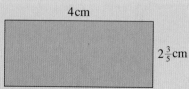

4 cm

$2\frac{3}{5}$ cm

8 Calculate:

a $\frac{3}{20}$ of 4 m, giving your answer in cm

b $\frac{7}{10}$ of 12 cm, giving your answer in mm

c $\frac{5}{6}$ of 2 days, giving your answer in days and hours

d $\frac{3}{8}$ of 7 kg, giving your answer in g

9 What are the missing numbers?

a $\frac{1}{4}$ of $\square$ = 7   **b** $\frac{3}{7}$ of $\square$ = 18

c $\frac{2}{5}$ of $\square$ = 120   **d** $\frac{1}{\square}$ of 800 = 160

e $\frac{2}{\square}$ of 45 = 10   **f** $\frac{3}{\square}$ of 160 = 12

g $\frac{\square}{4}$ of 120 = 30   **h** $\frac{\square}{7}$ of 35 = 15

i $\frac{\square}{\square}$ of 16 = 14

10 Fill in the missing numbers:

a $\frac{1}{4}$ of 40 = $\frac{2}{5}$ of $\square$ **b** $\frac{3}{8}$ of 200 = $\frac{3}{7}$ of $\square$

c $\frac{1}{5}$ of $\square$ = $\frac{1}{10}$ of $\square$

d There are many possible answers to part **c**. Find some more answer pairs. Write down the connection between the pairs of answers.

⟫ INVESTIGATION

Some things look wrong, even though they are correct.
For example: $2 - \frac{2}{3} = 2 \times \frac{2}{3}$

a Is this correct?
b How about $4 - \frac{4}{5}$ and $4 \times \frac{4}{5}$?
c Can you find other fractions that work? Any rules?

7.3 Division of fractions

▶ ACTIVITY

You will need paper and scissors. Use something circular, like the rim of a cup, to draw six identical circles.

Cut out each circle carefully. Fold it in half, then fold it in half again. Cut along the fold lines. This will give you quarter-circles.

Dividing whole numbers by fractions

You can divide a number by a fraction.

For example:

what is $2 \div \frac{1}{4}$?

One way of working this out is to ask how many quarter-circles make 2 circles.

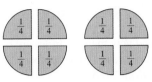

8 quarters = 2, so $2 \div \frac{1}{4} = 8$

Exercise 7G

1 **a** Using quarter-circles, make up 6 circles. How many quarter-circles are in 6 circles?
Copy and complete: $6 \div \frac{1}{4} = \square$

   b From your 6 circles make $\frac{3}{4}$ circles. How many $\frac{3}{4}$ circles can you make from 6 circles?
Copy and complete: $6 \div \frac{3}{4} = \square$

2 **a** Using quarter-circles, make up 9 circles. How many quarter-circles are in 9 circles?
Copy and complete: $9 \div \frac{1}{4} = \square$

   b From your 9 circles make up $\frac{3}{4}$ circles. How many $\frac{3}{4}$ circles can you make from 9 circles?
Copy and complete: $9 \div \frac{3}{4} = \square$

3 Look at your answers to Question **1**.

$6 \div \frac{1}{4} = 24$

$6 \div \frac{3}{4} = 8$

   a There are 24 quarter-circles in 6 circles. What number could you have *multiplied* 6 by, to get 24?
Copy and complete:

$6 \div \frac{1}{4} = 6 \times \square = 24$

   b What number should you *divide* 24 by, to find how many $\frac{3}{4}$ circles there are in 24 quarter-circles?
Copy and complete:

$6 \div \frac{3}{4} = \frac{6 \times 4}{\square} = 8$

4 Look at your answers to Question **2**.

$$9 \div \frac{1}{4} = 36 \qquad 9 \div \frac{3}{4} = 12$$

a There are 36 quarter-circles in 9 circles. What number could you have *multiplied* 9 by, to give 36?
Copy and complete:

$$9 \div \frac{1}{4} = 9 \times \square = 36$$

b To find how many $\frac{3}{4}$ circles are in 36 quarter-circles, what number should you *divide* 36 by?
Copy and complete:

$$9 \div \frac{3}{4} = \frac{9 \times 4}{\square} = 12$$

5 Look at some of the answers to Questions **3** and **4**:

$$6 \div \frac{1}{4} = 6 \times 4 = 24 \qquad 6 \div \frac{3}{4} = \frac{6 \times 4}{3} = 8$$

$$9 \div \frac{1}{4} = 9 \times 4 = 36 \qquad 9 \div \frac{3}{4} = \frac{9 \times 4}{3} = 12$$

Can you see the pattern?
Can you see a way to divide a whole number by a fraction? Explain.

- To divide a whole number by a fraction you turn the fraction upside down and multiply.

EXAMPLE 7

Work out:

a $2 \div \frac{3}{4}$ **b** $6 \div 1\frac{1}{2}$ **c** $\frac{4}{9} \div \frac{2}{5}$

....

a $2 \div \frac{3}{4} = 2 \times \frac{4}{3}$

Division by $\frac{3}{4}$ = multiplication by $\frac{4}{3}$

$$= \frac{8}{3}$$

$$= 2\frac{2}{3}$$

b $6 \div 1\frac{1}{2} = 6 \div \frac{3}{2}$

Change $1\frac{1}{2}$ to an improper fraction
$\div \frac{3}{2} = \times \frac{2}{3}$
Cancel!

$$= 6 \times \frac{2}{3}$$

$$= \frac{{}^{2}6}{1} \times \frac{2}{3_{1}}$$

$$= 4$$

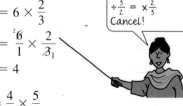

c $\frac{4}{9} \div \frac{2}{5} = \frac{4}{9} \times \frac{5}{2}$

$$= \frac{{}^{2}4}{9} \times \frac{5}{2_{1}} = \frac{10}{9} = 1\frac{1}{9}$$

Exercise 7H

1 Copy and complete:

a $4 \div \frac{2}{3} = 4 \times \frac{3}{2} = \frac{12}{2} = \square$

b $10 \div \frac{2}{5} = 10 \times \frac{5}{2} = \frac{\square}{\square} = \square$

c $10 \div \frac{4}{5} = 10 \times \frac{5}{4} = \frac{\square}{\square} = \square$

d $\frac{5}{9} \div \frac{3}{4} = \frac{5}{9} \times \frac{4}{3} = \frac{\square}{\square}$

e $8 \div 1\frac{3}{4} = 8 \div \frac{7}{4} = 8 \times \frac{4}{7} = \frac{\square}{\square} = \square$

2 Do the division and write the answer in its simplest form.

a $5 \div \frac{1}{9}$ **b** $4 \div \frac{3}{4}$ **c** $6 \div \frac{2}{3}$

d $8 \div \frac{5}{6}$ **e** $9 \div \frac{3}{4}$ **f** $7 \div \frac{7}{9}$

g $^-3 \div \frac{2}{5}$ **h** $^-2 \div \frac{5}{18}$ **i** $^-10 \div \frac{5}{7}$

j $\frac{7}{10} \div \frac{3}{4}$ **k** $\frac{4}{5} \div \frac{8}{15}$ **l** $\frac{3}{10} \div \frac{7}{9}$

m $\frac{5}{8} \div \frac{3}{4}$

3 Change the mixed number to an improper fraction, then do the division.

a $6 \div 1\frac{1}{4}$ **b** $4 \div 1\frac{1}{2}$ **c** $9 \div 1\frac{5}{6}$

d $7 \div 2\frac{3}{8}$ **e** $10 \div 3\frac{4}{5}$ **f** $8 \div 2\frac{5}{11}$

g $^-6 \div 2\frac{3}{5}$ **h** $^-4 \div 1\frac{1}{7}$ **i** $^-10 \div 4\frac{2}{3}$

4 How many half-litre bottles of juice can you get from a 10-litre container?

5 I have 12 oranges.
How many people can I give $\frac{2}{3}$ of an orange to?

6 Andrew uses $\frac{2}{5}$ of a bag of fertiliser each week. How long will one bag of fertiliser last?

7 Which is better value:

$\frac{3}{4}$ kg of soap powder for \$9,

or $\frac{2}{3}$ kg of soap powder for \$8?

7.4 Using known facts and laws of arithmetic to simplify calculations

- The **commutative law**: when adding two or more numbers or multiplying two or more numbers, the order of adding or multiplying doesn't matter.

  You know that $3 + 2 + 1 = 2 + 3 + 1$ and that $4 \times 5 \times 3 = 3 \times 4 \times 5$. The same applies to fractions.

EXAMPLE 8

Calculate:

a $\dfrac{3}{5} + 2\dfrac{1}{4} + \dfrac{2}{5}$ **b** $\dfrac{2}{5} \times \dfrac{3}{8} \times 10$

....

a $\dfrac{3}{5} + 2\dfrac{1}{4} + \dfrac{2}{5} = \dfrac{3}{5} + \dfrac{2}{5} + 2\dfrac{1}{4}$

$= 1 + 2\dfrac{1}{4} = 3\dfrac{1}{4}$

> By changing the order you add these, you avoid having to find a common denominator.

b $\dfrac{2}{5} \times \dfrac{3}{8} \times 10 = \dfrac{2}{5} \times 10 \times \dfrac{3}{8} = \dfrac{20}{5} \times \dfrac{3}{8}$

$= 4 \times \dfrac{3}{8} = \dfrac{12}{8} = \dfrac{3}{2} = 1\dfrac{1}{2}$

> By changing the order you multiply these, you avoid having to multiply a fraction by a fraction.

- The **distributive law** tells us that when a sum is being multiplied by a number, each number in the sum can be multiplied by the number first, then the products added.

You know that $5 \times (4 + 2) = 5 \times 6 = 30$ and $5 \times 4 \times 5 \times 2 = 20 + 10 = 30$.
The same applies to fractions.

EXAMPLE 9

Calculate $5 \times 12\dfrac{4}{23}$

....

$5 \times 12\dfrac{4}{23} = 5 \times \left(12 + \dfrac{4}{23}\right) = 5 \times 12 + 5 \times \dfrac{4}{23}$

$= 60 + \dfrac{20}{23} = 60\dfrac{20}{23}$

> By using the distributive law, there is no need to change the fraction into an improper fraction, avoiding a long multiplication sum.

Exercise 7I

1 Change the order so that you do not need to find a common denominator, then work out:

   a $\dfrac{1}{8} + 5\dfrac{1}{3} + \dfrac{7}{8}$ **b** $\dfrac{2}{9} + 3\dfrac{1}{5} + \dfrac{7}{9}$

2 Change the order so that you do not need to multiply a fraction by a fraction, then work out:

   a $\dfrac{3}{5} \times \dfrac{3}{4} \times 15$ **b** $\dfrac{5}{8} \times 18 \times \dfrac{7}{9}$

3 As in Example 9, use the distributive law so that you do not need to do a long multiplication sum, then work out:

   a $4 \times 15\dfrac{4}{21}$ **b** $25\dfrac{3}{19} \times 6$

4 Ahmad and Tarek were going for a walk to their friend's house. They walked $2\dfrac{3}{8}$ km then stopped for a rest. Then they walked $4\dfrac{1}{3}$ km and stopped for lunch. Finally they walked $2\dfrac{5}{8}$ km until they reached their friend's house. How far had they walked altogether?

5 This cuboid has side lengths $\dfrac{5}{8}$ cm, $\dfrac{2}{5}$ cm and 24 cm. What is its volume?
(**Remember:** volume = length $\times$ width $\times$ height)

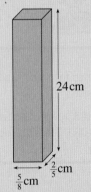

24 cm

$\dfrac{2}{5}$ cm

$\dfrac{5}{8}$ cm

6 Water flows through a hosepipe at a rate of $2250\dfrac{7}{25}$ litres every hour. Sue is using this pipe to fill a water tank which holds 6760 litres. If she leaves the water running for 3 hours, will the water tank be full?

- Inverse operations: we can use inverse operations to help us solve problems.

Multiplying and dividing are inverses of each other.

You know that if $4 \times 3 = 12$ then $12 \div 4 = 3$ and $12 \div 3 = 4$. The same applies to fractions.

Adding and subtracting are inverses of each other.

You know that if $2 + 5 = 7$ then $7 - 5 = 2$ and $7 - 2 = 5$. The same applies to fractions.

EXAMPLE 10

Find the missing numbers:

a $\square + 3\frac{2}{5} = 5\frac{7}{15}$

b $\square \div 4\frac{1}{5} = 3$

....

a $\square = 5\frac{7}{15} - 3\frac{2}{5} = 5\frac{7}{15} - 3\frac{6}{15} = 2\frac{1}{15}$

The inverse of $+ 3\frac{2}{5}$ is $- 3\frac{2}{5}$

b $\square = 3 \times 4\frac{1}{5} = 3 \times \frac{21}{5} = \frac{63}{5} = 12\frac{3}{5}$

The inverse of $\div 4\frac{1}{5}$ is $\times 4\frac{1}{5}$

• We can use known facts and place value to make calculations easier.

You know that if $3 \times 8 = 24$ then $30 \times 800 = 24\,000$. The same applies to fractions.

EXAMPLE 11

a If $240 \times \frac{3}{4} = 180$, what is

  i $240 \times \frac{3}{8}$ **ii** $24 \times \frac{3}{8}$?

b If $16\,000 \times \frac{7}{8} = 14\,000$, what is $16\,000 \times \frac{7}{800}$?

a **i** $\frac{3}{8}$ is half of $\frac{3}{4}$

so $240 \times \frac{3}{8}$ is half of $180 = 90$.

  ii $240 \times \frac{3}{8} = 90$

24 is 10 times smaller than 240,

so $24 \times \frac{3}{8} = 9$.

b $\frac{7}{800}$ is one hundredth of $\frac{7}{8}$

so $16\,000 \times \frac{7}{800} = 14\,000 \div 100 = 140$

• Multiplying fractions can be made easier by cancelling first.

If, when multiplying two numbers, we multiply one of the numbers by something and divide the other number by the same thing, the answer to the new calculation will be the same as the original. For example:

$$3 \times 60 = 180$$
$$\times 5 \downarrow \qquad \downarrow \div 5$$
$$15 \times 12 = 180$$

We use this idea for cancelling fractions.

EXAMPLE 12

Calculate $\frac{23}{25} \times 35$

....

$\frac{23}{25} \times 35^{7} = \frac{23 \times 7}{5}$

$= \frac{161}{5} = 32\frac{1}{5}$

23×7 is easier than 23×35

Exercise 7J

1 Find the missing numbers:

  a $3\frac{4}{5} + \square = 6\frac{2}{15}$ **b** $\square - 1\frac{7}{8} = 2\frac{3}{5}$

  c $\square \div 1\frac{3}{8} = 3$ **d** $\square \times 2\frac{3}{4} = {}^-3$

2 If $16 \times \frac{5}{8} = 10$, what is

 a $1600 \times \frac{5}{8}$ **b** $1600 \times \frac{5}{80}$

 c $16000 \times \frac{5}{800}$?

Explain how you worked these out without doing the calculation.

3 Cancel, then work out:

 a $\frac{11}{25} \times 45$ **b** $16 \times \frac{13}{24}$

 c $\frac{17}{18} \times 27$

4 What number added to $4\frac{1}{3}$ makes $7\frac{1}{2}$?

5 Kafele makes a mistake in his homework.
The question is: $1\frac{4}{35} \times 25$
Kafele writes:

$$1\frac{4}{\cancel{35}_{7}} \times \cancel{25}^{5} = 1\frac{4}{7} \times 5 = \frac{55}{7} = 7\frac{6}{7}$$

What mistake has he made?

6 If $2200 \times 1\frac{3}{5} = 3520$, what is

 a $220 \times 1\frac{3}{5}$ **b** $2200 \times \frac{4}{5}$?

Explain how you worked these out without doing the calculation.

7 Janine says that if $400 \times 1\frac{3}{4} = 700$, then $400 \times 1\frac{3}{8} = 350$. What mistake has she made?

8 $\frac{2}{3}$ of a number is 30. What is $\frac{2}{9}$ of the same number?

7.5 Applying order of operations rules to fractions

Previously you learned about the order of operations and how to apply BIDMAS:

Brackets first
Then Indices
Then Division and Multiplication
Then Addition and Subtraction

BIDMAS also applies to questions with fractions in them.

EXAMPLE 13

Work out:

 a $7 \times \left(2\frac{1}{5} - 1\frac{3}{4}\right)$ **b** $\frac{3}{4} + 2 \times 1\frac{2}{3} + \frac{5}{12}$

....

 a $7 \times \left(2\frac{1}{5} - 1\frac{3}{4}\right) = 7 \times \left(\frac{11}{5} - \frac{7}{4}\right)$

$$= 7 \times \left(\frac{44}{20} - \frac{35}{20}\right)$$

$$= 7 \times \frac{9}{20}$$

$$= \frac{63}{20}$$

$$= 3\frac{3}{20}$$

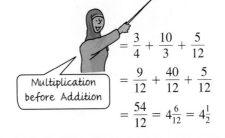 Brackets first

 b $\frac{3}{4} + 2 \times 1\frac{2}{3} + \frac{5}{12} = \frac{3}{4} + 2 \times \frac{5}{3} + \frac{5}{12}$

$$= \frac{3}{4} + \frac{10}{3} + \frac{5}{12}$$

$$= \frac{9}{12} + \frac{40}{12} + \frac{5}{12}$$

$$= \frac{54}{12} = 4\frac{6}{12} = 4\frac{1}{2}$$

Multiplication before Addition

Exercise 7K

1 Work out:

 a $2 \times \left(2\frac{3}{4} + 1\frac{1}{3}\right)$

 b $\left(1\frac{1}{8} - \frac{3}{4}\right) \times 4$

 c $2 \times \left(1\frac{7}{8} + 2\frac{2}{5}\right) + 5$

2 Work out:

 a $8\frac{7}{8} - 3 \times 2\frac{1}{6}$ **b** $5 \div \left(2\frac{3}{5} - 1\frac{1}{3}\right)$

 c $16 + \frac{17}{35} \times 25$

3 Work out:

 a $1\frac{7}{8} \times 5 + 3 \times 2\frac{1}{4} - 6\frac{1}{2}$

 b $4\frac{2}{3} + 3 \times 1\frac{3}{10} + \frac{2}{5}$

 c $\left(\frac{3}{5} - \frac{1}{20}\right) \times \left(\frac{2}{3} + \frac{4}{5} + \frac{8}{15}\right)$

d $\dfrac{3}{4} + \dfrac{1}{5} \times 4 - 3 \div 1\dfrac{1}{4}$

e $6\dfrac{9}{10} - 2 \times \dfrac{3}{8} + 1\dfrac{3}{5}$

7.6 Problem solving

Exercise 7L

1 In a mixed school, $\dfrac{3}{5}$ of the students are boys. What fraction are girls?

2 A bag of sugar has a mass of $\dfrac{3}{8}$ kg. A second bag has a mass of $\dfrac{9}{16}$ kg. What mass of sugar is there altogether?

3 A mouse eats $\dfrac{2}{15}$ kg of cheese each night.

How much cheese does it eat in:
a 30 nights **b** 16 nights?

4 A man walks $3\dfrac{1}{2}$ km each hour. How far does he walk in:
a 4 hours **b** 7 hours

5 Three oranges are divided equally between five people.

What fraction of an orange does each person get?

6 A length of string is 41 cm long. It is cut into 6 equal pieces. How long is each piece?

7 A barrel of juice holds 50 litres. How many bottles, each holding $\dfrac{2}{3}$ litre, can be filled from the barrel?

8 A plank of wood is $2\dfrac{1}{2}$ cm thick. How thick is a pile of 25 planks?

9 Mr Clow has no children so he decides to leave all his money to his three nephews, Ben, Bob and Billy. Ben gets $\dfrac{3}{5}$ of the money. Bob gets $\dfrac{1}{4}$ of the money. What fraction does Billy get?

10 The table shows how Form 3 students in Radley High School come to school every day.

Means of transport	Number of students
Bus	28
Car	12
Bicycle	15
Walking	45

a What fraction of the total number of students:
i come by bus
ii come by car
iii come by bicycle
iv walk to school?
Write each fraction in its simplest form.

b What number do you get when you add the four fractions?

11 The bar chart shows how Mr Damon spends his monthly salary.

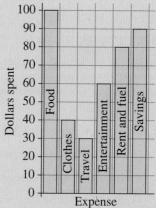

a How much does Mr Damon earn a month?
b Write down the fraction of his salary he uses for:
i food **ii** clothes
iii travel **iv** entertainment
v rent and fuel **vi** savings.
c When you add the six fractions, do you get 1 as your answer?

The money that Mr Damon spends on food is used in the following way:

$\dfrac{2}{5}$ on meat $\qquad \dfrac{1}{10}$ on fruit

$\dfrac{1}{4}$ on drinks $\qquad \dfrac{1}{10}$ on vegetables

d What fraction is used for other foods?
e What fraction of his total salary is spent on:
i meat **ii** drinks
iii fruit **iv** vegetables?

Consolidation

Example 1

Work out:

a $\dfrac{3}{5} + \dfrac{1}{3}$ **b** $2\dfrac{3}{4} - 1\dfrac{2}{3}$

a $\dfrac{3}{5} + \dfrac{1}{3} = \dfrac{9}{15} + \dfrac{5}{15}$

$= \dfrac{14}{15}$

b $2\dfrac{3}{4} - 1\dfrac{2}{3} = \dfrac{11}{4} - \dfrac{5}{3}$

$= \dfrac{33}{12} - \dfrac{20}{12}$

$= \dfrac{13}{12} = 1\dfrac{1}{12}$

Example 2

Calculate: $6 \times 3\dfrac{2}{5}$

$6 \times 3\dfrac{2}{5} = \dfrac{6}{1} \times \dfrac{17}{5}$

$= \dfrac{102}{5}$

$= 20\dfrac{2}{5}$

Example 3

Calculate: $^-5 \div 3\dfrac{1}{4}$

$\div \dfrac{13}{4}$ is the same as $\times \dfrac{4}{13}$

$^-5 \div 3\dfrac{1}{4} = {}^-5 \div \dfrac{13}{4}$

$= {}^-5 \times \dfrac{4}{13}$

$= \dfrac{^-20}{13}$

$= {}^-1\dfrac{7}{13}$

Example 4

Using the distributive law, calculate $6 \times 15\dfrac{5}{24}$

$6 \times 15\dfrac{5}{24} = 6 \times 15 + 6 \times \dfrac{5}{24}$

$= 90 + \dfrac{30}{24}$

$= 90 + 1\dfrac{6}{24}$

$= 91\dfrac{1}{4}$

Example 5

Find the missing number: $\square \times 3\dfrac{1}{4} = 8$

$8 \div 3\dfrac{1}{4} = 8 \div \dfrac{13}{4}$

$= 8 \times \dfrac{4}{13}$

$= \dfrac{32}{12}$

$= 2\dfrac{6}{13}$

The inverse of $\times 3\dfrac{1}{4}$ is $\div 3\dfrac{1}{4}$

Example 6

If $32 \times 1\dfrac{3}{4} = 56$, what is $32\,000 \times 1\dfrac{3}{4}$?

$32\,000 \times 1\dfrac{3}{4} = 56 \times 1000 = 56\,000$

Example 7

Calculate $\dfrac{11}{45} \times 55$

$\dfrac{11}{\cancel{45}_9} \times \cancel{55}^{11}$

$= \dfrac{11 \times 11}{9} = \dfrac{121}{9} = 13\dfrac{4}{9}$

Exercise 7

1 Work out:

a $\dfrac{1}{4} + \dfrac{1}{4}$ **b** $\dfrac{3}{5} + \dfrac{2}{5}$

c $2\dfrac{1}{4} + 1\dfrac{1}{4}$ **d** $3\dfrac{2}{5} + 1\dfrac{2}{5}$

e $4\dfrac{5}{9} + 2\dfrac{7}{9}$ **f** $\dfrac{1}{3} + \dfrac{1}{5}$

g $\dfrac{3}{4} + \dfrac{1}{3}$ **h** $\dfrac{3}{5} + \dfrac{3}{4}$

i $\dfrac{1}{6} + \dfrac{2}{3}$ **j** $\dfrac{3}{8} + \dfrac{5}{12}$

2 Calculate:

a $2\dfrac{3}{4} + 1\dfrac{2}{3}$ **b** $3\dfrac{1}{2} + 4\dfrac{3}{5}$

c $2\dfrac{1}{4} + 3\dfrac{4}{5}$ **d** $6\dfrac{1}{8} + 4\dfrac{2}{5}$

e $4\dfrac{5}{6} + 2\dfrac{1}{2} + 1\dfrac{3}{4}$

3 Work out:

a $\dfrac{3}{5} - \dfrac{1}{5}$ **b** $\dfrac{5}{6} - \dfrac{1}{6}$ **c** $\dfrac{7}{9} - \dfrac{5}{9}$

d $2\dfrac{4}{7} - 1\dfrac{3}{7}$ **e** $3\dfrac{2}{5} - 1\dfrac{4}{5}$ **f** $\dfrac{3}{4} - \dfrac{1}{3}$

g $\dfrac{5}{8} - \dfrac{1}{2}$ **h** $\dfrac{4}{7} - \dfrac{2}{9}$ **i** $\dfrac{2}{3} - \dfrac{3}{5}$

j $\dfrac{7}{12} - \dfrac{3}{8}$

4 Calculate:

a $3\dfrac{3}{4} - 1\dfrac{3}{5}$ **b** $2\dfrac{1}{2} - 1\dfrac{3}{4}$ **c** $3\dfrac{1}{3} - \dfrac{3}{4}$

d $2\dfrac{2}{5} - 1\dfrac{3}{4}$ **e** $4\dfrac{2}{7} - 1\dfrac{7}{8}$

5 Calculate:

a $\dfrac{3}{8} \times \dfrac{16}{27}$ **b** $12 \times \dfrac{2}{5}$ **c** $^-5 \times \dfrac{2}{3}$

d $4\dfrac{3}{7} \times {}^-3$ **e** $\dfrac{2}{9} \times \dfrac{3}{7}$ **f** $6 \times 2\dfrac{1}{5}$

g $^-2 \times 3\dfrac{7}{10}$ **h** $2\dfrac{6}{7} \times 34$

6 Calculate:

a $\dfrac{5}{8} \div \dfrac{2}{3}$ **b** $6 \div 3\dfrac{1}{5}$ **c** $^-4 \div \dfrac{3}{4}$

d $10 \div 2\dfrac{1}{2}$ **e** $^-2 \div 1\dfrac{5}{6}$ **f** $\dfrac{4}{5} \div \dfrac{3}{8}$

g $^-4 \div 2\dfrac{2}{9}$ **h** $3 \div 5\dfrac{1}{3}$

7 Using the distributive law, calculate:

a $2 \times 12\dfrac{5}{36}$ **b** $3 \times 25\dfrac{7}{24}$

c $8 \times 11\dfrac{3}{31}$

8 Find the missing numbers:

a $\square \times 5\dfrac{1}{2} = 10$ **b** $\square + 3\dfrac{1}{3} = 10$

c $\square \div 2\dfrac{3}{5} = 4$

9 If $470 \times \dfrac{3}{5} = 282$, what is

a $47000 \times \dfrac{3}{5}$ **b** $470 \times \dfrac{3}{10}$

c $4700 \times \dfrac{6}{5}$?

10 Calculate:

a $\dfrac{17}{54} \times 18$ **b** $12 \times \dfrac{19}{30}$

c $\dfrac{13}{56} \times 21$

11 The drawing shows the fuel gauge in Omar's car. When full, the fuel tank holds 60 litres.

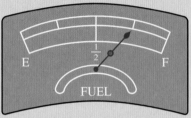

a How much fuel is in the tank now?

b After travelling a further 80 km, Omar noticed his tank was half full. How much fuel did he use?

12 There are 864 students at Greatfield High School. The number is expected to increase by $\dfrac{1}{12}$ next year.

a How many students are expected at Greatfield next year?

b One third of the new intake will be boys. How many new girls are expected?

Summary

You should know ...

1 To add or subtract fractions their denominators must be the same.
For example:

$\dfrac{3}{4} + \dfrac{2}{5}$

$= \dfrac{15}{20} + \dfrac{8}{20}$

$= \dfrac{23}{20}$

$= 1\dfrac{3}{20}$

> You can't add 3 fourths to 2 fifths

> but you can add 15 twentieths to 8 twentieths

Check out

1 a Work out:

i $\dfrac{1}{2} + \dfrac{1}{4}$ **ii** $\dfrac{2}{3} + \dfrac{1}{4}$

iii $3\dfrac{1}{2} + \dfrac{2}{3}$ **iv** $1\dfrac{4}{7} + \dfrac{2}{9}$

v $\dfrac{2}{3} - \dfrac{1}{2}$ **vi** $3\dfrac{3}{4} - 1\dfrac{7}{8}$

b Beverly drinks $\dfrac{2}{3}$ litre of milk from a $1\dfrac{1}{2}$ litre bottle. How much milk is left in the bottle?

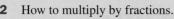

2 How to multiply by fractions.
For example:

a $6 \times \dfrac{3}{4} = \dfrac{6 \times 3}{4} = \dfrac{18}{4} = 4\dfrac{2}{4} = 4\dfrac{1}{2}$

b $^-3 \times 3\dfrac{3}{8} = {}^-3 \times \dfrac{27}{8} = \dfrac{^-81}{8} = {}^-10\dfrac{1}{8}$

c $\dfrac{7}{8} \times \dfrac{2}{5} = \dfrac{7}{_48} \times \dfrac{\overset{1}{2}}{5} = \dfrac{7}{20}$

---

3 To divide an integer by a fraction, you turn the fraction over and then multiply.

For example:

a $3 \div 2\dfrac{1}{8} = 3 \div \dfrac{17}{8} = 3 \times \dfrac{8}{17} = \dfrac{24}{17} = 1\dfrac{7}{17}$

b $\dfrac{3}{5} \div \dfrac{4}{15} = \dfrac{3}{5} \times \dfrac{15}{4} = \dfrac{3}{\cancel{5}_1} \times \dfrac{\cancel{15}^3}{4} = \dfrac{9}{4} = 2\dfrac{1}{4}$

---

4 How to use known facts and laws of arithmetic to simplify calculations.
For example:

a $5 \times 12\dfrac{3}{17} = 5 \times 12\dfrac{3}{17} = 5 \times 12 + 5 \times \dfrac{3}{17}$

$\qquad = 60 + \dfrac{15}{17} = 60\dfrac{15}{17}$

b $34 \times \dfrac{19}{51} = {}^2\cancel{34} \times \dfrac{19}{\cancel{51}_3} = 2 \times \dfrac{19}{3} = \dfrac{38}{3} = 12\dfrac{2}{3}$

2 a Work out:

   i $3 \times \dfrac{1}{2}$ **ii** $\dfrac{4}{5} \times \dfrac{2}{3}$

   iii $\dfrac{3}{4} \times 8$ **iv** $4 \times 2\dfrac{3}{5}$

   v $3\dfrac{1}{6} \times {}^-5$ **vi** $4 \times 5\dfrac{2}{3}$

b Kathy ate $\dfrac{3}{4}$ of a box of cornflakes. The box holds 152 g. What mass of cornflakes remains?

3 a Work out:

   i $\dfrac{7}{10} \div \dfrac{3}{8}$ **ii** $^-7 \div \dfrac{2}{3}$

   iii $5 \div 1\dfrac{3}{5}$ **iv** $4 \div 3\dfrac{2}{7}$

b How many $1\dfrac{3}{8}$ m pieces of material can be cut from a piece of material 11 m long?

4 a Work out:

   i $4 \times 15\dfrac{3}{19}$ **ii** $14 \times \dfrac{17}{49}$

b Find the missing number:
   $\square \div 3\dfrac{4}{5} = 7$

8 Expressions, equations and formulae

Objectives

- Construct and solve linear equations with integer coefficients (unknown on either or both sides, with or without brackets).

- Substitute positive and negative integers into formulae, linear expressions and expressions involving small powers, e.g. $3x^2 + 4$ or $2x^3$, including examples that lead to an equation to solve.

- Derive and use simple formulae, e.g. to convert degrees Celsius (°C) to degrees Fahrenheit (°F).

What's the point?

Different parts of the world use different currencies and different measuring scales. It is sometimes necessary to convert between currencies, for example when travelling or in business. Some parts of the world measure temperature in degrees Celsius, while some measure it in degrees Fahrenheit (both scales are named after the person who invented them). You have also learned about metric and imperial measurements. It is important to know which measuring scale is being used and how to convert between them − algebra and formulae help with this.

⊕ CANADA	CAD	0.9512	0.8883
CHINA	CNY	7.3169	6.0910
EURO	EUR	0.6644	0.6100
JAPAN	JPY	109.00	102.00
SINGAPORE	SGD	1.3712	1.2630
HONG KONG	HKD	7.0043	6.4072
NEW ZEALAND	NZD	1.1646	1.0675
MYR	3.2536	2.7818	

Before you start

You should know ...

1. About the order of operations, BIDMAS.

$(10 \times 6) - 5 \times 3^2$ Brackets first
$= 60 - 5 \times 3^2$ Then Indices
$= 60 - 5 \times 9$ Then Multiplication
$= 60 - 45$ Then Subtraction
$= 15$

Check in

1. Work out:
   a $7 \times 3^2 - (5 \times 3)$
   b $2 \times 10 - 12 \div (1 + 2)$
   c $(3^3 + 4) - 2^2 \times 5$
   d $\dfrac{4^2 + 5}{7}$

2 How to expand brackets and use the order of operations with algebra.
For example:

$$6 - 2(3x - 4) + 10x$$
$$= 6 - 6x + 8 + 10x$$
$$= 14 + 4x$$

3 How to multiply an integer by a fraction.
For example:

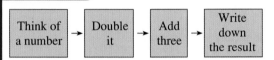

$$25 \times \frac{8}{5} = (25 \div 5) \times 8$$
$$= 5 \times 8 = 40$$

Find $\frac{1}{5}$ of 25 first, then $\times$ 8

2 a Expand:
  i $4(2p - 5)$
  ii $^-5(3 - 2x)$
 b Expand and simplify:
  i $4(3p - 5m)$
   $+ 5(2p + 7m)$
  ii $60 - 10(3x + 2)$
   $+ 80x$

3 Work out:

 a $64 \times \dfrac{5}{8}$

 b $55 \times \dfrac{4}{5}$

 c $60 \times \dfrac{5}{12}$

8.1 Solving linear equations

Using number machines

EXAMPLE 1

Think of a number	→	Double it	→	Add three	→	Write down the result

Using the given number machine:
a find the result if you start with 5
b find the result if the starting number is x
c find the starting number if the result is 15.

....

a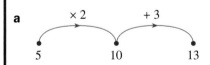

$\times 2 \qquad + 3$

5   10   13

The result is 13.

b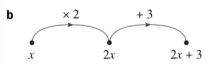

$\times 2 \qquad + 3$

$x \qquad 2x \qquad 2x + 3$

The result is $2x + 3$.

c To do this, use the machine backwards:
 -3 is the inverse of $+3$
 $\div 2$ is the inverse of $\times 2$
 So subtract 3 and then divide by 2:

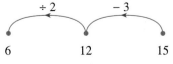

$\div 2 \qquad\qquad - 3$

6   12   15

The starting number was 6.

Exercise 8A

1 For each machine, write down the result when the starting number is

 i 5
 ii 12

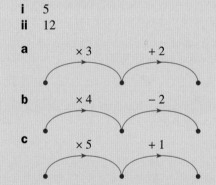

 a $\times 3$ $+ 2$

 b $\times 4$ $- 2$

 c $\times 5$ $+ 1$

2 Draw a diagram to show the reverse machine for each part of Question **1**.

3 For each machine in Question **1**, find what starting number will give 26 as the result. Use your machines from Question **2** to help you.

4 For each machine in Question **1**, write down the result when the starting number is x.

Simple equations can be solved using number machines.

EXAMPLE 2

Solve the equation $3x - 5 = 13$

First draw the machine which gives the result $3x - 5$ when you start with x.

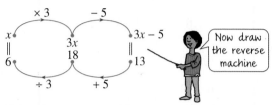

Now draw the reverse machine

Using the reverse machine you can see that, when $3x - 5 = 13$, $x = 6$.

Exercise 8B

1 Use the machines in Question **1** of Exercise 8A and the method in Example 2 to solve the equations:

   a $3x + 2 = 26$ **b** $4x - 2 = 26$
   c $5x + 1 = 26$

Do your results agree with your answers for Question **3** of Exercise 8A?

2 Draw a number machine which gives this result when you start with x:

   a $5x + 2$ **b** $7x - 4$ **c** $4x + 9$

3 Use the machines you drew in Question **2** to help you solve:

   a $5x + 2 = 27$
   b $7x - 4 = 24$
   c $4x + 9 = 33$

4 Solve:

   a $3x - 5 = 19$ **b** $4x + 1 = 21$
   c $7x - 8 = 25$ **d** $2x + 12 = 13$
   e $\frac{1}{2}x + 3 = 5$ **f** $6x - 9 = {}^-5$

5 Solve:

   a $3p - 5 = 4$
   b $26 = 8x - 6$
   c $7n + 8 = {}^-20$
   d $8 = 10 - y$
   e $9 = 7x + 16$
   f $2t + 0.5 = 12.5$
   g $r + r + 8 = 16$
   h $18 = 8m + 6 - 4m$
   i $12 = 18 - 3x$
   j $12 = 77 - 5r$

Balancing equations

Can you use a number machine to solve

   $2x = x + 5?$

You cannot because there are xs on both sides of the equation.

The idea of a balance can be used to solve equations like these.

Look at the balance. On one side it has 2 tins of paint. On the other side it has 1 tin of paint and a 5 kg weight.

Each tin of paint weighs x kg.

The masses on each side of the balance are equal.

If you remove one tin of paint from each side of the balance…

… you will see that $x = 5$.

Algebraically you can write:
   $2x = x + 5$

so, subtracting x from each side:
   $x = 5$

EXAMPLE 3

Use the balance idea to solve:

   a $w - 5 = 9$ **b** $4w = w + 3$

   c $\frac{x}{3} + 4 = 8$

Add 5 to both sides!

   a $w - 5 = 9$
      Add 5 to both sides:
      $w - 5 + 5 = 9 + 5$
           $w = 14$

   b $4w = w + 3$
      Subtract w from both sides:
      $4w - w = w + 3 - w$
         $3w = 3$

      Divide both sides by 3:
      $\dfrac{3w}{3} = \dfrac{3}{3}$
         $w = 1$

c $\dfrac{x}{3} + 4 = 8$

Subtract 4 from both sides:

$$\dfrac{x}{3} + 4 - 4 = 8 - 4$$

$$\dfrac{x}{3} = 4$$

Multiply each side by 3:

$$3\left(\dfrac{x}{3}\right) = 3(4)$$

$$x = 12$$

Exercise 8C

1 a Look at the diagram below.

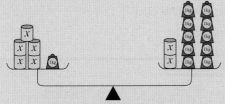

Do you agree that it shows the equation
$5x + 1 = 2x + 10$?

b Can you explain how the diagram below
was obtained from the diagram in part **a**?
Do the two sides still balance?

c Write an equation to show the information
in the diagram in part **b**.

d Draw a diagram like the one in part **b**, but
this time remove 1 kg from each side. Do
the two sides still balance?

e Work out the mass of one tin, x.

2 Solve the following equations:

a $7x + 3 = 3x + 11$
(**Hint:** Subtract $3x$ from both sides)

b $2x + 19 = 5x + 4$
(**Hint:** Subtract $2x$ from both sides)

c $5x + 1 = 8 - 2x$
(**Hint:** add $2x$ to both sides)

d $6x + 2 = 16 - x$
(**Hint:** add x to both sides)

3 Use the balance idea to solve for x:

a $9x + 3 = 4x + 8$

b $3x + 7 = 5x + 1$

c $2x + 3 = 28 - 3x$

d $4x - 7 = 13 - x$

e $6x - 7 = 2x + 3$

f $5x - 12 = 3 + 2x$

g $9x - 11 = 3x + 13$

h $8x - 12 = 5x - 3$

4 Use the balance idea to solve the following
equations. Write down what you are doing
to both sides of the equation, as in Example 3.

a $x + 3 = 5$ **b** $x - 3 = 7$

c $6 = 2x$ **d** $\dfrac{x}{3} = 4$

e $2x + 3 = 9$ **f** $2x - 3 = 9$

g $5p - 7 = 8$ **h** $\dfrac{x}{3} + 4 = 8$

i $7 + 6x = 7x$ **j** $4 - 2x = 2x$

k $\dfrac{x}{5} - 7 = 2$ **l** $7x = 20 - 3x$

m $3 - x = 2$ **n** $3x = x + 6$

5 Kevin tried to solve the equation
$5x - 4 = 3x + 6$
Here is his working:

	$5x - 4 = 3x + 6$
[− 4]	$5x = 3x + 2$
[− 3x]	$2x = 2$
[÷ 2]	$x = 1$

a What was Kevin's mistake?

b Solve the equation correctly for Kevin.

Sometimes you have to simplify equations by first
removing brackets.

EXAMPLE 4

Solve $5(x + 3) = 2x + 27$

Expand brackets:
$$5x + 15 = 2x + 27$$

Subtract $2x$ from both sides:
$$5x + 15 - 2x = 2x + 27 - 2x$$
$$3x + 15 = 27$$

Subtract 15 from both sides:
$$3x + 15 - 15 = 27 - 15$$
$$3x = 12$$

Divide both sides by 3:
$$x = 4$$

Sometimes you need to expand brackets and collect terms.

EXAMPLE 5

Solve $7(2p + 3) + 6(p - 2) = 10p + 2$

Expand brackets:
$$14p + 21 + 6p - 12 = 10p + 29$$

Simplify by collecting like terms:
$$20p + 9 = 10p + 29$$
$$[-10p] \quad 10p + 9 = 29$$
$$[-9] \quad\quad 10p = 20$$
$$[\div 10] \quad\quad p = 2$$

Exercise 8D

1 Solve:
  a $3(x + 3) = 21$
  b $10(t - 3) = 40$
  c $60 = 4(2x - 5)$
  d $42 = 3(3m - 4)$
  e $3(4 + x) = 24$
  f $2(x - 3) = 1$
  g $3(x + 4) = 2(x + 7)$
  h $5(x + 2) = 3(x + 10)$
  i $4(x - 5) = 2(x + 3)$
  j $7(p + 3) = 4(p + 6)$
  k $4(2x + 1) = 6(x + 3)$
  l $7d = 5(3 + d)$

2 Solve the following equations:
  a $4x + 2(x + 1) = 16$
  b $3(x + 1) + 2(x + 2) = 17$
  c $2(2x - 7) + x = 2x + 6$
  d $4 + 3(x - 5) = 10$
  e $x + (x + 1) + (x + 2) = 63$
  f $2(x + 7) - 6 = x + 15$

3 Solve the equations:
  a $\dfrac{x}{3} + 4 = 8$
  b $\dfrac{x}{5} - 6 = 1$
  c $\dfrac{x}{4} + 8 = 17$
  d $\dfrac{(x + 2)}{3} = 10$
  e $\dfrac{(x + 3)}{2} + 3 = 7$
  f $\dfrac{(x - 3)}{2} = 42$

Solving problems

Creating an equation and solving it is a powerful way of working on many types of problem.

There are four steps to solving a problem:

1 Understand the problem

2 Devise a plan

3 Carry out your plan

4 Look back.

EXAMPLE 6

10.5 cm

A rectangle with length 10.5 cm has perimeter 36 cm. What is its width?

Understand the problem
What is to be found?
The width, call it w.

Devise a plan
Form an equation:
$$w + 10.5 + w + 10.5 = 36$$

Carry out the plan
Solve the equation.
First, collect like terms:
$$w + w + 10.5 + 10.5 = 36$$

Simplify:
$$2w + 21 = 36$$

Subtract 21 from both sides:
$$2w = 15$$

Divide both sides by 2:
$$w = 7.5$$

Look back
The width of the rectangle is 7.5 cm.
This is correct, since:
$$10.5 + 7.5 + 10.5 + 7.5 = 36$$

Exercise 8E

For all of these questions, first write down an equation, then solve it.

1 The result when adding a number to 6 is 41. What is the number?

2 A certain number when multiplied by 2 and then added to 5 gives a result of 97. What is the number?

3 The sum of two consecutive whole numbers is 91. What are the numbers?

4 The perimeter of this isosceles triangle is 29 cm. What is the missing side length, t?

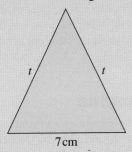

7 cm

5 Find the width of these rectangles.
   a length 5 cm, perimeter 24 cm
   b length 8.1 cm, perimeter 28.5 cm
   c length 17.3 m, perimeter 41.7 m

6 The perimeter of a triangle is 34 cm. What are the lengths of the sides if the first side is twice the length of the second side and the third side is 2 cm longer than the second side?

7 Janet is 6 years younger than Safiya. How old is each girl if the sum of their ages is 26?

8 Anton has three times as many marbles as Kamil. Kamil has 4 more marbles than Abdul. How many marbles does each boy have if there are 96 marbles altogether?

9 In 16 years' time Jim will be three times his current age. How old will Jim be in 4 years' time?

10 The sum of two consecutive even numbers is 214. What are the numbers?

11 The sum of three consecutive odd numbers is 243. What are the numbers?

TECHNOLOGY

Found this last section tough going?
Go over the 'Intro to Solving Equations' course at www.coolmath.com/prealgebra
Make sure you do the questions!

8.2 Substitution into expressions

You have already learned about expressions such as $3x + 4$, $2a + b$ and $7ty$. Notice there is no equals sign in an expression. When you know the values of the letters, you can find the value of the expression. You have already learned how to do this with positive numbers in Book 1. The next exercise extends that to using negative integers and expressions involving small powers.

Substituting negative values into expressions is just like substituting positive values.

EXAMPLE 7

If $x = {}^-3$, find the value of
a x^2 **b** $5x^2$
....
a $x^2 = ({}^-3)^2 = {}^-3 \times {}^-3 = 9$
b $5x^2 = 5 \times ({}^-3)^2 = 5 \times 9 = 45$

You can also work out more complex expressions.

EXAMPLE 8

If $x = {}^-2$, find the value of $2x^3 - x^2 + 10x + 6$
....
$$= 2({}^-2)^3 - ({}^-2)^2 + 10({}^-2) + 6$$
$$= 2({}^-8) - (4) + 10({}^-2) + 6$$
$$= {}^-16 - 4 + {}^-20 + 6$$
$$= {}^-40 + 6$$
$$= {}^-34$$

Exercise 8F

1 If $m = 2$, $n = {}^-3$, $p = {}^-5$ and $t = 10$, find the value of:
   a $m + n - t$ **b** $2t - m + 3p$
   c $n(m + t)$ **d** $2p + 10t - 3n$
   e $6t(p + n)$ **f** $3mp + 30$
   g $mp - tn$ **h** $5(m + n + p)$
   i ${}^-p - n - m - t$ **j** $4np - 6t$
   k $\dfrac{mnp}{t}$ **l** $\dfrac{7m + 2n}{t + p + 3}$

2 Find the value of each of the following. The first has been worked out for you.

  a $3x^2$ when $x = {}^-2$

  $3x^2 = 3 \times ({}^-2)^2$
  $ = 3 \times ({}^-2) \times ({}^-2)$
  $ = 12$

  b x^3 when $x = {}^-1$ **c** $4x^3$ when $x = 2$
  d $3x^2$ when $x = 5$ **e** $4t^2$ when $t = {}^-3$
  f $2p^3$ when $p = {}^-5$

3 Find the value of:

  a $x^2 + 4$ when $x = {}^-3$
  b $2x^2 + 5$ when $x = 2$
  c $3 + t^3$ when $t = {}^-1$
  d $4y^2 + y - 45$ when $y = {}^-5$
  e $24 + 3p^3$ when $p = {}^-2$
  f $100 - 2x^2$ when $x = 7$
  g $1 - m^3$ when $m = {}^-3$
  h $2w^2 + 3v^3 - r$ when $w = {}^-5$, $v = {}^-2$ and $r = {}^-8$

4 Find the value of each expression when
  i $x = 2$ **ii** $x = {}^-2$

  a $3x^2 - x$ **b** $2x^3 + x - 10$
  c $x^4 + 2x^2 + 7$ **d** $4x^3 + 3x^2 + 2x + 1$
  e $x^3 - 5x^2 - 4x - 15$

5 Find the value of each expression:

  a $x^2 - x$ when $x = {}^-3$
  b $x + x^2 - x^3$ when $x = {}^-4$
  c $2x^2 - x - 1$ when $x = {}^-5$
  d $x^3 + 4x^2 - 16x + 19$ when $x = {}^-1$
  e $x^3 - 3x$ when $x = {}^-3$

6 Copy these expressions.

  $\boxed{2x + y^2}$ $\boxed{4xy}$ $\boxed{x^2 + y}$ $\boxed{4y^2 - x^2 + x}$

  Tick (✓) which expressions have the same value when $x = 3$ and $y = {}^-2$.

7 a Find the value of each of the following:
  i $({}^-1)^3$ **ii** $({}^-1)^6$
  iii $({}^-1)^4$ **iv** $({}^-1)^7$

  b Can you guess the value of $({}^-1)^{100}$ and $({}^-1)^{101}$?
  c What is the value of $({}^-1)^n$ if n is an odd number?

8 Find the value of each expression when
  i $x = 2$ **ii** $x = {}^-2$

  a $5x^2$ **b** $x^4 + x^2$
  c $3x^4 - 2x^2$ **d** ${}^-5x^2 + x^4$

  What do you notice about the answers?

9 Find the value of each expression when:
  i $x = 2$ **ii** $x = {}^-2$

  a $7x^3$ **b** $x^3 - x$
  c $2x - 3x^3$ **d** $x^5 - x^3 + x$
  What do you notice about the answers?

10 Look at the expressions in Questions **8** and **9**.

  Can you see a pattern in the answers?
  Make up an expression in terms of x so that:
  a it has the same value for $x = 3$ and $x = {}^-3$
  b it has the same value but opposite sign for $x = 3$ and $x = {}^-3$.

8.3   Formulae

You have already learned about formulae such as $t = 3p + 4$, $C = 2a$ and $M = 3fd$. Notice there is an equals sign in a formula. A formula describes the relationship between different variables. When you know the values of some of the letters, you can find the value of the unknown variable by using substitution.

Substituting into formulae

EXAMPLE 9

In electricity, Ohm's law states that $V = IR$ where V = voltage, I = current and R = resistance.

Find V when $I = 0.5$ and $R = 6$

....

$V = 0.5 \times 6$
$V = 3$

When something moves in a straight line with constant acceleration, the following formulae apply:

$$v = u + at \qquad \text{and} \qquad s = ut + \tfrac{1}{2}at^2$$

where

a = acceleration u = initial velocity

t = time taken s = displacement (distance travelled from a point)

These are called equations of motion. They are used in mechanics and physics.

EXAMPLE 10

a Using $v = u + at$, find v when $u = 10$, $a = 2$ and $t = 15$

b Using $s = ut + \frac{1}{2}at^2$, find s when $u = 15$, $a = 3$ and $t = 20$

. . . .

a $v = u + at$
$v = 10 + 2 \times 15$ BIDMAS says Multiply before Adding
$v = 10 + 30$
$v = 40$

b $s = ut + \frac{1}{2}at^2$

$s = 15 \times 20 + \frac{1}{2} \times 3 \times 20^2$ BIDMAS says Indices first

$s = 15 \times 20 + \frac{1}{2} \times 3 \times 400$ Then Multiply

$s = 300 + 600$ Then Add

$s = 900$

Sometimes when you substitute into formulae you need to solve an equation to find the unknown letter because the unknown letter is not on its own.

EXAMPLE 11

a Using $v = u + at$, find u when $v = 50$, $a = 3$ and $t = 15$

b Using $v = u + at$, find t when $v = 40$, $u = 10$ and $a = 2$

. . . .

a $\qquad v = u + at$
$\qquad 50 = u + 3 \times 15$
$\qquad 50 = u + 45$

Subtract 45 from both sides:
$\qquad 50 - 45 = u + 45 - 45$
$\qquad\qquad 5 = u$

b $\qquad v = u + at$
$\qquad 40 = 10 + 2t$

Subtract 10 from both sides:
$\qquad 40 - 10 = 10 - 10 + 2t$
$\qquad\qquad 30 = 2t$

Divide both sides by 2:
$\qquad \dfrac{30}{2} = \dfrac{2t}{2}$
$\qquad\qquad 15 = t$

Note that all of the formulae used in the following exercise are used in real life.

Exercise 8G

1 Using $v = u + at$, find v when
  a $u = 20$, $a = 2$ and $t = 25$
  b $u = 50$, $a = {}^-3$ and $t = 10$

2 Using $V = IR$, find V when
  a $I = 2$ and $R = 5$
  b $I = 0.5$ and $R = 4$

3 Using $v = u + at$, find u when
  a $v = 100$, $a = 5$ and $t = 10$
  b $v = 20$, $a = {}^-5$ and $t = 9$

4 Using $V = IR$, find I when
  a $V = 8$ and $R = 4$
  b $V = 5$ and $R = 10$

5 Using $v = u + at$, find t when
  a $v = 60$, $a = 5$ and $u = 20$
  b $v = 40$, $a = {}^-2$ and $u = 68$

6 Using $F = ma$, find m when
  a $F = 24$ and $a = 8$
  b $F = 15$ and $a = 6$

7 Using $v = u + at$, find a when
  a $v = 70$, $t = 10$ and $u = 20$
  b $v = 30$, $t = 2$ and $u = 36$

8 Using $Ft = mv - mu$, find v when $F = 42$, $u = 10$, $t = 5$ and $m = 7$

9 Using $s = ut + \frac{1}{2}at^2$, find s when
  a $u = 20$, $t = 15$ and $a = 2$
  b $u = 0$, $t = 12$ and $a = 7$

10 Using $s = ut + \frac{1}{2}at^2$, find u when $s = 78$, $t = 3$ and $a = 4$

11 Using $v^2 - u^2 = 2as$, find
  a s when $v = 20$, $u = 8$ and $a = 4$
  b a when $v = 15$, $u = 5$ and $s = 200$

12 Hooke's Law states that $F = kx$ where F = force, x = displacement and k is the spring's constant (which tells you how powerful the spring is). Find k when $F = 20$ and $x = 0.1$

13 Using $v^2 - u^2 = 2as$, find v when $s = 21$, $u = 4$ and $a = 2$

14 Using $s = ut + \frac{1}{2}at^2$, find t when $s = 96$, $u = 0$ and $a = 3$

15 Using $v^2 - u^2 = 2as$, find u when $v = 35$, $a = 3$ and $s = 200$

Deriving formulae

In Chapter 4 you learned that some countries measure distances in miles and that one kilometre is about $\frac{5}{8}$ of a mile. You can derive a formula to convert between miles and kilometres.

To convert 8 km into miles do $8 \times \dfrac{5}{8} = 5$

To convert 24 km into miles do $24 \times \dfrac{5}{8} = 15$

To convert 56 km into miles do $56 \times \dfrac{5}{8} = 35$

To convert k km into miles do $k \times \dfrac{5}{8} = \dfrac{5k}{8}$

$\frac{5k}{8}$ is an expression for the number of miles in k km.

To turn it into a formula you need to have an equals sign. If m = the number of miles then the formula is $m = \dfrac{5k}{8}$

Some countries measure temperature using degrees Fahrenheit and some use degrees Celsius. To compare the two scales, consider the boiling point and freezing point of water:

	Celsius	Fahrenheit
Freezing temperature of water	0 °C	32 °F
Boiling temperature of water	100 °C	212 °F

You can see that when the temperature in Celsius increases by 100 °C the temperature in Fahrenheit increases by 180 °F.

So an increase of 1 °C will be an increase of 1.8 °F.

Since the freezing point of water in Fahrenheit is 32 °F we need to add 32 °F as well.

If F = the temperature in Fahrenheit and C = the temperature in Celsius, then

$F = 1.8C + 32$

You could write this using fractions to make calculations easier to do without a calculator:

$F = \dfrac{9}{5}C + 32$

Exercise 8H

1 Use the formula $m = \frac{5k}{8}$ to convert the following into miles:
  a 48 km   **b** 80 km   **c** 32 km

2 **a** Derive a formula to convert m miles into k kilometres.
  b Use your formula to convert the following into kilometres:
    i 50 miles   **ii** 75 miles

3 Leilah has x and Myesha has y. Derive an equation for each of these statements:
  a Leilah and Myesha have a total of $300.
  b Myesha has 3 times as many dollars as Leilah.
  c If Leilah gave Myesha $400 they would both have the same amount.

4 **a** Use the formula $F = \frac{9}{5}C + 32$, without a calculator, to convert the following into Fahrenheit:
    i 30 °C   **ii** 80 °C   **iii** 12 °C
  b Some people find this formula hard to remember and use. Did you find part **a iii** difficult without a calculator? Could you do it in your head or did you have to write down working?
    Instead of using the formula, people can find an approximation by doubling the degrees in Celsius then adding 32 to find the degrees in Fahrenheit. Derive the formula for this.
  c Without using a calculator, use your formula from part **b** to convert the following into Fahrenheit:
    i 30 °C   **ii** 80 °C   **iii** 12 °C
    Was part **iii** easier this time?

5 n cubes are stacked in a tower like the one below, then placed on a table.

Faces that touch each other or the table when the cubes are stacked are hidden from view and can't be seen.

a Derive a formula for the number of faces, f, that can be seen.

b Derive a formula for the number of faces, c, that can't be seen.

c Amira thinks that this formula shows the total number of faces: $f + c = 6n$
Do you think she is right? Use your answers to parts **a** and **b** to check.

6 José bought n apples and some oranges. He had 4 times as many oranges as he had apples.

a If José bought 3 apples how many oranges did he buy?

b Write an expression showing how many oranges José bought.

c If José bought 3 apples how many pieces of fruit did he buy altogether?

d Derive a formula showing t, the total number of pieces of fruit José bought.

e If José bought 30 pieces of fruit in total how many were oranges?

7 a Derive a formula for the perimeter, P, of this rectangle

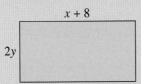
$x + 8$
$2y$

b Derive a formula for the area, A, of this rectangle

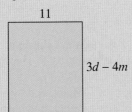
11
$3d - 4m$

c Use your answer to part **a** to find the value of y when $P = 76$ and $x = 20$

d Use your answer to part **b** to find the value of d when $A = 110$ and $m = 2$

8 I think of a number, n. I square it, multiply the answer by 4 and add 7. The answer is A.

a Derive the formula for A.

b Using your formula, find A when $n = {}^-3$

9 Cubes are stacked in a tower like the one below, with 4 cubes in each layer. The tower is placed on a table. It is n cubes high.

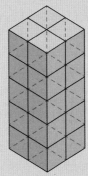

a Derive a formula for t, the total number of cubes used in the tower.

b Derive a formula for f, the number of faces that can be seen

c Use your formula from part **b** to find the total number of faces that can be seen when the tower is 12 cubes high.

d Use your formula from part **b** to find the height of the tower when the number of faces that can be seen is 76.

10 Derive a formula to convert degrees Fahrenheit into degrees Celsius.

⇒ INVESTIGATION

To convert from Celsius to Fahrenheit we use the formula $F = \frac{9}{5}C + 32$

This is not so easy to remember or use so many people use the approximation $F \approx 2C + 32$

Try a few conversions using both formulae. When is the approximation most accurate?

Consolidation

Example 1

Solve:

a $\dfrac{x}{3} - 4 = 7$ **b** $8(x - 5) = 2(3x + 4)$

....

a $\dfrac{x}{3} - 4 = 7$

   [+ 4] $\dfrac{x}{3} = 11$

   [× 3] $x = 33$

b $8(x - 5) = 2(3x + 4)$

   [Expand] $8x - 40 = 6x + 8$

   [− 6x] $2x - 40 = 8$

   [+ 40] $2x = 48$

   [÷ 2] $x = 24$

Example 2

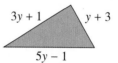

The perimeter of this triangle is 21 cm. Find the side lengths.

....

[Derive a formula] $P = 3y + 1 + y + 3 + 5y - 1$

[Simplify] $P = 9y + 3$

[Substitute 21 for P] $21 = 9y + 3$

[− 3] $18 = 9y$

[÷ 9] $2 = y$

Use substitution to find the three side lengths:

$3y + 1 = 3 \times 2 + 1 = 7$

$y + 3 = 2 + 3 = 5$

$5y - 1 = 5 \times 2 - 1 = 9$

Example 3

If $d = {}^-3$ find the value of $4d^2 + 7d$

....

    $4d^2 + 7d$

$= 4 \times ({}^-3)^2 + 7 \times {}^-3$ BIDMAS

$= 4 \times 9 + 7 \times {}^-3$ BIDMAS

$= 36 + {}^-21$ BIDMAS

$= 36 - 21$

$= 15$

Exercise 8

1   Solve:

   a $5x - 8 = 32$

   b $x + 8 = 4x - 4$

   c $7(x + 1) = 28$

   d $11y - 24 = 3y$

   e $5(2x - 4) = 4x + 10$

   f $\dfrac{d}{4} - 1 = 1$

   g $6(2x - 7) = 2(4x - 1)$

   h $3(2x + 7) + 1 = 5x + 2(7x - 2)$

2   The sum of three consecutive numbers is 54.

   a   Write an equation to show this.

   b   Solve the equation.

3   If $t = 3$, $v = {}^-2$ and $r = {}^-5$, find the value of:

   a $v + 2 - r$ **c** $v(r + t)$

   b $2r - v + 3t$ **d** $3rv + 5t$

4   Using $s = ut + \frac{1}{2}at^2$, find s when $u = 10$, $t = 5$ and $a = 4$.

5   If the area of this rectangle is 55, write an equation and solve it to find the value of x.

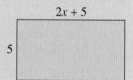

6   Find the value of

   a $x^2 + 3$ when $x = {}^-3$

   b $3x^2 + x + 1$ when $x = 2$

   c $3v + v^3$ when $v = {}^-2$

   d $2y^2 + 3y - 4$ when $y = {}^-4$

   e $125 + m^3$ when $m = {}^-5$

   f $100 - 5x^2$ when $x = {}^-4$

7   Using $v^2 - u^2 = 2as$ find s when $v = 17$, $u = 9$ and $a = 4$.

8   I think of a number, n. I square it, multiply the answer by 5 and subtract 3. The answer is t.

   a   Derive a formula for t.

   b   Using your formula, find t when $n = {}^-2$

   c   Using your formula, find n when $t = 42$

9   A rectangle is 6 cm longer than it is wide. It has a perimeter of 90 cm.

   a   Write down an equation to show this information.

   b   Solve the equation to find the length and width of the rectangle.

10   Three times Fitzroy's age 4 years ago will be the same as twice his age in two years' time. What is his current age?

11 These two rectangles have the same area.

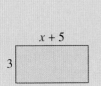

Find the area.

12 The formula for the nth triangle number is
$T = \frac{1}{2}n(n + 1)$

a What is
  i the 8th triangle number
  ii the 19th triangle number?

b Which triangle number is
  i 10   **ii** 78?

(**Hint:** you can use trial and improvement.)

Summary

You should know ...

1 How to construct and use formulae.
For example:

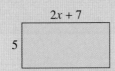

Write a formula for the perimeter, P, of the rectangle.
$P = 5 + 2x + 7 + 5 + 2x + 7$
$P = 4x + 24$
If the perimeter of the rectangle is 36, find its length.
$$36 = 4x + 24$$
$[- 24] \quad 12 = 4x$
$[\div 4] \quad\quad 3 = x$

The length of the rectangle is $2x + 7$, or
$2 \times 3 + 7 = 13$

2 How to solve equations.
For example:
$$3(4d - 6) = 2(5d + 1)$$
$[\text{Expand}] \quad\quad 12d - 18 = 10d + 2$
$[- 10d] \quad\quad 2d - 18 = 2$
$[+ 18] \quad\quad\quad 2d = 20$
$[\div 2] \quad\quad\quad\quad d = 10$

3 How to substitute integers into formulae.
For example:
$D = x^2 + 3p - 7$
Find p when $D = 12$ and $x = {}^-2$.
$$12 = ({}^-2)^2 + 3p - 7$$
$[\text{Simplify}] \quad 12 = 3p - 3$
$[+ 3] \quad\quad\quad 15 = 3p$
$[\div 3] \quad\quad\quad\; 5 = p$

Check out

1 a Write a formula for the area, A, of this rectangle.

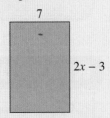

b If the area of this rectangle is 63, work out x.

2 Solve:
a $7x - 3 = 2x + 42$
b $5y + 8 = 8(y - 2)$
c $4(3m - 1) = 2(4m + 6)$
d $6(2x - 1) - 5x$
    $= 2(5x - 4) - 7$

3 $T = 2B^2 + 3n - 38$
Find n when $T = 1$ and
$B = {}^-3$

Geometry

Objectives

- Identify alternate angles and corresponding angles.
- Understand a proof for:
  – the angle sum of a triangle is 180° and that of a quadrilateral is 360°
  – the exterior angle of a triangle is equal to the sum of the two interior opposite angles.

- Solve geometric problems using properties of angles, of parallel and intersecting lines, and of triangles and special quadrilaterals, explaining reasoning with diagrams and text.
- Find the midpoint of a line segment AB given the coordinates of points A and B.

What's the point?

Cycling is a popular sport across the world. Top cyclists practise on special cycle tracks called velodromes. The velodromes are usually banked at an angle of 42° to prevent cyclists from falling off at the bends.

Before you start

You should know ...

1. How to use letters to name an angle.

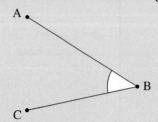

This is angle ABC.

Check in

1. Name the angles marked.

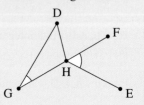

2 The symmetry properties of triangles and quadrilaterals and how to describe their angles and diagonals.

For example:

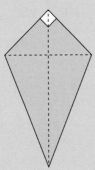

A kite is a quadrilateral with one line of symmetry and no rotational symmetry. This example has two equal obtuse angles, a right angle and an acute angle. One diagonal bisects the other.

2 Describe the symmetry properties, angles and diagonals of these shapes:
  a square
  b rectangle
  c isosceles triangle
  d parallelogram
  e rhombus
  f equilateral triangle
  g isosceles trapezium

9.1 All about angles

In Book 1 you learned that an angle is formed when two straight lines meet at a point.

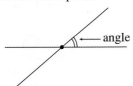

Angles are measured in degrees.

A complete turn is 360°.

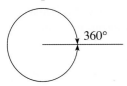

That is, angles at a point add up to 360°.

A half turn is 180°.

That is, angles on a straight line add up to 180°.

A quarter turn is 90°.

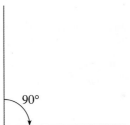

Vertically opposite angles are equal.

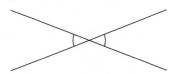

You can use these facts to calculate angles.

EXAMPLE 1

Find the angle x.

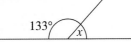

The angle on a straight line is 180°, so
$$133° + x = 180°$$
$$x = 180° - 133°$$
$$= 47°$$

129

Exercise 9A

1 Calculate the missing angles a, b, c and d.

a

b

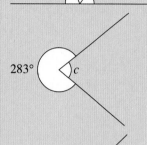

c

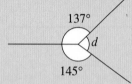

d

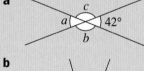

2 Find the missing angles a, b, c and d.

a

b

3 Work out the missing angles a, b, c, d and e.

a

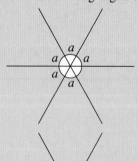

b

c

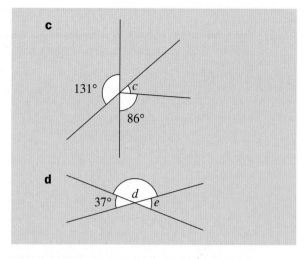

d

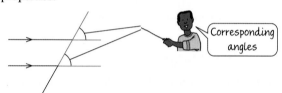

Parallel lines

When a line crosses two parallel lines you get other properties.

Corresponding angles are equal (look for an F shape).

Alternate angles are equal (look for a Z shape).

You can use the ideas of corresponding and alternate angles to solve more complex problems involving missing angles.

EXAMPLE 2

Find the missing angles a, b and c.

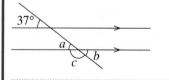

Write the reason for the answer.

$a = 37°$ (corresponding angles)
$a = b = 37°$ (vertically opposite angles)
$b + c = 180°$ (angles on a straight line)
So $37° + c = 180°$
hence $c = 180° - 37° = 143°$

Exercise 9B

1 Find the missing angles *a–i*. Give reasons for your answers.

a

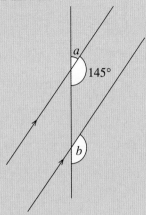

b

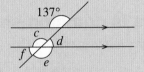

c

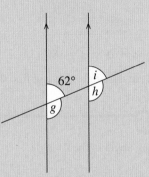

2

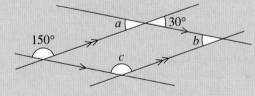

Copy and complete these sentences

a Angle *a* is ____° because

_____.

b Angle *b* is ____° because

_____.

c Angle *c* is ____° because

_____.

3 Calculate angles *a–l*. Give reasons for your answers.

a

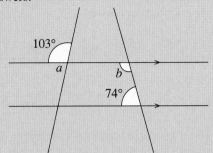

b

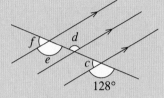

c

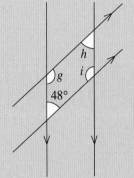

d

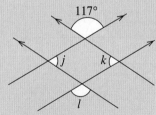

4 Find angles *a–f*.

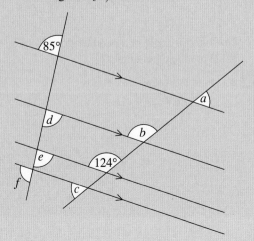

131

5 Without measuring any angles, say which diagrams below are drawn correctly. Give a reason for your answer.

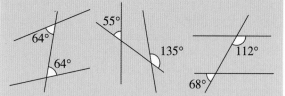

6 Find the marked angles *a–i* in these parallelograms.

a

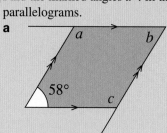

b

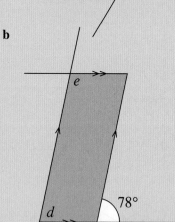

c

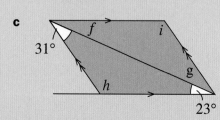

7 One interior angle of a parallelogram is 142°. What are the other angles?

8 a Find the values of *x* and *y*.

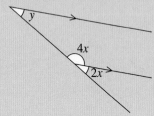

b Find the values of *m*, *n* and *p*.

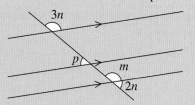

c Find the values of *t*, *u*, *v* and *w*.

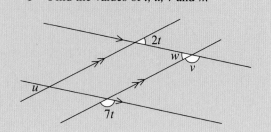

9.2 Angles in common shapes

The sum of the angles in a triangle is 180°.

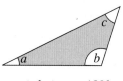

$$a + b + c = 180°$$

Given two angles in a triangle you can find the third.

EXAMPLE 3

Find the missing angle in the triangle.

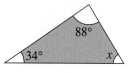

As the angle sum is 180°,
$$34° + 88° + x = 180°$$
$$122° + x = 180°$$
$$x = 180° - 122°$$
$$= 58°$$

You should understand the proof that the angle sum of a triangle is 180°.

Start with a triangle with angles labelled a, b and c:

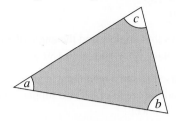

Draw a line touching the **apex** (top) of this triangle and parallel to the base of the triangle, marking the angles formed x and y, as shown below:

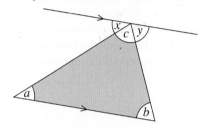

$x + y + c = 180°$ because angles on a straight line add up to 180°
$x = a$ because alternate angles are equal
$y = b$ because alternate angles are equal
so $x + y + c = a + b + c = 180°$

Therefore the angles of a triangle always add up to 180°.

Quadrilaterals are made up of two triangles.

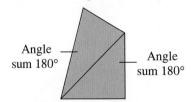

Angle sum 180°

Angle sum 180°

Hence the angle sum in a quadrilateral is
$180° + 180° = 360°$

You should understand the proof that the angle sum of a quadrilateral is 360°.

Start with a quadrilateral with angles labelled w, x, y and z:

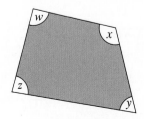

Split the quadrilateral into two triangles with angles labelled a, b, c, d, e and f.

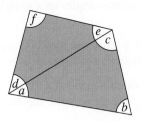

$a + b + c = 180°$ because the angles in a triangle add up to 180°
$d + e + f = 180°$ because the angles in a triangle add up to 180°
$a + b + c + d + e + f = 180° + 180° = 360°$

$a + d = z$, $e + c = x$, $f = w$ and $b = y$
$a + d + e + c + f + b = 360°$
so $z + x + w + y = 360°$

Therefore the angles of a quadrilateral always add up to 360°.

You can use this fact to calculate missing angles in a quadrilateral.

EXAMPLE 4

Find the missing angle in the quadrilateral.

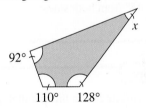

The angle sum is 360°.
Hence, $92° + 110° + 128° + x = 360°$
$330° + x = 360°$
$x = 30°$

An **exterior angle** of a triangle is found by extending one of the sides of the triangle. The exterior angle x is marked in the diagram below.

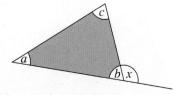

The exterior angle is related to the interior angles as shown in the proof opposite.

$a + b + c = 180°$ because the angles in a triangle add up to $180°$

$b + x = 180°$ because angles on a straight line add up to $180°$

$a + b + c = b + x$ because they both equal $180°$
subtracting b from both sides gives us $a + c = x$

Therefore the exterior angle of a triangle is equal to the sum of the two interior opposite angles.

EXAMPLE 5

Find the size of angle a.

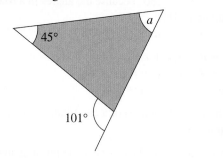

The exterior angle of a triangle is equal to the sum of the two interior opposite angles
so $101° = 45° + a$

Subtracting $45°$ from both sides:
$a = 56°$

EXAMPLE 6

Find the missing angles in this parallelogram.

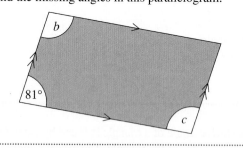

To find angle b, extend one side, as shown:

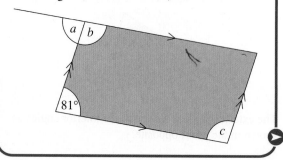

Angle $a = 81°$ (alternate angles)
$a + b = 180°$ (angles on a straight line)
$81° + b = 180°$
So $b = 99°$

This shows that angle b and the angle of $81°$ are supplementary (they add up to $180°$).

You can use the same technique to show that angle c and the angle of $81°$ are supplementary, so $c = 99°$ as well.

Don't forget to use common sense. In most questions to do with finding angles, diagrams will not be drawn to scale, so you cannot measure them. It is worth remembering that acute angles should be less than $90°$ and obtuse angles should be between $90°$ and $180°$ when you work them out: do a common sense check of your answer.

Exercise 9C

1 Find the missing angles in these shapes.

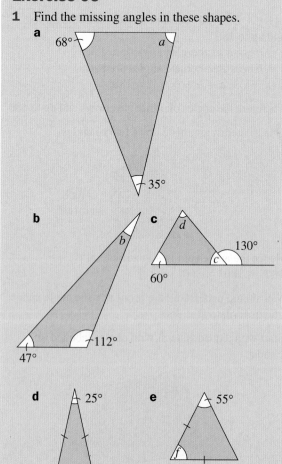

2 Calculate the missing angles.

a

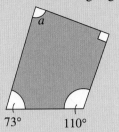

73° 110°

b

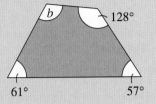

b 128°

61° 57°

c

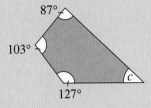

87°
103°
c
127°

3 Find the missing angles.

a

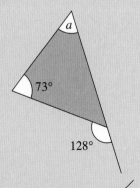

a

73°

128°

b

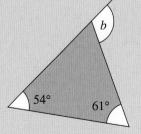

b

54° 61°

c

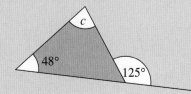

c

48° 125°

d

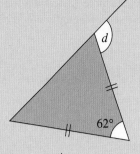

d

62°

e

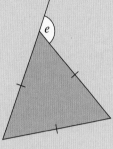

e

f

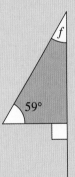

f

59°

4 Find the missing angles.

a

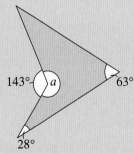

143° *a* 63°

28°

b

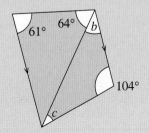

61° 64° *b*

104°

c

135

5 Sam measures the three angles of a triangle. Which of these sets of 3 angles could be for his triangle?

Set A: 64°   44°   62°

Set B: 57°   48°   74°

Set C: 25°   77°   78°

Set D: 97°   41°   52°

6 Work out angle *w*.

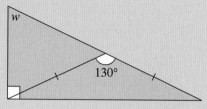

7 One of the angles in this quadrilateral has been labelled wrongly. How can you tell?

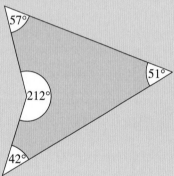

8 Find the missing angles in these parallelograms.

a

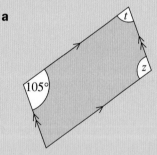

b

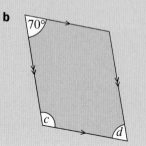

9 Here is a quadrilateral:

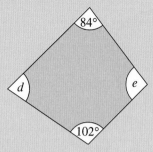

Which two of the following angles could be the values of *d* and *e*?

78°   136°   68°   116°   96°   48°

10 Find the missing angles in these diagrams.

a

b

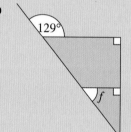

c

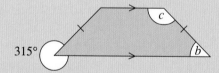

d

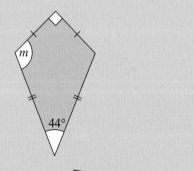

e

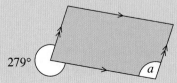

11 Find all of the angles inside each of these triangles and quadrilaterals.

a

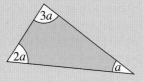

b

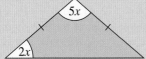

c

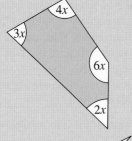

d

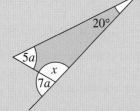

e

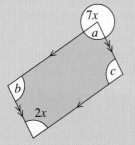

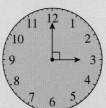
TECHNOLOGY

For further revision on angles, including videos, visit

www.onlinemathlearning.com/geometry-help.html

Click on the relevant sections under the heading 'Angles'. It has lots of information!

INVESTIGATION

Trace onto card the two *identical* right-angled triangles below. Cut them out.

Which of the following shapes can you make from them?

a A parallelogram   **b** A square
c A rectangle   **d** A rhombus

Repeat for two identical isosceles triangles of your own.

Repeat for two identical scalene triangles of your own.

9.3 Geometry problems using coordinate axes

Sometimes geometry problems will involve a coordinate grid. You should remember how to plot coordinates in all four quadrants. The first number tells you how far to move across, the second number tells you how far to move up or down.

EXAMPLE 7

Plot the points A (0, 5), B (8, 3), C (1, 1) and D (9, ⁻1). Join them to make a quadrilateral.

What sort of quadrilateral is it?

Measure angle ABD. Work out the sizes of the other three angles without measuring.

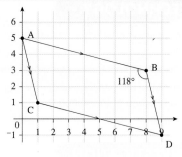

ABCD is a parallelogram.

Angle ABD is 118°. The other three angles are 118°, 62° and 62°.

Sometimes you need to find midpoints of line segments. If you look at the line AB in Example 7, you can see by counting squares that the midpoint, M, is at (4, 4).

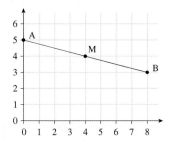

You don't have to draw the line segment to find the midpoint – there is another way.

The coordinates of the midpoint of a line segment AB, where A is the point (x_1, y_1) and B is the point (x_2, y_2), are

$$\left(\frac{x_1 + x_2}{2}, \frac{y_1 + y_2}{2} \right)$$

Can you see why this is the case?

EXAMPLE 8

Find the midpoint of the line segment XY where X is the point $(^-3, ^-2)$ and Y is the point $(4, ^-4)$.

....

Midpoint $= \left(\dfrac{^-3 + 4}{2}, \dfrac{^-2 + ^-4}{2} \right)$

$= \left(\dfrac{1}{2}, \dfrac{^-6}{2} \right)$

$= \left(\dfrac{1}{2}, ^-3 \right)$

Exercise 9D

1 Plot the points A (1, 2), B (3, 4), C (5, 2) and D (3, ⁻5). Join them to make a quadrilateral. What sort of quadrilateral is it? Measure angles ABC and ADC. Work out the other 2 angles without measuring.

2 **a** Find the midpoint of the line segment AB where A = (⁻2, 4) and B = (10, 2)

b Find the midpoint of the line segment CD where C = (⁻2, 0) and D = (10, 6)

c Draw the line segments AB and CD from parts **a** and **b** on the same axes. Mark the point of intersection as point M. What are the coordinates of M? What do you notice?

d What sort of quadrilateral is ADBC?

e Measure angle AMC.

f Without measuring, what are angles BMD, AMD and BMC?

3 Find the midpoints of the line segments with end points:

a A(4, 3), B(2, 7)   **b** C(1, ⁻2), D(⁻1, 6)
c E(2, ⁻5), F(⁻2, 3)   **d** G(0, ⁻6), H(⁻3, ⁻3)

4 X is the midpoint of the line segment AB. The coordinates of X are (2, 4) and of A are $(^-1, 3)$. Find the coordinates of B.

5 The midpoint of a line segment AB is (6, 2). Write down some possible coordinates for A and B.

6 The diagram below shows a rectangle ABCD. What are the coordinates of vertices A, B and C?

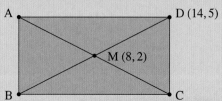

7 The diagram below shows a square ABCD. What are the coordinates of the vertices B, C and D?

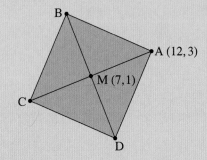

Consolidation

Example 1

Find the sizes of the missing angles a, b and c.

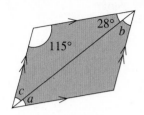

$a = 28°$ (alternate angles)
$28° + 115° + c = 180°$ (angles in a triangle add up to 180°)
So $\quad\quad\quad c = 37°$
$\quad\quad\quad b = c$ (alternate angles)
$\quad\quad\quad b = 37°$

Example 2

Find angles x and y in parallelogram WXYZ.

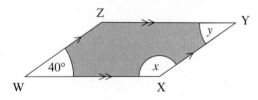

The angles W and X are supplementary.
$40° + x = 180°$
so $\quad x = 140°$

The angles X and Y are also supplementary.
$\quad\quad x + y = 180°$
or $\quad 140° + y = 180°$
so $\quad\quad y = 40°$

Example 3

Find angle a.

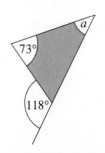

The exterior angle of a triangle is equal to the sum of the two interior opposite angles.
$118° = 73° + a$
so $a = 45°$

Example 4

Find the midpoint of the line segment AB if A = (7, 2) and B = (5, ¯4).

....

Midpoint is at $\left(\dfrac{x_1 + x_2}{2}, \dfrac{y_1 + y_2}{2}\right)$

$= \left(\dfrac{7 + 5}{2}, \dfrac{2 + {}^-4}{2}\right) = \left(\dfrac{12}{2}, \dfrac{{}^-2}{2}\right) = (6, {}^-1)$

Exercise 9

1 Calculate angles a–d.

a

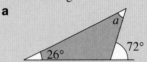

b

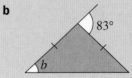

c

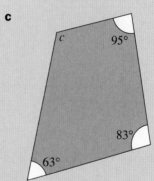

d

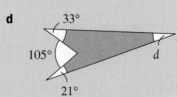

2 Without measuring, find the value of each angle, a–f.

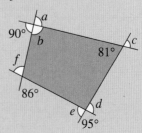

3

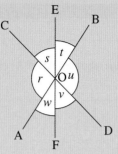

AB, CD and EF are straight lines that cross at O.
a If $s = 35°$ find v.
b If $r = 60°$ and $t = 65°$ find v.
c If $r + s = 120°$ and $v = 50°$ find u.

4

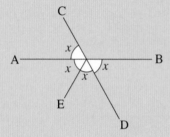

Find the size of x.

5

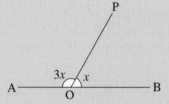

Find the value of x.

6 Calculate angles a–f. Give reasons for your answers.

a

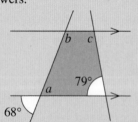

b

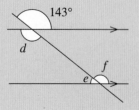

7 Calculate the sizes of angles a–e.

a

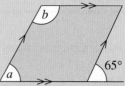

b

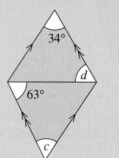

c

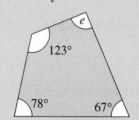

8 Copy and complete these proofs.

a

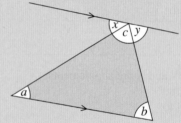

$x + y + c = 180°$ because _____

_____ because alternate angles are equal

$y = b$ because _____

so $x + y + c = a + b + c = $ _____°

Therefore the angles of a triangle always add up to _____°

b

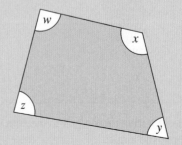

Split the quadrilateral into two triangles with angles labelled a, b, c, d, e and f:

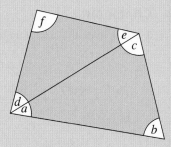

$a + b + c = 180°$ because _____

_____ = 180° because
the angles in a triangle add up to 180°

$a + b + c + d + e + f =$ ____° + ____°
= ____°

$a + d = z$, $e + c = x$, $f = w$ and $b = y$

$a + d + e + c + f + b =$ ____°

so _____ = 360°

Therefore the angles of a quadrilateral always add up to ____°

9 Write down the size of the angles marked by letters. Give reasons for your answers.

a

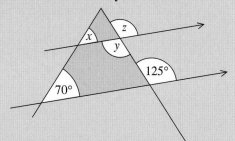

b

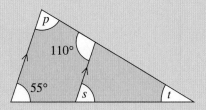

c

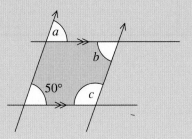

d

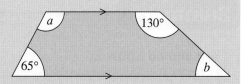

10 Find the missing angles marked by letters in these parallelograms.

a

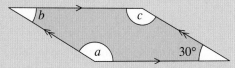

b

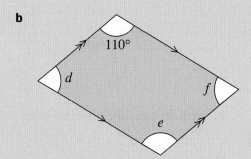

c

11 a Find the midpoint of the line segment CD
    if C = (3, 4) and D = (¯1, 8)

b Find the midpoint of the line segment EF
    if E = (6, 9) and F = (3, ¯5)

12 a Find the midpoint of the line segment AC
    where A = (¯4, 2) and C = (4, 0)

b Find the midpoint of the line segment BD
    where B = (1, 5) and D = (¯1, ¯3)

c Plot points A, B, C and D from parts **a**
    and **b** on the same axes. What sort of
    quadrilateral is ABCD?

d Draw in the diagonals of this quadrilateral.
    Where do they intersect? What do you
    notice?

Summary

You should know ...

1 a Angles on a straight line add up to 180°.

$a + b + c = 180°$

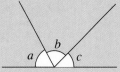

b Angles at a point add up to 360°.

$a + b + c + d = 360$

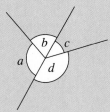

c Vertically opposite angles are equal.

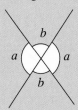

2 a Alternate and corresponding angles are equal.

a and *b* are alternate angles so $a = b$

b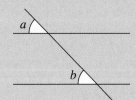

a and *b* are corresponding angles so $a = b$

Check out

1 Calculate the size of angles *a*–*c*.

a

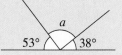

b

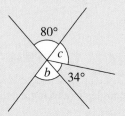

2 Find the size of angles *a*–*d*.

a

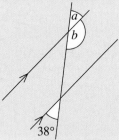

b

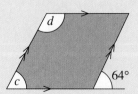

3 The exterior angle of a triangle is equal to the sum of the two interior opposite angles.

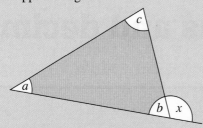

$a + c = x$

3 Complete this proof for the triangle shown on the left.

$a + b + c = 180°$ because

_____

_____ because angles on a straight line add up to 180°.

$a + b + c = b + x$ because they both equal 180°.

Subtracting b from both sides gives us _____

4 How to solve geometric problems using properties of angles.
For example:

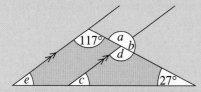

Angle $a = 117°$ (alternate angles)
Angle $b = 63°$ (angles on a straight line add up to 180°)
Angle $d = 117°$ (vertically opposite angles)
Angle $c = 36°$ (angles in a triangle add up to 180°)
Angle $e = 36°$ (corresponding angles)

4 Find the missing angles *a–e*, giving reasons for your answers.

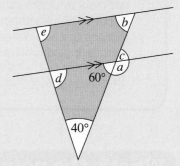

5 How to find the midpoint of a line segment.
For example:
F = (3, ⁻2) and G = (8, 10)

Midpoint is at $\left(\dfrac{x_1 + x_2}{2}, \dfrac{y_1 + y_2}{2} \right)$

$= \left(\dfrac{3 + 8}{2}, \dfrac{⁻2 + 10}{2} \right) = \left(\dfrac{11}{2}, \dfrac{8}{2} \right) = (5.5, 4)$

5 Find the midpoint of line segment XY if
  a X = (7, 7) and Y = (⁻1, 3)
  b X = (2, ⁻4) and Y = (7, 10)

10 Fractions and decimals

Objectives

- Consolidate adding and subtracting integers and decimals, including numbers with differing numbers of decimal places.

- Multiply and divide integers and decimals by decimals such as 0.6 or 0.06, understanding where to place the decimal point by considering equivalent calculations, e.g. $4.37 \times 0.3 = (4.37 \times 3) \div 10$, $92.4 \div 0.06 = (92.4 \times 100) \div 6$.

- Convert a fraction to a decimal using division; know that a recurring decimal is a fraction.

- Order fractions by writing with common denominators or by dividing and converting to decimals.

- Recall simple equivalent fractions and decimals.

What's the point?

Doctors and nurses are just two professions that work with decimals on a daily basis. Have you gained weight? How tall are you? Do you have a temperature?

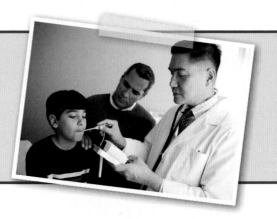

Before you start

You should know ...

1   A decimal is a way of writing a number using place values of tenths, hundredths, etc.

For example:

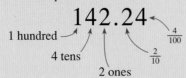

$$142.24$$

1 hundred
4 tens
2 ones
$\frac{4}{100}$
$\frac{2}{10}$

Check in

1   Write down the value of each digit in these decimals.

a	62	**b**	6.2
c	2.6	**d**	25.6
e	5.06	**f**	621.4
g	26.41	**h**	2.416
i	2.461		

2 A fraction with a denominator of 10 or 100 can be written as a decimal.

For example:

$$\frac{78}{100} = \boxed{\begin{array}{c|c|c} U & \frac{1}{10} & \frac{1}{100} \\ \hline 0 & 7 & 8 \end{array}} = 0.78$$

3 How to cancel fractions to their simplest form.

For example:

$$\frac{18}{42} = \frac{3}{7} \quad \begin{array}{c} \div 6 \\ \div 6 \end{array}$$

2 Write as decimals:

   a $\frac{3}{10}$ **b** $\frac{13}{100}$

   c $\frac{3}{100}$ **d** $1\frac{13}{100}$

   e $12\frac{2}{100}$

3 Cancel these fractions to their simplest form.

   a $\frac{42}{56}$ **b** $\frac{24}{32}$

   c $\frac{35}{49}$ **d** $\frac{70}{77}$

   e $\frac{26}{39}$

10.1 Ordering decimals

You can use a number line to help you order decimals.

EXAMPLE 1

Which is larger, 0.2 or 0.18?

On a number line:

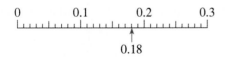

So 0.2 is larger than 0.18

Exercise 10A

1 Draw number lines to help you find the larger of each pair of numbers.

   a 0.3, 0.26   **b** 0.53, 0.7
   c 0.04, 0.3   **d** 1.4, 0.86
   e 2.3, 2.04   **f** 5.07, 4.89
   g 16.1, 6.98   **h** 3.25, 3.3

2 a Copy this number line:

    0   1.0   2.0

   b On your line mark these points:
     A = 0.3   B = 0.6   C = 1.1
     D = 1.7   E = 0.45   F = 1.65

3 Estimate the value of the points marked on this number line.

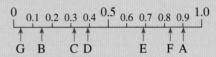

4 Write in order of size, smallest first:
   a 0.3, 0.21, 0.46, 0.18, 2.1
   b 4.8, 0.5, 0.46, 2.3, 0.41
   c 0.9, 0.81, 1, 1.01, 0.68
   d 2.4, 2.04, 4.2, 2, 2.13

5 Write down three numbers between 0.3 and 0.4.

6 This line is 4.2 cm long.

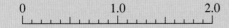

   What are the lengths of these lines, in centimetres?
   a _____
   b _____
   c _____

7 Which is longer, 4 cm or 4.0 cm? Explain.

8 What is the value of each of the points shown on this line?

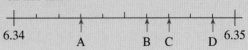

10.2 Adding and subtracting integers and decimals

To add and subtract decimals you must first line up the decimal points.

EXAMPLE 2

Work out $2.40 + 0.61 + 9$

....

It helps to write in the column headings:

Remember
$9 = 9.00$

T	U	.	$\frac{1}{10}$	$\frac{1}{100}$
	2	.	4	0
	0	.	6	1
+	9	.	0	0
1	2	.	0	1

↳ line up the decimal points

EXAMPLE 3

Work out $6 - 0.23$

....

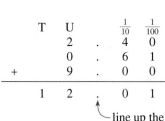

$6 = 6.00$

U	.	$\frac{1}{10}$	$\frac{1}{100}$
6	.	0	0
− 0	.	2	3
5	.	7	7

line up the decimal points

Exercise 10B

1 Work out:
  a $2.3 + 1.4$ **b** $2.7 + 2.83$
  c $4.2 + 5$ **d** $0.68 + 4.3$
  e $1.98 + 2.1$ **f** $1.3 + 5 + 0.67$
  g $16 + 2.5 + 0.9$ **h** $3 + 0.42 + 6.3$
  i $102 + 0.3 + 6.25$ **j** $0.103 + 4.1 + 38$

2 Use your calculator to check your answers to Question **1**.

3 Work out:
  a $2.4 - 1.3$ **b** $2.4 - 1.65$
  c $2 - 1.46$ **d** $5.3 - 2.24$
  e $6.8 - 3.92$ **f** $9 - 0.41$
  g $12 - 2.67$ **h** $103 - 7.963$
  i $31.3 - 0.04$ **j** $11 - 0.94$

4 Use your calculator to check your answers to Question **3**.

---

5 Teresa is 1.58 m tall. Her brother Tony is 2 m tall. How much taller is Tony than Teresa?

6 To install his new bath, Ashton requires two pieces of pipe.

  The first piece needs to be 3.45 m long and the second 4 m long.
  a What is the total length of pipe that Ashton requires?
  b If pipe is sold in 6 m lengths, how many lengths should he buy?
  c How much pipe would be left over from the job?

7 When James was well his body temperature was 36.8°C. When he was ill it went up to 38°C. How much did his body temperature go up by?

8 A builder estimated these costs for a job:
  Material: $498.75
  Transport: $38.65
  Profit: $350
  Labour: $835.50

  What was his total estimate for the job?

9 Nadia has $5. She buys a pen for $1.79 and two pencils costing $0.68 each. Will she have enough money left to buy a ruler for $1.95? Explain your answer.

10 Find the perimeter of the following two shapes.
  a

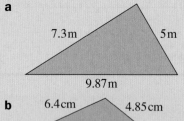

  b

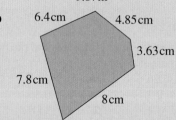

11 Yemi earns $1539 a month in his job. Maduka earns $1384.76 a month in his job.

  What is the difference in their pay?

12 The mass of my hand luggage is 5.25 kg. The mass of my suitcase is 18 kg.
  a What is the total mass of my hand luggage and suitcase?
  b What is the difference between the two masses?

13 Sami spent $39.68 on new shoes. She paid with a $50 note. How much change did she receive?

14 Copy and complete the tables below using the rules given. The first row in each table is done for you.

a **Rule:** Add 2.84

Input	Output
7	9.84
3.6	
4.78	
3.935	
⁻6	

b **Rule:** Subtract 2.67

Input	Output
7.98	5.31
4.89	
⁻12.6	
8.3	
7	

15 Expenses for the Hague's family vacation were $1283.50 for accommodation, $483 for food, $876.45 for flights and $285.72 for gifts and souvenirs.
   a How much did they spend in total?
   b How much change did they have from $3000?

16 Victoria cycled 78.4 km one day, 87.65 km the next day and 69 km the day after. How far did she cycle altogether?

17 In July 1999, Hicham El Guerrouj of Morocco set a world record by running the mile in 3 minutes 43.13 seconds. By how much did he beat the 4-minute mile barrier?

10.3 Multiplying and dividing decimals

Multiplying by 10, 100 and 1000

When you multiply a number by 10, 100 or 1000 you get a pattern:

$$36 \times 10 \quad = 360$$
$$36 \times 100 \quad = 3600$$
$$36 \times 1000 \quad = 36\,000$$
$$36 \times 10\,000 = 360\,000$$

The same is true when you multiply a decimal by 10, 100 or 1000:

$$3.64 \times 10 \quad = 36.4$$
$$3.64 \times 100 \quad = 364$$
$$3.64 \times 1000 \quad = 3640$$
$$3.64 \times 10\,000 = 36\,400$$

Dividing by 10, 100 and 1000

As with multiplication, there is a simple pattern when you divide a number by 10, 100 or 1000.

$$42 \div 10 \quad = 4.2$$
$$42 \div 100 \quad = 0.42$$
$$42 \div 1000 \quad = 0.042$$
$$42 \div 10\,000 = 0.0042$$

Exercise 10C

1 Work out:
   a 10×3.65 **b** 10×4.3
   c 10×9.1 **d** 10×2.06
   e 10×0.63 **f** 10×0.124

2 Work out:
   a 100×3.65 **b** 100×4.3
   c 100×9.1 **d** 100×2.06
   e 100×0.63 **f** 100×0.124

3 Copy and complete these divisions:
   a $74.5 \div 10 = \square$ **b** $89 \div 10 = \square$
   c $3.4 \div 10 = \square$ **d** $30.4 \div 10 = \square$
   e $0.6 \div 10 = \square$ **f** $10.06 \div 10 = \square$
   g $184 \div 10 = \square$ **h** $216.3 \div 10 = \square$

4 Copy and complete these divisions:
   a $745 \div 100 = \square$
   b $890 \div 100 = \square$
   c $34 \div 100 = \square$
   d $304 \div 100 = \square$
   e $6 \div 100 = \square$
   f $100.6 \div 100 = \square$
   g $1840 \div 100 = \square$
   h $2163 \div 100 = \square$

5 Work out:
   a 6.253×1000 **b** 0.125×1000
   c 6.48×1000 **d** 6.4×1000
   e 0.4×1000 **f** 0.003×1000

6 Copy and complete these sentences.
   a When a decimal is multiplied by 10, each number in the decimal moves ... place to the .... .

147

b When a decimal is multiplied by 100, each number in the decimal moves ... places to the ... .

c When a decimal is multiplied by 1000, each number in the decimal moves ... places to the ... .

7 Copy and complete this table.

Number	Number ÷ 10	Number ÷ 100	Number ÷ 1000
74.5	7.45	0.745	0.0745
89	8.9	0.89	
3.4			
30.4			
0.6			
485			
1024			
102.4			
10.24			
6			

Use your calculator to check your answers.

8 Copy and complete these sentences.

a When a decimal is divided by 10, each digit in the decimal moves ... place to the ... .

b When a decimal is divided by 100, each digit in the decimal moves ... places to the ... .

c When a decimal is divided by 1000, each digit in the decimal moves ... places to the ... .

9 Copy and complete:

a $3.8 \times \square = 380$

b $\square \times 1000 = 120$

c $\square \div 100 = 0.43$

d $7.9 \div \square = 0.079$

e $0.03 \times \square = 30$

f $0.2 \times \square = 200$

g $1.72 \div \square = 0.172$

h $3.9 \div \square = 0.039$

i $43.5 \times \square = 43\,500$

j $\square \times 100 = 135.7$

k $\square \div 1000 = 0.0543$

l $31.2 \div \square = 0.0312$

10 Another way to multiply a decimal by 10 or 100 is to change it back into a fraction. Look at these examples:

$$10 \times 2.68 = 10 \times \left(2 + \frac{6}{10} + \frac{8}{100}\right)$$

$$= 10 \times 2 + 10 \times \frac{6}{10} + 10 \times \frac{8}{100}$$

$$= 20 + 6 + \frac{8}{10}$$

$$= 26.8$$

$$100 \times 0.17 = 100 \times \left(\frac{1}{10} + \frac{7}{100}\right)$$

$$= 100 \times \frac{1}{10} + 100 \times \frac{7}{100}$$

$$= 10 + 7$$

$$= 17$$

Do these multiplications in the same way.

a $10 \times (20 + 4)$

b $10 \times \left(20 + 2 + \frac{3}{10}\right)$

c $10 \times \left(1 + \frac{3}{10} + \frac{5}{100}\right)$

d $10 \times \left(6 + \frac{7}{10} + \frac{4}{100} + \frac{9}{1000}\right)$

e $100 \times \left(3 + \frac{2}{10}\right)$

f $100 \times \left(5 + \frac{3}{10} + \frac{7}{100}\right)$

g $100 \times \left(\frac{6}{10} + \frac{9}{100} + \frac{8}{1000}\right)$

h 10×1.2

i 10×0.25

j 10×3.289

k 100×4.9

l 100×13.45

m 1000×7.163

Multiplying decimals by decimals

Multiplying decimals is like multiplying whole numbers. The only difference is you have to put the decimal point in the right place. You can do this from an estimate of the answer.

EXAMPLE 4

Work out 7.4×8

....

The estimate is $7 \times 8 = 56$

$$
\begin{array}{r}
7.4 \\
\times\ 8 \\
\hline
59_3.2
\end{array}
$$

The digits in the answer are 592

The estimate is 56

Place the decimal point to give 59.2

That is, $7.4 \times 8 = 59.2$

The method is the same even if both numbers are decimals.

EXAMPLE 5

Work out 2.3×0.6

....

The estimate is $2 \times 1 = 2$

$$
\begin{array}{r}
2.3 \\
\times\ 0.6 \\
\hline
1.3_18
\end{array}
$$

The digits are 138

The estimate is 2

Place the decimal point to give 1.38

That is, $2.3 \times 0.6 = 1.38$

Example 5 can be thought of like this:
Do the sum 2.3×0.6 as if there are no decimal places, as 23×6.

Notice that

$$
\begin{array}{ccc}
23 & \times\ 6 & =\ 138 \\
\downarrow & \downarrow & \downarrow \\
2.3 & \times\ 0.6 & =\ 1.38
\end{array}
$$

> 2.3 is 10 times smaller than 23
> 0.6 is 10 times smaller than 6
> so 1.38 is 100 times smaller than 138.

If there are two decimal places in the product, there must be two decimal places in the answer:

> The product and the answer both have two decimal places.

$$2.\underline{3} \times 0.\underline{6} = 1.\underline{38}$$

Exercise 10D

1 Copy and complete:

Calculation	Estimate	Answer
3.2×4	12	
6.9×9		
32.5×8		
1.68×3		
12.4×13		

2 Work out:
- **a** 0.4×8 **b** 0.9×6
- **c** 2.3×5 **d** 4.7×8
- **e** 6×1.4 **f** 12×0.6
- **g** 11×0.42 **h** 4×0.65
- **i** 3.24×8 **j** 4.85×12

3 Find the cost of 12 pencils each selling for $0.45.

4 If a packet of rice holds 1.4 kg, what is the mass of 6 packets?

5 Given that $45 \times 19 = 855$, find:
- **a** 45×1.9 **b** 4.5×19
- **c** 45×0.19 **d** 4.5×1.9
- **e** 0.45×19 **f** 0.45×1.9

6 Work out:
- **a** 3.4×0.5 **b** 7.12×0.3
- **c** 0.4^2 **d** 7.15×0.7
- **e** 8.2×0.02 **f** 4.8×0.06
- **g** 62.3×0.4 **h** 0.34×0.08
- **i** 0.7×4.52 **j** 0.03×6.4
- **k** 0.09×53.1 **l** 0.06^2
- **m** 0.23×1.4 **n** 0.38×2.6
- **o** 2.3^2 **p** $^-0.7 \times 3.7$
- **q** $^-2.61 \times\ ^-0.04$ **r** $(^-0.2)^2$

7 What is the total length of 20 pieces of pipe each 3.25 m long?

8 A tyre costs $153.94.

What would be the cost of 6 tyres?

9 Another way to multiply decimals is to consider equivalent calculations. Look at the example below.

$$
\begin{aligned}
4.23 \times 0.4 &= (4.23 \times 4) \div 10 \\
&= 16.92 \div 10 = 1.692
\end{aligned}
$$

Use this method to copy and complete:
- **a** $3.17 \times 0.3 = (\square \times \square) \div 10$
  $$= \square \div 10 = \square$$

b $21.8 \times 0.04 = (\square \times \square) \div 100$
$= \square \div 100 = \square$

c $9.87 \times 0.8 = (\square \times \square) \div \square$
$= \square \div \square = \square$

d $16.5 \times 0.03 = (\square \times \square) \div \square$
$= \square \div \square = \square$

e $458.3 \times 0.002 = (\square \times \square) \div \square$
$= \square \div \square = \square$

f $0.3 \times 17.4 = (\square \times \square) \div \square$
$= \square \div \square = \square$

g $0.6 \times 9.84 = (\square \times \square) \div \square$
$= \square \div \square = \square$

10 Another way to multiply decimals is to change them to fractions. Look at the example below.

$$6 \times 4.32 = 6 \times 4\frac{32}{100}$$

$$= 6 \times \frac{432}{100}$$

$$= \frac{2592}{100}$$

$$= 25\frac{92}{100}$$

$$= 25.92$$

Use this method to work out:

a 3×2.4 **b** 6.7×8
c 4×0.3 **d** 12×0.7
e 2.48×9 **f** 13×0.07
g 1.3×1.6 **h** 4.6×0.7
i 0.3×0.4 **j** 0.32×0.8

Dividing decimals by decimals

When you divide a decimal by a decimal, turn the divisor into a whole number.

EXAMPLE 6

Work out $3.66 \div 0.6$

Multiply by 10 to turn 0.6 into a whole number.

$$3.66 \div 0.6 = \frac{3.66}{0.6}$$

$$= \frac{3.66 \times 10}{0.6 \times 10}$$

$$= \frac{36.6}{6}$$

$$= 6.1$$

$6\overline{)36.6}$ → 6.1

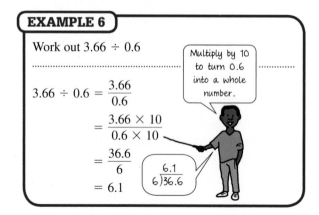

In Example 6 both the numerator and denominator were multiplied by the same number. You had to do this to keep the answer the same, using an equivalent fraction method. Sometimes you will need to multiply by 100.

EXAMPLE 7

Work out $0.028 \div 0.04$

Multiply by 100 to turn 0.04 into a whole number.

$$0.028 \div 0.04 = \frac{0.028}{0.04}$$

$$= \frac{0.028 \times 100}{0.04 \times 100}$$

$$= \frac{2.8}{4}$$

$$= 0.7$$

$4\overline{)2.8}$ → 0.7

Exercise 10E

1 Look at this division:

$$72 \div 0.9 = \frac{72}{0.9}$$

a Is it easier to divide by 0.9 or by 9?
b By what must you multiply 0.9 to give 9?
c So by what must you multiply the 72?
d Copy and complete:

$$\frac{72}{0.9} = \frac{72 \times 10}{0.9 \times 10} = \frac{\square}{\square} = \square$$

2 Look at this division:

$$6.182 \div 0.02 = \frac{6.182}{0.02}$$

a Is it easier to divide by 0.02 or by 2?
b By what must you multiply 0.02 to give 2?
c So by what must you multiply the 6.182?
d Copy and complete:

$$\frac{6.182}{0.02} = \frac{6.182 \,\square\, 100}{0.02 \,\square\, 100} = \frac{\square}{\square} = \square$$

3 Work out:

a **i** $6.4 \div 0.8$ **ii** $64 \div 8$
b **i** $3.5 \div 0.5$ **ii** $35 \div 5$
c **i** $4 \div 0.5$ **ii** $40 \div 5$
d **i** $4.8 \div 0.03$ **ii** $480 \div 3$
e **i** $3.64 \div 0.4$ **ii** $36.4 \div 4$

Compare your answers to parts **i** and **ii** in each case. What do you notice?

4 Work out:

a $5.6 \div 0.4$ **b** $6.4 \div 0.04$
c $57 \div 0.03$ **d** $1.25 \div 0.5$
e $69 \div 0.3$ **f** $34.65 \div 0.07$
g $2.5 \div 0.05$ **h** $14.4 \div 0.6$
i $1.38 \div 0.6$ **j** $1.34 \div 0.08$

5 Another way to divide decimals is to consider equivalent calculations:

$$2.36 ÷ 0.4 = (2.36 × 10) ÷ 4$$
$$= 23.6 ÷ 4 = 5.9$$

Use this method to copy and complete:

a $7.35 ÷ 0.3 = (□ × 10) ÷ □$
  $= 73.5 ÷ □ = □$

b $6.88 ÷ 0.04 = (□ × 100) ÷ □$
  $= □ ÷ □ = □$

c $10.08 ÷ 0.8 = (□ × □) ÷ □$
  $= □ ÷ □ = □$

d $30.1 ÷ 0.07 = (□ × □) ÷ □$
  $= □ ÷ □ = □$

e $1.284 ÷ 0.006 = (□ × □) ÷ □$
  $= □ ÷ □ = □$

6

a Measure the thickness of a coin, in millimetres.

b How many coins, placed on top of one another, would make a 10 cm tall pile?

c How many would make a 100 m tall pile?

7 There are 2.2 lb in 1 kg.

How many kilograms are there in packets of rice marked:

a 1 lb   **b** 2 lb   **c** 7.5 lb?

Give your answers to 2 d.p.

8

A centipede is timed as travelling 4.65 m in 7.5 s. What is its speed in metres per second?

9

1.3 kg of Ultra Tide soap powder costs $10.66. How much does:

a 1 kg of powder cost

b 0.75 kg of powder cost?

---

10 Work out:

a	$5.28 ÷ 2.2$	**b**	$19.52 ÷ 6.1$
c	$29.93 ÷ 4.1$	**d**	$0.441 ÷ 1.4$
e	$2.52 ÷ 0.126$	**f**	$0.336 ÷ 2.4$
g	$9.64 ÷ 0.16$	**h**	$0.042 ÷ 0.14$
i	$628 ÷ 3.14$	**j**	$1.75 ÷ 0.005$

TECHNOLOGY

That's quite a bit on decimals. You may need to review what you've done in your own time.

Visit

www.aaastudy.com/dec.htm

or

www.coolmath.com/prealgebra

for a complete series of lessons on all decimal operations.

10.4 Writing fractions as decimals

Any fraction can be written as a decimal.

For example,

$$\frac{1}{2} = 1 ÷ 2 = 1.0 ÷ 2 = 0.5$$

$$\frac{3}{4} = 3 ÷ 4 = 3.00 ÷ 4 = 0.75$$

To convert fractions to decimals you will need to divide.

EXAMPLE 8

Change these fractions to decimals.

a $\frac{1}{4}$ **b** $\frac{3}{8}$

....

a $\frac{1}{4} = 1 ÷ 4 = 1.00 ÷ 4$

$$\begin{array}{r} 0.25 \\ 4\overline{)1.0^20} \end{array}$$

So $\frac{1}{4} = 0.25$

b $\frac{3}{8} = 3 ÷ 8 = 3.000 ÷ 8$

$$\begin{array}{r} 0.375 \\ 8\overline{)3.0^60^40} \end{array}$$

So $\frac{3}{8} = 0.375$

Exercise 10F

1 Change these fractions to decimals.

a $\dfrac{2}{5}$ **b** $\dfrac{3}{5}$ **c** $\dfrac{4}{5}$ **d** $\dfrac{1}{4}$

e $\dfrac{3}{4}$ **f** $\dfrac{3}{8}$ **g** $\dfrac{5}{8}$ **h** $\dfrac{7}{8}$

i $\dfrac{1}{16}$ **j** $\dfrac{5}{16}$

2 Change these fractions to decimals.

a $\dfrac{1}{3}$ **b** $\dfrac{2}{3}$ **c** $\dfrac{1}{6}$ **d** $\dfrac{1}{7}$

What do you notice?
These decimals are called **non-terminating decimals** because they go on for ever.

3 The fraction $\frac{1}{3} = 0.333\ldots = 0.\dot{3}$
The digits in the decimal repeat themselves.
Decimals like this are called **recurring decimals**. The dot above the 3 shows it repeats.
Find four other recurring decimals, using your calculator to help you.

4 Look at this diagram. It shows ten identical strips of paper. A fraction of each is shaded.

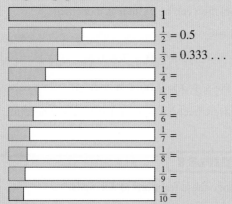

The fractions are written on the right-hand side.
The first two have been converted to decimals.

a Why are there dots after the fraction for $\frac{1}{3}$?
b Copy the list of fractions and convert the rest of them to decimals.
c Which of these decimals are
  i terminating **ii** recurring?

Use your calculator for these questions.

5 Convert the fraction to a decimal. If the decimal is non-terminating, find out whether it recurs:

a $\dfrac{5}{6}$ **b** $\dfrac{2}{7}$ **c** $\dfrac{4}{7}$ **d** $\dfrac{2}{9}$

e $\dfrac{5}{9}$ **f** $\dfrac{2}{11}$ **g** $\dfrac{1}{14}$ **h** $\dfrac{1}{12}$

6 In Question **6**, did you find that all the decimals were recurring? When you change a fraction to a decimal, does the decimal always either terminate or recur?

7 Change these fractions to decimals. Check each answer on your calculator.

a $\dfrac{1}{16}$ **b** $\dfrac{11}{16}$ **c** $\dfrac{3}{20}$ **d** $\dfrac{7}{20}$

e $\dfrac{11}{20}$ **f** $\dfrac{3}{200}$ **g** $\dfrac{7}{200}$ **h** $\dfrac{1}{250}$

8 a Change these fractions to decimals.

  i $3\frac{2}{5}$ **ii** $7\frac{3}{8}$ **iii** $2\frac{1}{3}$ **iv** $12\frac{3}{16}$

b Change these decimals to fractions, giving your answers as mixed numbers.
  i 2.8 **ii** 7.625
  iii $2.\dot{4}$ **iv** 15.4375

9 The decimal

$$0.333\ldots = \dfrac{1}{3}$$

Find fractions that represent these decimals.
a 0.111... **b** 0.555... **c** 0.1818...

All recurring decimals can be written as fractions.

Look at the examples below. Try to spot the pattern.

$$0.\dot{4} = \dfrac{4}{9} \qquad 0.\dot{6} = \dfrac{6}{9} = \dfrac{2}{3}$$

$$0.\dot{5}\dot{6} = \dfrac{56}{99} \qquad 0.\dot{1}\dot{2} = \dfrac{12}{99} = \dfrac{4}{33} \qquad 0.\dot{0}\dot{8} = \dfrac{8}{99}$$

$$0.\dot{1}7\dot{5} = \dfrac{175}{999} \qquad 0.\dot{0}6\dot{1} = \dfrac{61}{999}$$

You should know some simple equivalent fractions and decimals. Question **1** in the next exercise expects you to know some fractions and decimal equivalents without having to do any calculations.

However, at stage 8 you do not need to know *how* to convert recurring decimals to fractions. Questions **2** to **6** and the Challenge on the following page are included as extenion material for you to try if you have time.

Exercise 10G

1 Copy and complete this table of equivalent fractions and decimals. Cancel each fraction to its simplest form.

Decimal	Fraction
0.6	
	$\dfrac{3}{4}$
0.6̇ (or 0.666 . . .)	
	$\dfrac{7}{10}$
0.25	
0.03	
	$\dfrac{27}{100}$
	$\dfrac{2}{5}$
0.8	

Questions **2** to **6** are extension material.

2 Change these recurring decimals into fractions. Cancel them to their simplest form if necessary.

a $0.\dot{2}$ b $0.\dot{3}$
c $0.\dot{2}\dot{8}$ d $0.\dot{0}\dot{9}$
e $0.0\dot{7}$ f $0.3\dot{6}$
g $0.\dot{1}4\dot{3}$ h $0.00\dot{7}$
i $0.\dot{0}1\dot{2}$

3 Make up five recurring decimals of your own and change them to fractions.

4 Dhanesh wrote this working:

$$\frac{2}{9} = 0.222$$

$$\frac{7}{9} = 0.777$$

$$0.222 + 0.777 = 0.999$$

So $\frac{2}{9} + \frac{7}{9} = 0.999$

What mistake has he made?

5 Look at the pattern:

$$0.0\dot{7} = \frac{7}{90}$$

$$0.0\dot{6} = \frac{6}{90} = \frac{1}{15}$$

$$0.00\dot{6} = \frac{6}{900} = \frac{1}{150}$$

$$0.01\dot{3} = \frac{13}{990}$$

Notice that not all numbers after the decimal point repeat. For example, $0.01\dot{3}$ means 0.013131313 … and not 0.013013013013 … . Look at how this affects the denominator of the fraction.

Change these recurring decimals into fractions. Cancel them to their simplest form if necessary.

a $0.0\dot{2}$ b $0.0\dot{5}$
c $0.0\dot{8}$ d $0.02\dot{9}$
e $0.00\dot{7}$

6 Change these recurring decimals into fractions. Cancel them to their simplest form if necessary.

a $0.001\dot{8}$
b $0.0\dot{2}0\dot{3}$
c $0.000\dot{7}$

Challenge

To convert a decimal that begins with a non-repeating part, such as 0.48888 . . . (or 0.48̇) to a fraction, follow these steps.

Step 1
Write it as the sum of the non-repeating part and the repeating part.
0.4 + 0.088888 . . .

Step 2
Convert each of these decimals to fractions.
(See the pattern in Exercise 10G, Question **5** to help you.)

$$\frac{4}{10} + \frac{8}{90}$$

Step 3
Write these as fractions with a common denominator.

$$\frac{36}{90} + \frac{8}{90}$$

Step 4
Add the fractions.

$$\frac{36}{90} + \frac{8}{90} = \frac{44}{90}$$

Step 5
Cancel the result to its simplest form (if necessary).

$$\frac{44}{90} = \frac{22}{45}$$

Your challenge is to convert these recurring decimals to fractions.
1 0.71̇
2 0.37̇
3 0.245̇
4 0.016̇
5 0.482̇3̇

Ordering fractions

One of the reasons for converting fractions to decimals is so that you can write fractions more easily in order of size. Another way of ordering fractions is to write them with common denominators. You will practise both methods in the next exercise.

EXAMPLE 9

Which fraction is larger:

a $\frac{4}{5}$ or $\frac{7}{9}$ **b** $\frac{1}{3}$ or $\frac{2}{7}$?

....

a $\frac{4}{5} = \frac{8}{10} = 0.8$

$\frac{7}{9} = 7.000 \div 9 = 9\overline{)7.000...}$ 0.777...

Make both decimals the same number of decimal places, e.g. 2 decimal places, to compare them more easily. 0.80 is larger than 0.78, so the larger fraction is $\frac{4}{5}$.

b $\frac{1}{3} = \frac{7}{21}$

$\frac{2}{7} = \frac{6}{21}$

> Write both fractions with a common denominator.

So $\frac{1}{3}$ is larger.

Exercise 10H

1 By converting to fractions with a common denominator, which is larger

  a $\frac{1}{4}$ or $\frac{3}{10}$ **b** $\frac{2}{5}$ or $\frac{3}{8}$ **c** $\frac{7}{20}$ or $\frac{2}{5}$?

2 By converting to decimals, which is smaller

  a $\frac{4}{9}$ or $\frac{3}{7}$ **b** $\frac{4}{25}$ or $\frac{3}{20}$ **c** $\frac{7}{50}$ or $\frac{1}{8}$?

3 Choose five pairs of fractions of your own, with different denominators. By writing equivalent fractions with a common denominator for each pair, write which fraction is larger.

4 Write down which of these is bigger than $\frac{3}{5}$.

  0.61 $\frac{59}{100}$ $\frac{5}{8}$ $\frac{6}{10}$ 0.597 $\frac{13}{20}$

5 Are these statements true or false?

  a $\frac{4}{15} < \frac{1}{5}$ **b** $\frac{14}{30} = \frac{21}{45}$ **c** $\frac{5}{9} > \frac{3}{5}$

6 Choose five pairs of fractions of your own, with different denominators. Convert each fraction to a decimal by division, then write which fraction is larger for each pair.

7 Choose from <, = or > to complete these:

  a $\frac{4}{7} \square \frac{5}{9}$ **b** $\frac{4}{3} \square 1\frac{2}{5}$ **c** $1.8 \square \frac{9}{5}$

Consolidation

Example 1

Work out:

a $6.2 + 0.08 + 12$ b $17 - 0.13$

a
$$
\begin{array}{r}
\text{T U} \;\; \frac{1}{10} \; \frac{1}{100} \\
6\,.\,2 \\
0\,.\,0\;8 \\
+\,1\,2\,.\,0\;0 \\
\hline
1\,8\,.\,2\;8
\end{array}
$$

b
$$
\begin{array}{r}
\text{T U} \;\; \frac{1}{10} \; \frac{1}{100} \\
1\,7\,.\,0\;0 \\
-\quad\;\; 0\,.\,1\;3 \\
\hline
1\,6\,.\,8\;7
\end{array}
$$

Example 2

What is

a 0.63×10 b 17.4×100
c $0.63 \div 10$ d $17.4 \div 100$?

a $0.63 \times 10 = 6.3$ b $17.4 \times 100 = 1740$
c $0.63 \div 10 = 0.063$ d $17.4 \div 100 = 0.174$

Example 3

What is 3.2×0.06?

$$
\begin{array}{r}
32 \\
\times\,6 \\
\hline
192
\end{array}
$$

The digits are 192.

There are 3 decimal places in $3.\underline{2} \times 0.0\underline{6}$ so there are 3 decimal places in the answer.

So $3.2 \times 0.06 = 0.192$

Example 4

Work out:

a $6.55 \div 5$ b $17.37 \div 0.03$

a
$$
\begin{array}{r}
1.31 \\
5)\overline{6.55}
\end{array}
$$

b
$$
\frac{17.37}{0.03} = \frac{17.37 \times 100}{0.03 \times 100}
$$
$$
= \frac{1737}{3}
$$
$$
\begin{array}{r}
579 \\
3)\overline{1737}
\end{array}
$$

So $17.37 \div 0.03 = 579$

Example 5

Write as decimals:

a $\dfrac{3}{5}$ b $\dfrac{3}{7}$ to 2 d.p.

a $\dfrac{3}{5} = 3 \div 5$

$$
\begin{array}{r}
0.6 \\
5)\overline{3.0}
\end{array}
$$

So $\dfrac{3}{5} = 0.6$

b $\dfrac{3}{7}$

$\dfrac{3}{7} = 3 \div 7$

$$
\begin{array}{r}
0.428 \\
7)\overline{3.000}
\end{array}
$$

So $\dfrac{3}{7} = 0.43$ (2 d.p.)

> You need to work out 3 d.p. so you know how to round to 2 d.p.

Example 6

Write these numbers in order, starting with the smallest.

$$\frac{2}{5}, \; 0.42, \; \frac{3}{10}, \; 0.\dot{6}, \; \frac{1}{3}$$

There are two ways to do this.

• **Method 1**

Convert all to fractions then write with a common denominator.

As fractions:

$$0.42 = \frac{42}{100} = \frac{21}{50}$$

$$0.\dot{6} = \frac{6}{9} = \frac{2}{3}$$

Change $\frac{2}{5}, \frac{21}{50}, \frac{3}{10}, \frac{2}{3}, \frac{1}{3}$ so they all have a common denominator of 150 (the LCM of all the denominators):

$$\frac{60}{150}, \frac{63}{150}, \frac{45}{150}, \frac{100}{150}, \frac{50}{150}$$

Put these in order:

$$\frac{45}{150}, \frac{50}{150}, \frac{60}{150}, \frac{63}{150}, \frac{100}{150}$$

List in order the numbers as they were given in the question:

$$\frac{3}{10}, \frac{1}{3}, \frac{2}{5}, \; 0.42, \; 0.\dot{6}$$

• **Method 2**

Convert all to decimals then write with the same number of decimal places to help compare.

$$\frac{2}{5}, \; 0.42, \; \frac{3}{10}, \; 0.\dot{6}, \; \frac{1}{3}$$

As decimals:

$$\frac{2}{5} = 2 \div 5 = 0.4$$

$$\frac{3}{10} = 3 \div 10 = 0.3$$

$$\frac{1}{3} = 1 \div 3 = 0.\dot{3}$$

Change 0.4, 0.42, 0.3, 0.$\dot{6}$, 0.$\dot{3}$ so they are all decimals with 2 d.p.:
0.40, 0.42, 0.30, 0.67, 0.33

Put these in order:
0.30, 0.33, 0.40, 0.42, 0.67

List in order the numbers as they were given in the question:
$\frac{3}{10}, \frac{1}{3}, \frac{2}{5}$, 0.42, 0.$\dot{6}$

Exercise 10

1   Work out:
    a   **i**   0.7 + 0.5   **ii**   5 + 0.6
        iii   12.3 + 0.14   **iv**   6 + 0.02
        v   3.8 + 0.04 + 2
    b   **i**   4 − 1.2   **ii**   16 − 0.1
        iii   99 − 0.99   **iv**   4.35 − 3.8
        v   101.1 − 0.011

2   Calculate:
    a   **i**   6.1 × 10   **ii**   6.15 × 10
        iii   23.4 × 10   **iv**   0.13 × 100
        v   2.784 × 100
    b   **i**   4.6 × 0.3   **ii**   0.38 × 0.04
        iii   13.68 × 0.7   **iv**   41.6 × 0.06
        v   1.3 × 2.4

3   Calculate:
    a   **i**   6.15 ÷ 10   **ii**   0.615 ÷ 10
        iii   47 ÷ 10   **iv**   0.3 ÷ 100
        v   36.2 ÷ 1000
    b   **i**   7.5 ÷ 5   **ii**   0.75 ÷ 5
        iii   36.3 ÷ 3   **iv**   15.3 ÷ 9
        v   0.688 ÷ 4
    c   **i**   4 ÷ 0.2   **ii**   36 ÷ 0.09
        iii   4.78 ÷ 0.02   **iv**   57.6 ÷ 0.04
        v   7.11 ÷ 0.9

4   Write these recurring decimals as fractions, cancelling where necessary.
    a   0.$\dot{7}$ **b**   0.$\dot{6}$
    c   0.$\dot{1}\dot{2}$ **d**   0.4$\dot{5}$

5   Write as decimals:
    a $\frac{3}{4}$ **b** $\frac{2}{5}$ **c** $\frac{7}{8}$ **d** $\frac{3}{8}$
    e $\frac{5}{16}$ **f** $\frac{7}{25}$ **g** $\frac{2}{7}$ **h** $\frac{2}{9}$

    i $\frac{5}{6}$ **j** $\frac{6}{13}$ **k** $\frac{13}{25}$ **l** $1\frac{2}{3}$
    m $2\frac{1}{8}$ **n** $3\frac{4}{5}$

6   The cost of 9 kg of pistachio nuts is \$111.24.
    a   What is the cost of 10 kg?
    b   What is the cost of 10.6 kg?

7   Write these numbers in order of size, starting with the smallest, by
    a   converting to fractions
    b   converting to decimals.
       i   0.$\dot{2}$, $\frac{1}{8}$, $\frac{3}{10}$, 0.25, $\frac{1}{5}$
       ii $\frac{2}{3}$, $\frac{7}{9}$, 0.8, $\frac{5}{8}$, 0.75

8   Find the perimeter of each shape.
    a   **b**

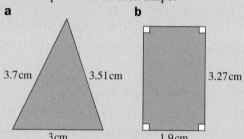

9   A pay packet of \$2980.11 is shared equally between 7 people. How much does each receive?

10 What is the area of this rectangle?

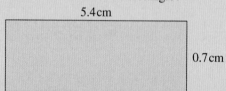

11 A scientist observes a snail move 1.2 cm in 8 seconds.
    a   How far does the snail move, on average, in one second?
    b   At this rate, how far could the snail go, in metres, in 100 seconds?

12 A square garden plot requires 63.4 m of fence to enclose it.
    a   What is the area of the plot?
    b   If the plot's area is increased by 10 m^2 but it still remains square, how much fencing is required to enclose it?

Give your answers to 1 d.p.

Summary

You should know ...

1 To add and subtract decimals you need to line up decimal points and write numbers with the same number of decimal places.

For example: $7 - 4.38$

U		$\frac{1}{10}$	$\frac{1}{100}$
7	.	0	0
− 4	.	3	8
2	.	6	2

2 You can multiply decimals by decimals.

For example: 2.8×0.3

$$\begin{array}{r} 28 \\ \times\ .3 \\ \hline 84 \end{array}$$

The digits in the answer are 84.

There are 2 decimal places in 2.8×0.3 so there are 2 decimal places in the answer.

So $2.8 \times 0.3 = 0.84$

3 To divide a decimal by a decimal you turn the divisor into a whole number.

For example: $3.8 \div 0.04 = \dfrac{3.8}{0.04} = \dfrac{3.8 \times 100}{0.04 \times 100} = \dfrac{380}{4} = 95$

4 You can change fractions to decimals.

For example:

$\dfrac{1}{3} = 1 \div 3 = 1.00 \div 3$

> This decimal goes on forever!

$\begin{array}{r} 0.33\ldots \\ 3\overline{)1.0^10\ldots} \end{array}$

So $\dfrac{1}{3} = 0.33\ldots$

$= 0.\dot{3}$

5 How to order fractions by writing them with common denominators or by dividing and converting them to decimals.

For example:

Write $\dfrac{3}{5}, \dfrac{2}{3}$ and $\dfrac{11}{20}$ in order of size, starting with the smallest.

$\dfrac{3}{5} = 3 \div 5 = 0.6 = 0.60$ to 2 d.p.

$\dfrac{2}{3} = 2 \div 3 = 0.\dot{6} = 0.67$ to 2 d.p.

$\dfrac{11}{20} = \dfrac{55}{100} = 55 \div 100 = 0.55$ to 2 d.p.

In size order they are: $\dfrac{11}{20}, \dfrac{3}{5}, \dfrac{2}{3}$

Check out

1 Work out:
  a $2.46 + 3.89 + 7$
  b $8 - 4.76$
  c $3.2 - 0.487$
  d $2.82 + 3.4 - 4.23$

2 Calculate:
  a 3.2×0.5
  b 0.4×7.4
  c 12.2×0.6
  d 8.4×0.03
  e 9.42×0.8
  f 1.36×0.42

3 Work out:
  a $8 \div 0.5$
  b $4 \div 0.2$
  c $3.6 \div 0.3$
  d $5.8 \div 0.02$
  e $0.4 \div 0.02$

4 Write as decimals.
  a $\dfrac{1}{4}$ b $\dfrac{2}{3}$
  c $\dfrac{3}{8}$ d $\dfrac{5}{6}$
  e $\dfrac{2}{9}$ f $\dfrac{3}{16}$

5 Write $\dfrac{7}{9}, \dfrac{17}{20}$ and $\dfrac{3}{4}$ in order of size, starting with the smallest.

11 Time and rates of change

Objective

- Draw and interpret graphs in real-life contexts involving more than one component, e.g. travel graphs for more than one person.

What's the point?

Drawing graphs helps you to identify patterns in numbers more easily. From graphs you can make simple predictions and use them to answer such questions as: how much rainfall is expected in September? Will sales increase in March? How much should my baby sister weigh next month?

Before you start

You should know ...

1 How to write down the coordinates of plotted points.
For example:

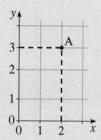

The coordinates of A are (2, 3). Coordinates are also called ordered pairs.

2 How to convert fractions to decimals.
For example:

$\frac{1}{10} = 0.1$

Check in

1

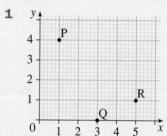

Write down the coordinates of P, Q and R.

2 Convert these fractions to decimals.

a $\frac{1}{8}$ b $\frac{7}{8}$

c $\frac{9}{11}$ d $\frac{2}{3}$

11.1 Plotting points

Before beginning work on graphs in real-life contexts it is important that you understand coordinates or ordered pairs. The first exercise revises these. You will need graph paper.

Scales and decimals

- The way in which numbers are marked on an axis is called the **scale**.

The divisions on the scale are not always whole numbers.

The interval between 0 and 5 on the *x*-axis is divided into 10 small squares so 2 squares represent 1 unit. The interval between 0 and 5 on the *y*-axis is divided into 5 small squares so 1 square represents 1 unit.

A represents $(4.5, 6)$

B represents $(2.5, 9.5)$

EXAMPLE 1

What are the coordinates of the points A, B, C and D on the graph below?

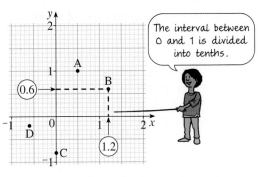

The interval between 0 and 1 is divided into tenths.

B is $(1.2, 0.6)$.

Each small division on both axes represents 0.1. Look carefully at the graph to see why B has coordinates $(1.2, 0.6)$. The coordinates for the other three points can then be matched.

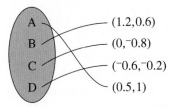

A — $(1.2, 0.6)$

B — $(0, {}^{-}0.8)$

C — $({}^{-}0.6, {}^{-}0.2)$

D — $(0.5, 1)$

EXAMPLE 2

What are the coordinates of the points A and B?

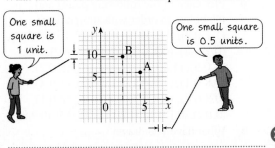

One small square is 1 unit.

One small square is 0.5 units.

Exercise 11A

1 Write down the coordinates of the eight vertices of this shape:

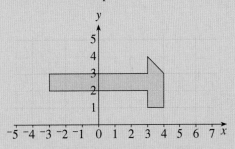

2 Make a copy of the graph below. Show the position of the points with coordinates:
   B (4, 1), C $(4, {}^{-}2)$, D (4, 6), E (2, 3), F (0, 3), G $({}^{-}2, 3)$

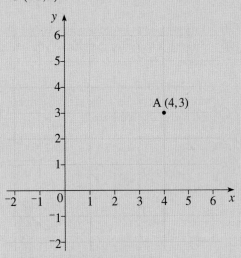

A (4, 3)

3 **a** On centimetre graph paper, what does each small division represent if
   i 1 cm represents 2 units
   ii 2 cm represents 1 unit
   iii 1 cm represents 5 units?

b How many small divisions will represent 0.1 if
  i 4 cm represents 1 unit
  ii 8 cm represents 1 unit
  iii 6 cm represents 1 unit?

4 Write down the coordinates of the points shown in the graph.

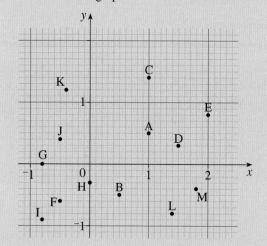

5 a Copy the axes in Question **4** on graph paper. Plot the following points:
  (0.2, 0.8), (0.2, 1.2), (0.6, 1.6), (1, 1.2), (1, 0.8), (1.4, 0.8), (1.4, 1.2), (1.8, 1.4), (2, 1.2), (1.8, 1.2), (1.8, 0.8), (1.6, 0.6), (1.6, 0), (1.4, 0.4), (1, 0), (1, 0.2), (0.6, 0.6), (0.6, 0.4), (0.2, 0.8)
  b Join the points in order with straight lines.
  c Suggest a name for the picture you have made.

6 a On the graph how many small divisions are there between 1 and 2? What does each division represent?

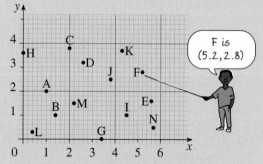

  b Write down the coordinates of the points shown.

7

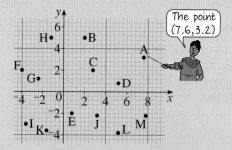

  a What does each small division on the y-axis represent?
  b What does each small division on the x-axis represent?
  c Write down the coordinates of the points shown.

8 a Copy on graph paper the axes in Question 7, using the same scale.
  b Plot these points:
  A (1, 4), B (3, 7), C (⁻3, 8), D (⁻5, ⁻5), E (1.5, 7), F (⁻4.5, ⁻9), G (7.8, ⁻2), H (⁻2.3, 6.8), I (⁻4.9, ⁻2.9), J (6.7, ⁻3.7)

9 a What does each small division on each axis represent in this graph?

  b Do you agree that the point marked A has coordinates (7.6, 3.2)?
  c Write down the coordinates of the other points.

10 Use a scale of 4 cm to represent 1 unit on both the x- and y-axes on your graph paper. Number both axes from 0 to 2.
  a Plot each of the following points and join the points in order with straight lines.
  (1.1, 0.9), (1.1, 1), (1, 1.1), (0.8, 1.1), (0.7, 1), (0.7, 0.8), (0.8, 0.7), (0.9, 0.7), (0.9, 0.5), (1, 0.5), (1.4, 0.4), (1.7, 0.3), (1.8, 0.2), (1.8, 0.1), (1.7, 0.2), (1.4, 0.3), (1, 0.4), (0.8, 0.4), (0.4, 0.3), (0.1, 0.2), (0, 0.1), (0, 0.2), (0.1, 0.3), (0.4, 0.4), (0.8, 0.5), (0.9, 0.5)
  b What have you drawn?

ACTIVITY

A coordinate game for the whole class

Arrange the desks in your class in rows and columns. Number the rows and columns so that each student is sitting at a coordinate (row, column).

The game

A student calls out an ordered pair. The student sitting in that position stands up and calls out another ordered pair. Any student who fails to stand when his or her ordered pair is called is out of the game. Also, any student calling out an ordered pair which is an empty or non-existent seat is out of the game.

Each of these graphs uses a different scale, but the figure looks the same.

In each case write down the coordinates of A, B, C and D.

(**Note:** the scale on the *x*-axis is not necessarily the same as the scale on the *y*-axis.)

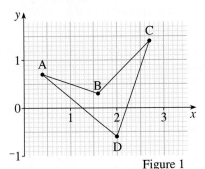

Figure 1

Figure 2

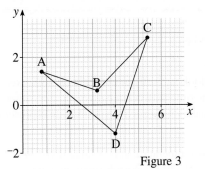

Figure 3

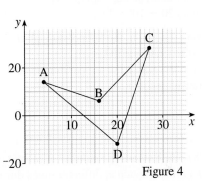

Figure 4

Are the coordinates of A, B, C and D the same in each figure? Explain.

11.2 Interpreting real-life graphs

If it makes sense, you can join the points on a graph to make a curve. Then you can read off values between the points.

EXAMPLE 3

The temperature of some hot water in a beaker is taken every 50 seconds. The results are shown in the graph.

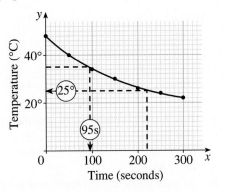

a What is the temperature of the water after 220 seconds?

b When is the temperature 35°C?

....

First look carefully at the scale on both axes.
On the time axis, a large division represents
100 seconds, so each small division represents
$100 \div 10 = 10$ seconds.

On the temperature axis, a large division
represents 20°C, so each small division
represents $20 \div 10 = 2$°C.

a From the graph, the temperature after
220 seconds is 25°C.

b From the graph, the temperature is 35°C
after 95 seconds.

Exercise 11B

1 The graph shows the number of people present
at a cricket match at different times during
the day.

a What does each small division on the
vertical axis represent?

b How much time does each small division
represent on the horizontal axis?

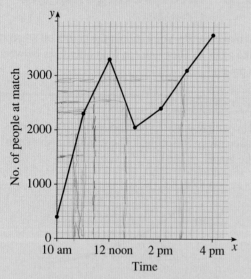

c How many people were there at the
match at:

i 11 am   **ii** 12.36 pm

iii 2.48 pm?

d At what times was the number of people
at the match:

i 1500   **ii** 1900

iii 2700?

2 The cost per unit of electricity is shown in
the graph.

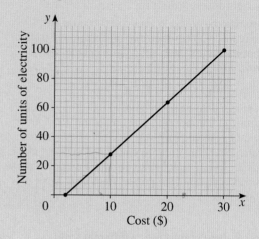

a What is the cost of:

i 20 units   **ii** 28 units

iii 76 units?

b How many units of electricity are used if
the total cost is:

i $5   **ii** $9.50   **iii** $29?

c What is the standing charge, that is, the
charge applied even if no electricity
is used?

3 The graph shows the number of US dollars
that can be obtained for a given number of
Emirati dirhams (AED).

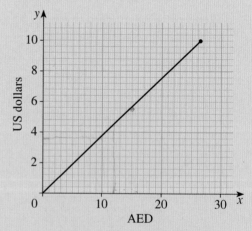

a How many US$ can be obtained for:

i AED15   **ii** AED20

iii AED8?

b How many AED can be obtained for:

i US$6   **ii** US$4.40

iii US$3.60?

4 The masses of different volumes of kerosene are shown in the graph.

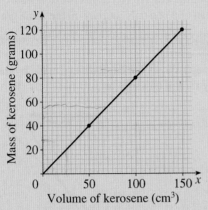

a What mass of kerosene has a volume of:
  i 50 cm³   **ii** 70 cm³
  iii 106 cm³?
b What volume of kerosene has a mass of:
  i 20 g   **ii** 32 g   **iii** 105 g?
c What is the mass of 1 cm³ of kerosene (its density)?

11.3 Drawing graphs

When you draw a graph you must pay special attention to the scale.

EXAMPLE 4

The temperature of water heated in a kettle is taken every minute. The table shows the results.

Time (min)	0	1	2	3	4	5	6	7
Temperature (°C)	24	29	36	46	58	71	80	88

a Plot a graph of the information in the table using a scale of 1 cm to 1 minute on the horizontal axis and 1 cm to 10°C on the vertical axis.
b Use your graph to find when the temperature was 62°C.

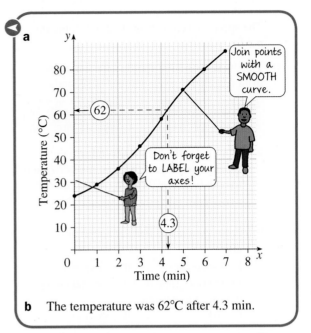

b The temperature was 62°C after 4.3 min.

Exercise 11C

1 The table shows the conversion from metres per second to kilometres per hour.

Speed (m/s)	0	10	20	30	40	50	60
Speed (km/h)	0	36	72	108	144	180	216

a Draw a graph to show this information. Use a scale of 2 cm to 10 m/s on the horizontal axis and 1 cm to 10 km/h on the vertical axis.
b Use your graph to convert these speeds to km/h:
  i 15 m/s   **ii** 45 m/s
c Use your graph to convert these speeds to m/s:
  i 90 km/h   **ii** 150 km/h

2 A piece of meat is taken out of a freezer. Its temperature rises steadily, as shown in the table.

Time (min)	0	10	20	30	40	50
Temp (°C)	⁻5	0	5	10	15	20

Draw a graph, using a scale of 1 cm to represent 5 minutes on the horizontal axis, and 2 cm to represent 5°C on the vertical axis.

Use the graph to estimate:
a the temperature of the meat after 17 minutes
b the time taken for the meat to reach a temperature of 12°C.

3 The speed of a car is measured in kilometres per hour (km/h).

This table shows the speed of the car starting from rest after a given number of seconds:

Time (s)	0	2	4	6	8	10	12
Speed (km/h)	0	4	16	30	44	56	60

a Draw a graph to show this information. Use a scale of 1 cm to 1 s on the horizontal axis and 2 cm to 10 km/h on the vertical axis. Make sure you join the points with a smooth curve.

b From your graph find the speed of the car after:

   i 5 seconds   **ii** $9\frac{1}{2}$ seconds

c From your graph find the time taken for the speed to become:

   i 20 km/h   **ii** 35 km/h

4 The volume of water that flows into a water tank over a period of 20 minutes is given in the table.

Time (min)	0	1	3	11	13.5	17	20
Water volume (ℓ)	0	13	39	143	175.5	221	260

a Draw a graph to show this information. Use a scale of 2 cm to represent 4 minutes on the horizontal axis and 2 cm to represent 50 litres on the vertical axis.

b From your graph, find how much water is in the tank after:

   i 10 min   **ii** 16.4 min

c From your graph, find when the volume of water in the tank is:

   i 20 litres   **ii** 155 litres

5 The cost of petrol for a given number of litres is shown in the table.

No. of litres	2	6	16	22	38	46
Cost ($)	3.70	11.10	29.60	40.70	70.30	85.10

a Draw a graph of this information. Use a scale of 1 cm to represent 5 litres on the horizontal axis and 1 cm to represent $10 on the vertical axis.

b From your graph, find the cost of the following quantities of petrol:

   i 5 litres   **ii** 33 litres

c From your graph, find how much petrol can be bought for:

   i $20   **ii** $55

11.4 Travel graphs

In Book 1 you learned how to draw and interpret distance–time graphs. Here, this is extended to showing more than one person's journey on the same graph.

EXAMPLE 5

The graph shows the car journeys made by the Astley family (A) and the Brown family (B).

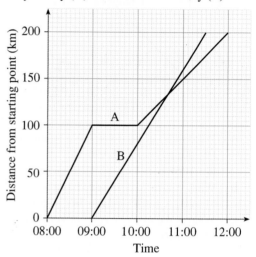

- What time does each family set off?
  The Astley family left home at 08:00, the Brown family at 09:00.
- Which family stops for a break in their journey?
  The Astley family stops for a break between 09:00 and 10:00 after driving 100 km, shown by the horizontal line in their graph.
- At what time have both families travelled the same distance?
  10:40 (the point where both lines intersect, or cross each other).
- Which family is driving faster at the start of their journey?
  The Astley family. Their line is steeper and they travel 100 km in the first hour of their journey (or at a speed of 100 km/hr), whereas in the first hour of their journey the Brown family travel 80 km (or at a speed of 80 km/hr).
- Which family is driving faster for the last 100 km of their journey?
  The Brown family has a steeper line for the last 100 km of the journey.

They travel 100 km in $1\frac{1}{4}$ hours, whereas the Astley family take 2 hours to travel 100 km.

Exercise 11D

1 The graph below shows two people making the same journey, starting at the same time, one cycling and the other walking.

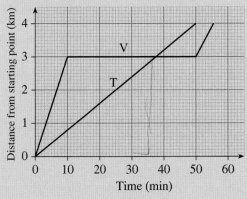

a Does graph V or T show the cyclist? Give a reason for your answer.
b For how long did person V stop?
c Describe what happened 37.5 minutes into the journey.
d What time did person V reach the destination?
e Who arrived at the destination first?
f What speed, in km/hr, was person V travelling
    i before they stopped
    ii after they stopped?
g What speed, in km/hr, was person T travelling?

2 Neema left the park and rode her bike home. The graph below shows her journey.

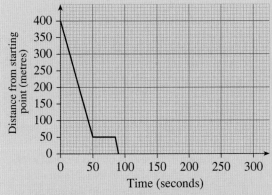

Starting at the same time as Neema, Neema's sister Samira walked from home to the park. The table opposite describes her journey.

Time (seconds)	Distance from home (metres)
0	0
50	50
80	50
300	400

a Copy the graph and draw Samira's journey on it.
b How far is the park from Neema and Samira's house?
c The two sisters met each other.
    i After how many seconds into their journey did they meet?
    ii How long did they spend talking to each other?
d Describe how the graph shows who is travelling faster in the first 50 seconds of the journeys.
e In m/s, how fast was Neema cycling for the first part of her journey?
f In m/s, how fast was Samira walking during the first part of her journey?

3 The graph below shows two friends, Devaj and Farhad, going for a run and coming home again.

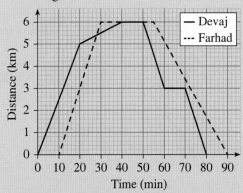

Are these statements true or false?

a Farhad stopped for half an hour.
b The slowest running was by Devaj, in the last kilometre he ran.
c After 70 minutes Devaj was nearer home than Farhad.
d Farhad began running at 15 km/hr.
e Devaj ran the last part of his journey at 18 km/hr.
f Devaj and Farhad were away from home for the same length of time.
g The friends met each other twice on their run.
h On the way home Devaj stopped for a 10 second break.

165

Consolidation

Example 1

The graph shows the speed of an athlete in metres per second over time.

a What is the athlete's speed after 2 seconds?

b When is the athlete's speed 2 m/s?

On the vertical (speed) axis each small square represents $1 \div 5 = 0.2$ m/s

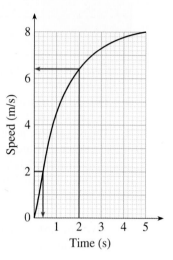

....

a After 2 seconds, speed is 6.4 m/s.

On the horizontal (time) axis, each small square represents $1 \div 5 = 0.2$ s.

b The athlete's speed is 2 m/s after 0.4 s.

Example 2

The graph below shows the journeys of Yasmina and Jadee as they went out for a run.

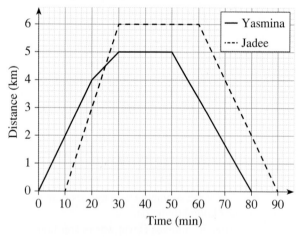

a Who travelled furthest?

b How long did each runner stop for?

c When did Jadee pass Yasmina?

....

a Jadee travelled furthest, as she ran 12 km in total while Yasmina only ran 10 km.

b Yasmina stopped for 20 minutes, Jadee stopped for $\frac{1}{2}$ hour.

c Jadee passed Yasmina after 25 minutes.

Exercise 11

1 **a** On graph paper with the x-axis numbered from $^-5$ to 5 and the y-axis numbered from $^-6$ to 6, plot the following points: $(2, 1)$, $(2, 5)$, $(3, 4)$, $(5, 4)$, $(4, 5)$, $(^-3, 5)$, $(^-1, 4)$, $(^-1, 2)$, $(3, 0)$, $(^-2, 1)$, $(^-4, 3)$, $(^-4, 6)$, $(^-2, 5)$, $(0, 6)$, $(0, 3)$, $(^-1, 2)$, $(2, ^-1)$.

b Join the points in order with straight lines.

c Suggest a name for your picture.

2 Starting at A and moving in a clockwise direction, list in order the coordinates of the vertices.

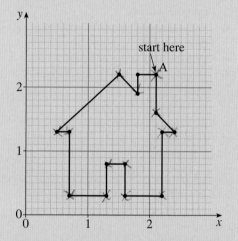

3 The population of a colony of insects increases as follows:

Week	1	2	3	4	5	6
No. of insects	3	9	27	81	243	729

a Using a scale of 2 cm to represent 1 week on the horizontal axis and 1 cm to represent 100 insects on the vertical axis, plot the graph of the above information.

b From your graph, estimate the number of insects after:

i $3\frac{1}{2}$ weeks **ii** $5\frac{1}{2}$ weeks

c From your graph, estimate how long it is before there are:

i 50 insects **ii** 620 insects

4   The graph shows the time it takes a car to complete a journey when travelling at different speeds.

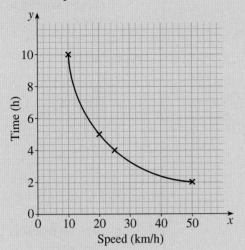

Speed (km/h)

a   How long does the journey take at a speed of
   i   30 km/h   **ii**   15 km/h?

b   At what speed is the car travelling when the journey takes
   i   5 hours   **ii**   8 hours?

c   Approximately how far is the journey?

5   One car travels from Town A to Town B. Starting at the same time, another car travels from Town B to Town A.

The journey of one car is shown on the graph below.

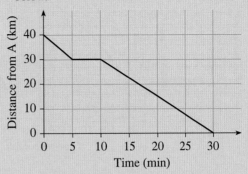

Time (min)

The other car started at Town A, drove 30 km in 15 minutes, stopped at the garage for 5 minutes for fuel, then it drove the rest of the way at a speed of 120 km/h.

a   Describe the journey of the car starting at Town B.

b   Copy the graph above and draw on it the journey of the car starting at Town A.

c   After how long did the cars pass each other? How far from Town A were they?

Summary

You should know ...

1   How to read and plot points with positive, negative and decimal coordinates.

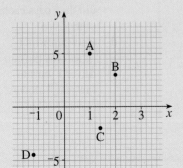

For example:
A is the point $(1, 5)$
B is the point $(2, 3)$
C is the point $(1.4, {}^-2)$
D is the point $({}^-1.2, {}^-4.5)$

Check out

1   **a**   Write down the coordinates of the points P, Q, R and S.

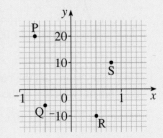

b   Copy the graph and plot T $(0.5, 4)$ and U $({}^-0.2, 20)$

167

2 How to read information from graphs.
For example:

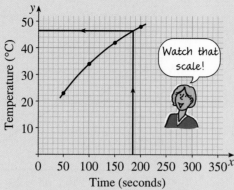

The temperature after 185 seconds is 46.5°C.

2 A baby's temperature was taken every 30 minutes starting at 5 o'clock.

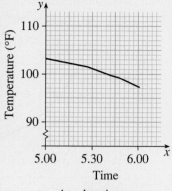

a At what time was the baby's temperature 100°F?

b What was the baby's temperature at 5.18?

....

3 How to draw and interpret graphs with more than one component.
For example:
This graph shows a journey to and from the town centre made by bus and by car, starting at the same time.

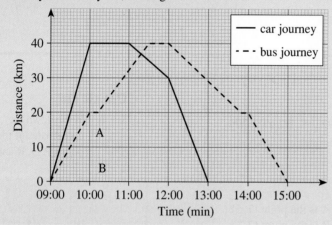

How many stops did the bus make on the way to the bus station (which is in the town centre)?

There was one stop at 10 am.

3 Using the graph on the left:

a How long did the car driver stop in the town centre?

b How long did the bus driver stop for at the bus station?

c The car driver was on the way home while the bus was still on its way into town. At what time did they pass each other?

d Between which times was the car driver's journey **i** fastest **ii** slowest?

12 Presenting data and interpreting results

Objectives

- Draw and interpret:
  - pie charts
  - frequency diagrams for discrete and continuous data
  - simple line graphs for time series
  - stem-and-leaf diagrams.

- Interpret tables, graphs and diagrams for discrete and continuous data, and draw conclusions, relating statistics and findings to the original question.

- Compare two distributions using the range and one or more of the mode, median and mean.

- Compare proportions in two pie charts that represent different totals.

What's the point?

Statistics are used all the time in the real world, in many different areas. For example, a good way to see the profits or losses of a company is by drawing a graph which shows visually whether the general trend is up or down, and which can be read far more quickly and easily than a page of numbers.

Before you start

You should know ...

1. How to calculate the mean, mode, median and range of a given set of data.

   For example:

   Carlene scored a total of 520 marks in 8 subjects in her last examination. Her mean score was $\frac{520}{8} = 65$.

Check in

1. Find the mode, median, mean and range of these numbers:

   15, 13, 13, 12, 11, 11, 10
   10, 10, 10, 9, 7, 8, 8, 7

2 How to draw a frequency table from given data.
For example:
The number of eggs laid each week by Kurt's hens during a 20-week period are:

4, 6, 4, 7, 3, 3, 2, 6, 3, 1
3, 7, 4, 6, 2, 7, 2, 7, 4, 3

This is the frequency table for the data:

No. of eggs	Frequency
1	1
2	3
3	5
4	4
5	0
6	3
7	4

2 The following are the numbers of pairs of shoes owned by 20 people.

7, 9, 6, 10, 8
8, 9, 11, 8, 7
9, 6, 8, 10, 9
8, 7, 7, 8, 9

Draw a frequency table for this information.

12.1 Frequency diagrams and pie charts for discrete data

Remember there are two basic types of data:
- **discrete** and
- **continuous**.

Discrete data can only take definite values.
For example:
   Clothes size   –   Small, Medium, Large etc.

Continuous data can take any value.
For example:
   Height
   Mass
   Time

Bar charts and pie charts are usually used to represent discrete sets of data.

A frequency table displays data as numbers. A frequency diagram represents these frequencies in a graph or chart. A bar chart is a frequency diagram for discrete data.

Pie charts

Pie charts are another way to show information.
For example, the pie chart on the right shows how Bob spends 24 hours.

He spends $\frac{1}{8}$ of his time doing homework.

The angle of the sector is $\frac{1}{8} \times 360° = 45°$

EXAMPLE 1

Alan has $180. He uses it to buy the following:

Item bought	Shoes	Books	Shirt	Umbrella
Amount spent	$60	$40	$50	$30

Show this data on a pie chart.

There are 360° in a circle.
Alan spends a total of $180.
This is represented by 360°.

That is, $1 is represented by $\frac{360°}{180} = 2°$

Angle representing $60 spent on shoes = 60 × 2°
                                                        = 120°

The other angles are:
Books: $40 \times 2° = 80°$
Shirt: $50 \times 2° = 100°$
Umbrella: $30 \times 2° = 60°$

Check the angle sum:
$120° + 80° + 100° + 60° = 360°$

Now draw a circle. With your protractor, carefully mark in the angles and label the sectors.

Pie chart showing how Alan spends his $180:

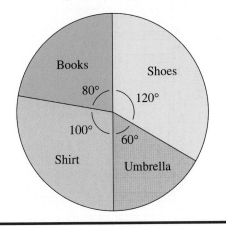

Angle representing girls in pie chart
$$= \frac{12}{30} \times 360°$$
$$= 144°$$

Pie chart showing boys and girls in class:

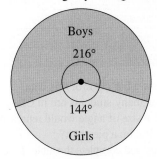

Notice the two different methods for drawing pie charts. In Example 1 the method used is to work out how many degrees for each $1 then multiply this by the amount of money spent on each item. In Example 2 the method used is to work this out in one step by multiplying the appropriate fraction by 360°.

EXAMPLE 2

In a class of 30 there are 18 boys and 12 girls. Show this data on: **a** a bar chart
                              b a pie chart.

....

a   Chart showing boys and girls in class

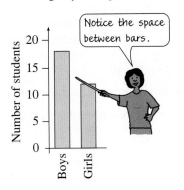

Notice the space between bars.

b   Angle representing boys in pie chart
$$= \frac{18}{30} \times 360°$$
$$= 216°$$

Exercise 12A

1   A family had 8 cows. Here is a bar graph of milk produced by each cow in 2012.

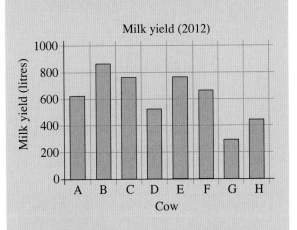

a   How much milk was produced by Cow G?
b   Which cow produced most milk?
c   Which cows produced the same amount?

2 The frequency table shows the favourite sports of Class 2C.

Favourite sport	Number of students
Cricket	14
Football	9
Badminton	5
Volleyball	4
Swimming	6
Tennis	2

a Draw a bar chart of this data.
b How many students are in Class 2C?
c What size of angle would represent one student on a pie chart?
d Copy and complete the table, for each sport.

Favourite sport	Number of students	Size of angle on pie chart
Cricket	14	
Football	9	
Badminton		
Volleyball		

e Now draw a circle of radius 6 cm. Using a protractor, draw an angle in your circle to represent each sport. Label your chart.

3 36 motorists together own four separate makes of car. The numbers who own each make is listed:

Toyota	15
Suzuki	9
Ford	2
Mitsubishi	10

a What size of angle would represent one motorist on a pie chart?
b Copy and complete the table for each make of car.

Car make	Number of motorists	Size of angle on pie chart
Toyota	15	
Suzuki	9	
Ford	2	
Mitsubishi	10	

c Now draw a circle of radius 6 cm. Using a protractor, draw an angle in your circle to represent each make of car. Label each sector.

4 The bar chart shows the rainfall statistics, for each month last year, for a South American country.

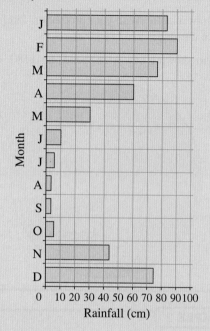

a What do, J, F, M, ... stand for?
b How much rain fell in July?
c In which month did most rain fall?
d When was the dry season?

5 In a government survey, 240 people were asked their occupations. Their replies were:

Management 10   Services 30
Clerical 36   Skilled Labour 72
Teaching 12   Unemployed 50
Agricultural 30

Show this information on a pie chart.

6 Here are the shoe sizes of 80 people:

Size 3   2   Size 7   12
Size 4   2   Size 8   28
Size 5   6   Size 9   16
Size 6   8   Size 10   6

Show this information on a pie chart.

7 Mr Pinder earns $400 a week. Here is what he did with his salary last week:

Food $120 Entertainment $40
Clothes $50 Rent and fuel $100
Travel $30 Savings $60

a In a pie chart to show this, $400 is represented by 360°.
What angle represents $1?

b Copy and complete the table.

	Amount	Angle
Food	$120	108°
Clothes	$50	
Travel	$30	

c Draw a pie chart to show the information in your table. Don't forget to label it.

8 a Make a tally chart to show how often each vowel appears in this poem:

A funny old beast is the yeti,
Mostly he eats just spaghetti,
Tomatoes, ham
And iced mulberry jam,
Then he showers himself with confetti.

> A vowel is *a, e, i, o* or *u*

b Draw a frequency table to show this data.
c Draw a bar chart to show this data.

9 Carry out a survey of 20 students. Find out their favourite:

a game **b** colour **c** singer.

Draw a pie chart to show the information in each case.

💻 TECHNOLOGY

You can draw pie charts using Microsoft Excel® (or other spreadsheet software).

Use the data from Question **3** of Exercise 12A.

First type the data into separate cells of a spreadsheet:

	A	B	C
1	Toyota	15	
2	Suzuki	9	
3	Ford	2	
4	Mitsubishi	10	
5			

Then, highlight your table and select the pie chart.

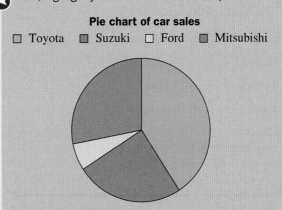

Pie chart of car sales

☐ Toyota ▨ Suzuki ☐ Ford ▨ Mitsubishi

Use a spreadsheet to draw pie charts for the other questions in Exercise 12A.

Reading pie charts

To read a pie chart you have to find the quantity or number that each angle actually represents.

> **EXAMPLE 3**
>
> There are 120 second-year students in St Mary's College. The pie chart shows their favourite colours.
>
>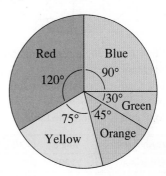
>
> Find out how many students chose each colour.
>
> ....
>
> 120 students are represented by 360°
>
> So 1 student is represented by $\frac{360}{120} = 3°$
>
> The colour blue, 90°, represents $\frac{90°}{3°} = 30$ students

The other colours are chosen by

Red $= \frac{120°}{3°} = 40$ students

Yellow $= \frac{75°}{3°} = 25$ students

Orange $= \frac{45°}{3°} = 15$ students

Green $= \frac{30°}{3°} = 10$ students

Check:
total number of students
$= 30 + 40 + 25 + 15 + 10$
$= 120$

Exercise 12B

1 This pie chart represents the flavours of the 180 ice creams sold by Joss in one weekend.

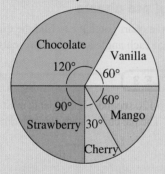

a What size of angle represents one ice cream?

b How many chocolate ice creams did Joss sell?

c Copy and complete the table.

Ice cream flavour	Angle	Number sold
Chocolate	120°	
Vanilla	60°	
Mango	60°	
Cherry	30°	
Strawberry	90°	

2 The pie chart shows how the workers at the National Cement Works travel to work.

Method of travel used by cement workers

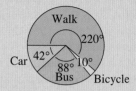

a If 5 people cycle to work, what angle represents 1 worker?

b Find out how many workers come by bus, by car, and by walking.

c How many workers are employed at the cement works?

3 **The favourite subjects of Class 2A**

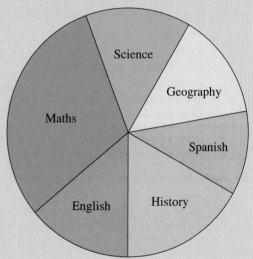

a From the pie chart, can you tell how many students are in Class 2A?

b Can you tell how many students prefer History?

c Measure the angle for History.

d If 6 students prefer History, what angle represents one student?

e Now measure the angle for each of the other subjects.

f Calculate the number of students preferring each of the other subjects.

g How many students are in Class 2A?

4 The bar chart shows Class 2B's favourite subjects.

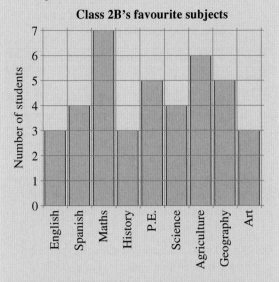

Class 2B's favourite subjects

a How many students are in Class 2B?
b Draw a frequency table for the graph.
c Do you think it is easier to read information from a bar chart than from a pie chart? Why?
d Compare this bar chart for Class 2B with the pie chart for Class 2A, in Question **3**. What can you find out from the bar chart that you cannot find out from the pie chart?

5 The way John Boyd spent his wages last month is shown in this pie chart.

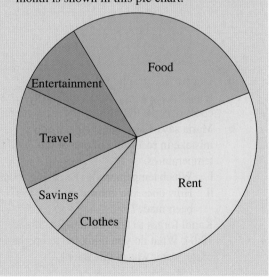

a About what fraction of his wages did John spend on rent?
b About what fraction did he save?
c Measure the angles at the centre. Write them down.
d Can you tell how much John earned last month?
e Last month John spent $200 on food. Work out how much money was used in each of the other ways.
f How much did John earn last month?
g If John's wages last month were $1080, calculate how much he spent on:
  i clothes   **ii** travel.

6 The pie chart shows how Farmer Errol spends a typical day.

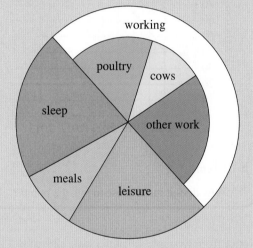

a What fraction of the day does Errol spend eating his meals?
b What percentage of the day does Errol have for leisure?
c How many degrees represent one hour?
d Work out how long Errol spends on each activity.

12.2 Line graphs

A **line graph** is another way of showing data.

EXAMPLE 4

a Draw a line graph to show Bently's height on 1 January each year, from his birth in 2009.

b Estimate his height in July 2011.

....

a

Year	2009	2010	2011	2012	2013
Height (cm)	40	65	80	90	100

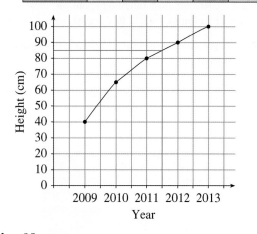

b 85 cm

Why does it make sense to join the points?

Notice from Example 4 that the scale chosen is important. A change of scale can make the graph look different:

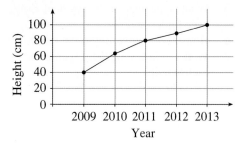

You need to choose scales and axes carefully when drawing charts. If you do not, your graphs could mislead!

Exercise 12C

1 a Draw a line graph to show the change of temperature over the course of a day. Start the temperature axis at 20°C.

Time	6 am	8 am	10 am	12 noon	2 pm	4 pm
Temperature (°C)	24	25	27	29	29	26

  b What do you think the temperature was at
  i 7 am **ii** 3 pm?

  c At what times do you think the temperature was 28°C?

  d Why does it make sense to join the points?

  e Redraw the graph with the temperature axis beginning at 0°C. Which graph is easier to read?

2 Kamil was conducting a science experiment to see how boiling water in a mug cools in 10 minutes. He took 9 measurements of the temperature, as shown in the table below.

Time	Temperature (°C)
10:00	100
10:01	96
10:02	91
10:03	87
10:04	
10:05	83
10:06	80
10:07	81
10:08	77
10:09	75
10:10	74

  a Draw a line graph to show the temperature of water in a mug as it cools over 10 minutes.

  b Maria says Kamil must have made a mistake in recording of one of the temperatures.
  i Which temperature is she talking about?
  ii How does she know a mistake has been made?

  c Kamil forgot to take the temperature at 10:04. What do you think the temperature of the water was at this time?

3 The graph shows temperatures during a day in July in Oxford, UK.

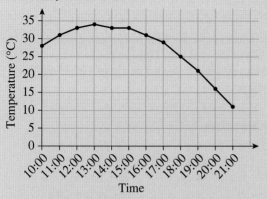

a At what time was the highest temperature recorded?

b Estimate the temperature at 17:30.

c The temperature stayed the same for an hour. When was this?

d Which hour showed the greatest rise in temperature?

e Which hour showed the greatest fall in temperature?

f Estimate the time when the temperature was 23°C.

4 A football match was due to start at 15:00. The sports stadium gates were opened at 14:00. The number of people inside the stadium was recorded every 10 minutes. The results are shown in the table.

Time	Number of people inside stadium
14:00	0
14:10	4000
14:20	15000
14:30	23000
14:40	30000
14:50	34000
15:00	35000

a Draw a line graph to show the number of people in the stadium between 14:00 and 15:00.

b Use your graph to estimate how many people were in the stadium at 14:25.

c In which 10-minute period did the most people enter the stadium?

d What have you assumed when answering part **b**?

5 Look at these two graphs of sales at an electronics store over a six-month period.

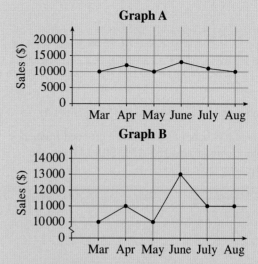

a Why do the graphs look different?

b Which graph is easier to read?

c Would it be fair to say that the sales in June were excellent? Why?

d Which graph would you show to the manager of the electronics store? Why?

TECHNOLOGY

View the page on misleading graphs and charts at www.bbc.co.uk/schools

(type 'misleading graphs' into the search bar).

Look for examples of misleading graphs and charts from magazines, newpapers or the internet.

Present some of these to your class.

12.3 Histograms

Histograms are very similar to bar charts. A histogram is usually used to show continuous data, while a bar chart is used to represent discrete data. Histograms are also often used for grouped data.

For example, look at the heights of 25 children measured to the nearest centimetre:

139, 141, 142, 142, 145, 146, 147, 147, 151, 151, 152, 152, 153, 153, 154, 156, 156, 157, 157, 158, 160, 161, 162, 162, 166

Drawing this information on a graph gives:

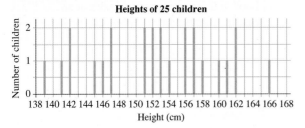

Heights of 25 children

Such a graph is not very useful – explain why.

Instead it is better to construct a grouped frequency table. Using groups 135–139, 140–144, etc. the frequency table is:

Interval	Tally	Frequency
135-139	l	1
140-144	lll	3
145-149	llll	4
150-154	llll ll	7
155-159	llll	5
160-164	llll	4
165-169	l	1

The histogram of this data is:

Histogram of heights of 25 children

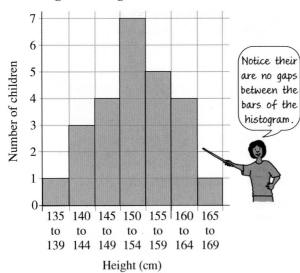

Notice their are no gaps between the bars of the histogram.

Note that it is more usual to label the horizontal axis as shown below, instead of using the '135 to 139' notation:

135 140 145 150 155 160 165 170
Height (cm)

You may wish to look back at Chapter 6, section 6.3 on frequency tables to see the different ways people use to label classes in tables. These can also be used to label the horizontal axis of a histogram.

The horizontal axis of the histogram above shows the height of the children. This is a continuous variable. It is quite possible to have a height of 138.43 cm, for example.

Exercise 12D

1 **a** Draw a histogram for the heights of the 25 children using intervals of:
  i 10 cm   **ii** 20 cm
  b Look at all the charts for this data. Which one shows best how heights vary?

2 The masses, in kilograms, of 25 children are:

Mass (kg)	10-19	20-29	30-39	40-49	50-59
No. of children	1	4	7	12	1

  a Draw a histogram to show this information.
  b What is the modal mass?

3 The heights of 30 plants, in centimetres, 6 weeks after planting were:

Height (cm)	Frequency
5-9	5
10-14	9
15-19	8
20-24	6
25-29	2

Draw a histogram to show the data.

4 The heights in centimetres of 30 different plants were measured 6 weeks after planting. The results are shown in this histogram:

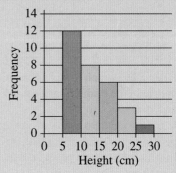

a Copy and complete the table.

Height ( )	
5-9	
	8
15-19	
20-24	
	1

b Compare the two histograms from Questions **3** and **4**. Do you think these plants were the same species? Justify your answer.

5 Vendra counted the words in the first 40 sentences of her book. Her findings were:

Number of words	Frequency
0-9	13
10-19	15
20-29	5
30-39	3
40-49	3
50-59	1

a Draw a histogram for this information.
b What is the modal interval?

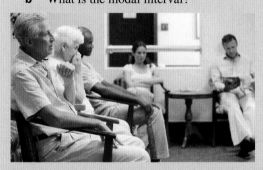

6 The waiting time, in minutes, for 25 patients to see a doctor was:
8, 35, 12, 45, 4, 15, 38, 28, 30, 23, 14, 38, 53, 26, 33, 32, 15, 18, 48, 37, 34, 34, 28, 51, 16
a Construct a frequency table using an interval of your choice to show the data.
b Draw a histogram to illustrate the data.
c How many patients waited less than 10 minutes?
d What percentage of patients waited half an hour or more?
e Was your interval choice the best?

7 The masses, in kilograms, of 30 items to be transported are shown below.

7, 13, 18, 24, 46, 29, 31, 34, 36, 37, 38, 38, 41, 43, 43, 44, 45, 46, 47, 48, 49, 52, 54, 55, 56, 59, 63, 72, 85, 97

a Jalad drew this histogram for the data.

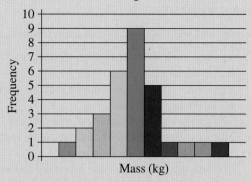

What class intervals did he use?

b Amy drew this histogram for the data.

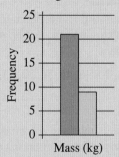

What class intervals did she use?

c Compare the two histograms. Which do you think is better? Why? Can you think of a reason why the other one might be used?

▶ ACTIVITY

Measure the heights and masses of the students in your class.
a Construct suitable frequency tables to show this data.
b Draw histograms to illustrate the distribution of heights and masses in your class.
c Repeat the process, but this time draw separate tables and histograms for boys and girls.
d Comment on your results.

12.4 Stem-and-leaf diagrams

A stem-and-leaf diagram is similar to a histogram in that the shape of the diagram gives an overall picture of the trend. The advantage of a stem-and-leaf diagram over a histogram is that you can still see all the data values. Bars in histograms are often drawn vertically, while in stem-and-leaf diagrams numbers are usually written horizontally.

A stem-and-leaf diagram has four main features:

- The key – which explains what the stem and leaves are worth.
- The stem – often the tens column (but not always – check the key to find out).
- The leaves – often the units column (but not always – check the key to find out). Note the leaves *must be in numerical order*.
- The spacing – numbers *must be evenly spaced* to maintain the shape clearly.

EXAMPLE 5

Draw a stem-and-leaf diagram for the data below, which shows the height, in centimetres, of 15 plants.

14, 34, 21, 20, 36, 45, 56, 52, 33, 44, 63, 49, 47, 49, 58

....

Look at the numbers to see what the stems should be. The tens columns for these numbers are 1, 2, 3, 4, 5 and 6. Write these in a vertical list with a vertical line to the right of them. Write the first two values, 14 cm and 34 cm, in the stem-and-leaf diagram and include a key. Make sure you label the diagram 'Unordered' so that you remember to put it in order later.

Unordered

Stem	Leaves
1	4
2	
3	4
4	
5	
6	

Key
1 | 4 means 14 cm

Beginning with an *unordered* stem-and-leaf diagram is the best way to make sure that you do not miss any values out, particularly when there

are a lot of values in a list. Starting from the beginning and working your way through the data, write in the other values. Don't worry too much about spacing at this point as this is just rough work.

Your stem-and-leaf diagram should look now look like this:

Unordered

Stem	Leaves
1	4
2	1 0
3	4 6 3
4	5 4 9 7 9
5	6 2 8
6	3

Key
1 | 4 means 14 cm

Now place all the leaves in numerical order and make sure that they are evenly spaced. Label the new diagram 'Ordered'.

Ordered

Stem	Leaves
1	4
2	0 1
3	3 4 6
4	4 5 7 9 9
5	2 6 8
6	3

Key
1 | 4 means 14 cm

Notice that if a bar chart or histogram was used, only the shape would be known, not the original values:

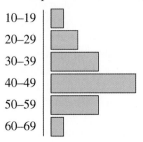

A useful feature of stem-and-leaf diagrams is that they are good for finding averages.

```
1 | 4
2 | 0 1
3 | 3 4 6
4 | 4 5 7 9 9
5 | 2 6 8
6 | 3
```

Key
1 | 4 means 14 cm

Using this stem-and-leaf diagram, you can see that the mode is 49 cm as that is in the only row with repeated digits. Because the numbers are in order you can easily find the middle number: the median is 45 cm. Finding the mean is also possible as you have all the original data. The mean for this data is:

$(14 + 20 + 21 + 33 + 34 + 36 + 44 + 45 + 47 + 49 + 49 + 52 + 56 + 58 + 63) \div 15 = 41.4$ cm to 1 d.p.

Exercise 12E

1 For a school project, Guntur surveyed 15 streets in his town to see how many houses there were in each street. These were the results:

17, 55, 48, 57, 61, 30, 26, 57, 46, 51, 40, 60, 58, 31, 42

His friends Kersen, Fatima and Eva drew these stem-and-leaf diagrams:

Kersen

```
1 | 7
2 | 6
3 | 0 1
4 | 0 2 8
5 | 1 5 7 7 8
6 | 0 1
```

Fatima

```
1 | 7
2 | 6
3 | 0 1
4 | 8 6 0
5 | 5 7 7 1 8
6 | 1 0
```

Key
1 | 7 means 17 houses

Eva

```
1 | 7
2 | 6
3 | 0 1
4 | 0 2 6 8
5 | 1 5 7 7 8
6 | 0 1
```

Key
1 | 7 means 17 houses

Explain what is wrong with each of these stem-and-leaf diagrams.

2 a Draw an ordered stem-and-leaf diagram from this unordered diagram about the masses of spare parts for a machine.

Unordered

```
2 | 7
3 | 3 1 9
4 | 6 8 8 1 5
5 | 3 5 2 5 0 5 9
6 | 5 8 6 1
7 | 2 3
```

Key
2 | 7 means 2.7 kg

b In the stem-and-leaf diagram in part **a**, which two numbers are represented by this row?

7 | 2 3

c What is the modal mass?
d What is the median mass?
e What is the smallest mass?
f What is the largest mass?

3 All of the students in a class were asked by their maths teacher to estimate the size of an angle that she drew on the board. These were the results:

45, 62, 38, 51, 44, 38, 36, 43, 50, 44, 40, 55, 48, 37, 32, 38, 29, 40, 41, 35, 46, 25, 42, 41, 39, 39, 45, 32, 43, 46

a Draw a stem-and-leaf diagram for this data.
b What is the modal guess?
c What is the median guess?
d Only one student guessed the angle correctly. Which angle do you think this was? Justify your answer.

4 The two data sets below show the heights of all the teachers in Yasmina's school, measured in centimetres.

Set A
162, 153, 157, 159, 165, 168, 190, 184, 181, 173, 174, 172, 166, 164, 155, 162, 171, 153, 164

Set B
192, 182, 176, 190, 168, 180, 179, 177, 174, 191, 185, 181, 183, 186, 167, 183, 177, 182, 158

a Copy and complete the two stem and leaf diagrams started below.

```
15 | 3
16 |
17 |
18 |
19 |
```

Key
15 | 3 means 153 cm

```
15 | 8
16 |
17 |
18 |
19 |
```

Key
15 | 8 means 158 cm

b One set of data is for the male teachers the other set is for the female teachers. Which set, A or B, do you think represents the male teachers? Why?

c Is it easier to answer part **b** by looking at the stem-and-leaf diagrams or by looking at the lists of numbers?

5 Draw a line on a piece of paper. Ask your classmates to estimate the length of the line, in millimetres. Draw a stem-and-leaf diagram for this data.

6 This stem-and-leaf diagram shows the number of watches sold by a jeweller each day, for a period of 20 days.

```
0 | 4  6  7
1 | 1  3  3  3  6  8  9
2 | 2  2  4  7  7  7
3 | 1  2  4
4 | 1
```

Key
4 | 1 means 41

a Work out the mean, median, mode and range for this data.

The next day 27 watches are sold.

b Without doing any calculations, what effect does this extra value have on the mean, median, mode and range? Why?

12.5 Interpreting and comparing data and diagrams

Comparing distributions means looking for what is similar and what is different about them. In particular, we usually look for which values are larger or smaller, for example which distribution has the larger mean, median, mode or range.

Interpreting tables, graphs and diagrams means understanding what these show. For example, a larger range shows that data is more spread out or more varied.

Often when you have interpreted and compared data you will be able to write some sort of **conclusion**. In Question **4** of Exercise 12E you interpreted the two stem-and-leaf diagrams and compared them to come to a conclusion based on female and male heights.

The next exercise is all about interpreting and comparing data. It is important that you think not just about the numbers, but also about how these numbers relate to the original question.

The dual bar chart below shows the number of ice creams and pizzas sold in a café during one week in September 2012.

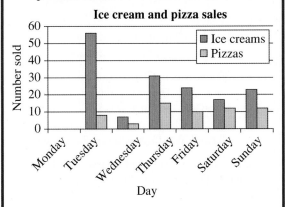

Ice cream and pizza sales

a The café closes one day a week. Which day was the café closed?

b Give a possible reason for the high level of ice cream sales on Tuesday.

c The café closed early one day. Which day do you think this was? Why?

....

a The café was closed on Monday, as it is the only day nothing was sold.

b The high level of ice cream sales on Tuesday could be because it was a much hotter day or because ice creams were being sold with a special discount that day. We cannot tell this information for certain from the chart – there may be other reasons. Did you think of any?

c The shop may have closed early on Wednesday, as sales are much lower that day, but we cannot know this for certain – we can only make a sensible guess.

The holiday destinations of 30 students during 2012 are recorded in the first pie chart. The second pie chart shows the holiday destinations of 60 students during 2013.

2012 holiday destinations

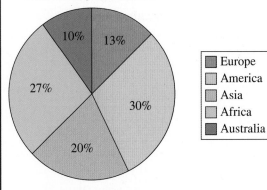

2013 holiday destinations

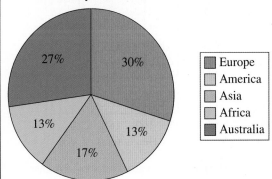

Compare the two pie charts by considering

a the most and least popular resorts

b the students travelling to Asia each year.

....

a The most popular two holiday destinations in 2012 were America and Africa but these were the least popular in 2013. The least popular destination in 2012 was Australia, however this was one of the most popular in 2013.

b 20% of students went to Asia in 2012. This was 6 students in total (0.2 × 30 = 6). 17% of students went to Asia in 2013. This was 10 students in total (0.17 × 60 = 10.2, rounded to the nearest whole number). So although a lower percentage of students went to Asia in 2013, this represented a higher number.

You will have seen in Example 7 that it is important to be careful when pie charts represent different total numbers. You must make sure you talk about percentages and numbers with care.

Exercise 12F

1 The histogram shows the heights of some men.

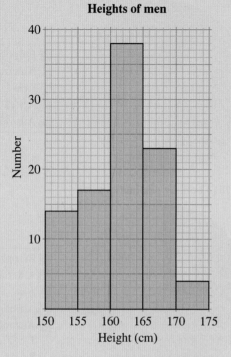

Heights of men

a How many men are represented in the histogram?
b How many men are shorter than 160 cm?
c How many men are 165 cm or taller?
d Is it possible to say how tall the tallest man is? Explain your answer.

2 30 students from Class 2 took part in a quiz. The bar chart shows their scores.

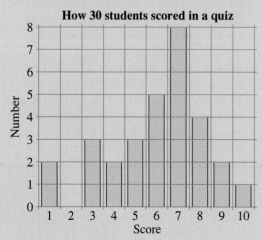

How 30 students scored in a quiz

a How many students scored
  i the lowest mark
  ii the highest mark?

b What mark did most students score?

c What is
  i the mode of the set of marks
  ii the range of the set of marks?

d 30 students in Class 3 took part in the same quiz. The modal score for Class 3 was 6 and the range was 5. The lowest score in Class 3 was 4. Compare the two Classes. What can you say about them?

3 Govinda, Jack, Billy and Chakor are playing cricket. Their runs for 10 overs are:
Govinda   10, 8, 5, 8, 5, 9, 4, 5, 9, 7
Jack   6, 7, 1, 4, 5, 2, 2, 4, 4, 5
Billy   12, 8, 9, 1, 1, 54, 5, 3, 6, 1
Chakor   15, 11, 14, 10, 9, 11, 8, 5, 5, 12
a For each boy, find the mean score per over.
b Find the range for each boy.
c Find the modal score for each boy.
d Which boy would you want on your team? Why?
e Which is the better average to use in this case, the mean or the mode? Why?

4 a In the graph below the solid line represents the number of baboons. The dotted line represents the number of cheetahs. Compare the shapes of the two graphs. What conclusions can you reach?

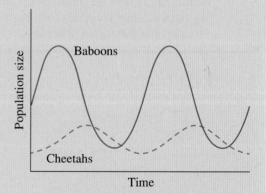

b Suggest a possible scale for the horizontal axis. Justify your choice.

5

The masses of adult Asian and African elephants are given in this table

Mass (kg)	Frequency	
	African	Asian
3000–4000	2	8
4000–5000	4	23
5000–6000	19	7
6000–7000	10	1
7000–8000	6	0

a Show this data on two frequency graphs.
b What is the modal mass of
   i Asian elephants
   ii African elephants?
c An elephant has a mass of
   i 7250 kg
   ii 3245 kg
   iii 5050 kg.

   Decide which species you think each of these elephants is.
d How confident can you be about your answers to part **c**?

6 An artist showed four paintings, A, B, C and D, to some adults and asked them which one they liked best. She then showed them to some children. The pie charts below show the adults' and children's choices.

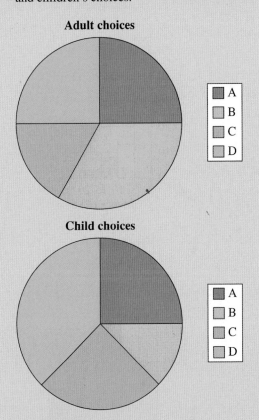

Adult choices

Child choices

Which of the following statements are
i true
ii false
iii impossible to say?
Give reasons for your answers.
a The same percentage of adults and children liked Picture A.
b More adults than children liked Picture B.
c More children liked Picture D than Picture A.
d A higher percentage of children than adults liked Picture D.
e Picture B was the most popular among the adults.
f The same number of adults liked Picture A and Picture D.
g A lower percentage of adults than children liked Picture C.
h Picture A was the most popular among the children.

7 Two samples of 80 people at a school were surveyed to see how far they lived from the school. The results are shown in the histograms below.

Histogram A

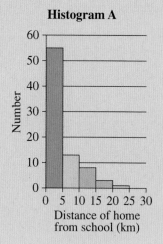

Histogram B

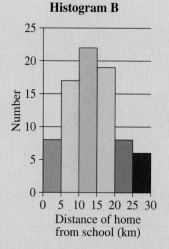

a In Histogram A, what was the modal distance from school?

b In Histogram B, what was the modal distance from school?

c Is this statement true, false or impossible to say: 'The range for histogram A was 24'? Give a reason for your answer.

d One histogram was for the students and one was for the teachers. Which do you think was for the teachers? Justify your answer.

e Write a sentence comparing the two histograms.

8 The table below shows the working time for two different laptops, Type A and Type B.

Number of laptops still working after	Type A	Type B
0 h	32	38
1 h	30	37
2 h	29	35
3 h	26	28
4 h	21	15
5 h	11	7
6 h	3	2

a How many laptops were there in the sample, altogether?

b Is it easy to see from the data above which laptop seems to last longer?

c Copy and complete the line graph started below.

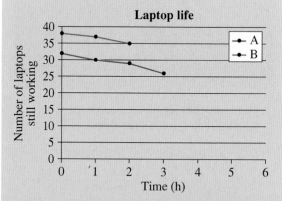

d After 3.5 hours how many laptops of
i Type A
ii Type B
   are still working?

e What percentage of the original laptops are still working after 3.5 hours?

f Is it easy to see from the graph which laptop seems to last longer?

9 The pie charts show the number of students passing or failing their driving test on the first attempt. The sample is taken from one year group in Cransley High School.

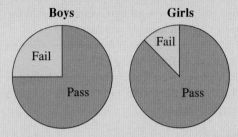

Boys   **Girls**

a Write a comparison between the percentage of girls and boys passing or failing.

The number of boys who failed the test first time is 20.

Twice as many girls as boys took the test.

b Copy and complete this two-way table.

	Boys	**Girls**	**Total**
Pass			
Fail	20		
Total			

c Write a comparison between the numbers of girls and boys passing or failing.

10 The dual bar chart below shows the ages at death to the nearest year for male elephants, based on a sample of 166 elephants in India.

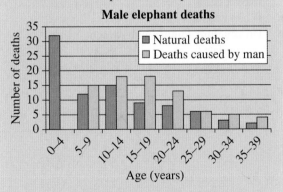

Male elephant deaths

■ Natural deaths
□ Deaths caused by man

Age (years)

a How do the bars for the 0–4 range compare to all the other age ranges? Why do you think this is the case?

b What is the modal age of death of these male elephants for
  i natural deaths
  ii deaths caused by man?

c Is it possible to work out the range for the ages of death? Give a reason for your answer.

d Can you suggest any possible problems with the data?

11 The two pie charts below show how students in Year 10 spend their free time.

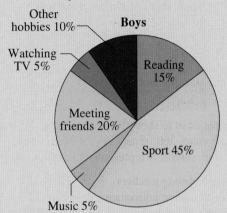

Boys

Other hobbies 10%
Watching TV 5%
Meeting friends 20%
Reading 15%
Sport 45%
Music 5%

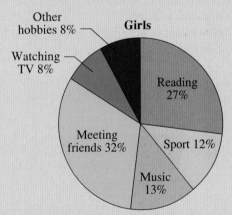

Girls

Other hobbies 8%
Watching TV 8%
Meeting friends 32%
Reading 27%
Sport 12%
Music 13%

a Compare the percentages in the two pie charts.

b Boys spent 10% of their time on 'Other hobbies' while girls only spent 8% of their time in this way. 8 boys spent their time on other hobbies. If there were 180 students surveyed in total, how many girls spent their time on other hobbies?

c Write a paragraph comparing the way Year 10 boys and girls spend their free time.

Consolidation

Example 1

A class of 30 seven-year-olds were asked what sort of job they would like when they grew up. Their answers were as follows:

Occupation	Number of students
Teacher	10
Policeman	8
Doctor/Nurse	4
Farmer	3
Other	5

Draw a pie chart to show this information.
In the pie chart, 360° represents the 30 children.
That is, one student is represented by 360° ÷ 30 = 12°

Angle representing teachers   = 10 × 12° = 120°
Angle representing policemen   = 8 × 12° = 96°
Angle representing doctors/nurses = 4 × 12° = 48°
Angle representing farmers   = 3 × 12° = 36°
Angle representing others   = 5 × 12° = 60°

Desired job

Example 2

The heights of 25 children, measured to the nearest centimetre, are shown in the histogram.

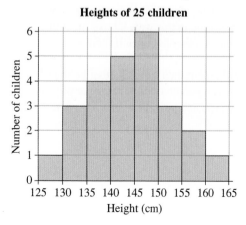

Heights of 25 children

a Which interval is the mode?
b How many children were 150 cm or more in height?
c Five children were all in the same interval for height. Which interval was that?

---

a The mode is from 145 to 150.
b Six children were 150 cm or more in height.
c The five children were all in the 140 to 145 cm height interval.

Example 3

The depth of water in a tidal river over a 12-hour period is shown in the table.

Time	Depth (m)
03:00	1.2
05:00	2.6
07:00	4.2
09:00	5.4
11:00	4.6
13:00	3.0
15:00	1.4

a Draw a line graph for this time series.
b Use your line graph to estimate the depth of the river at 06:00.
c When was high tide?
d Do you think joining with straight lines is appropriate?

---

a

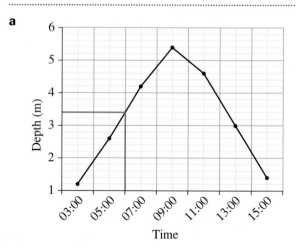

b At 06:00 the river was about 3.4 m deep.
c High tide is at around 09:00.
d Joining with straight lines assumes constant increase or decrease in water levels. This may not be the case but it will give a good estimate.

Example 4

a Draw a stem-and-leaf diagram for these grades of 11 students out of 50 marks in a recent chemistry test:

11, 47, 34, 21, 20, 36, 45, 48, 50, 34, 44

b What was the modal score in the test?

....

a

```
1 | 1
2 | 0 1
3 | 4 4 6
4 | 4 5 7 8
5 | 0
```

Key

1 | 1 means 11 marks out of 50

b The modal score in the test was 34.

Example 5

The two frequency diagrams below show the scores of 40 students in their biology and physics tests.

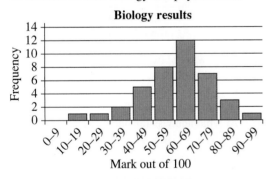

Biology results

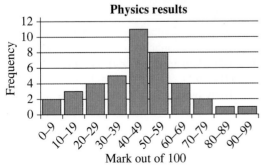

Physics results

Which was the hardest test? Justify your answer.

The modal score for biology is 60–69 and the modal score for physics is 40–49. More candidates scored lower in physics so it looks like physics was the harder test.

Exercise 12

1 Dave sells soft drinks at his stall in the market. His sales for this week are shown in the bar chart.

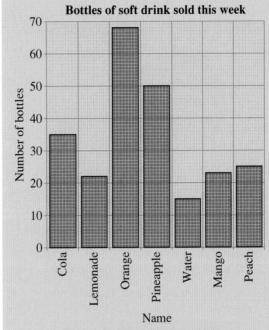

Bottles of soft drink sold this week

a How many bottles of soft drink has Dave sold altogether, this week?
b Which drink should Dave keep the largest stock of? Why?

2

Subject	Maths	Physics	Biology	Sociology	Accounting
No. of graduates	15	12	5	9	19

The table gives the number of graduates by subject from Seaview College in 2012.

a Calculate the number of students that graduated.
b What angle on a pie chart would represent one graduate?
c Draw a pie chart to show the information.

3 The table shows the number of people employed at a building firm.

Types of personnel	Number employed
Unskilled workers	30
Masons	25
Carpenters	25
Draughtsmen	7
Tilers	3

Draw a bar chart to represent this information.

4

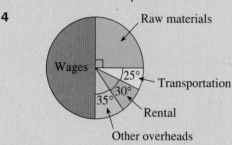

The pie chart illustrates how a firm spent its budget of $540 000 for a year on rental, raw materials, transportation, wages and other overheads.

Calculate the amount spent on each area.

5 The masses of 50 students in a school are shown in the table.

Mass (kg)	No. of students
40–44	8
45–49	12
50–54	17
55–59	9
60–64	4

a Draw a histogram of the data.
b What is the modal mass interval?
c How many students had a mass of 55 kg or more?

A different group of students had the following masses:

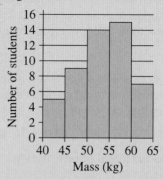

d Which group of students do you think was younger? Give a reason for your answer.

6 The mass of a baby during its first 24 months of life is shown in the graph.

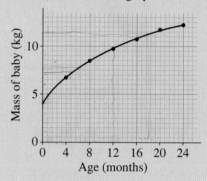

a What was the mass of the baby:
i at birth
ii after 6 months
iii after 19 months?
b At what age was the baby's mass:
i 5 kg   **ii** 9 kg   **iii** 7.3 kg?

7 Display each of these data sets on a pie chart.
a 60 students select sports options as follows:
20 choose football   10 choose cricket
15 choose rounders   15 choose netball
b 120 adults travel to work as follows:
60 walk   20 by car
10 by bicycle   30 by bus

8 Half of a class of 32 students completed a psychological test and were timed while doing so.

a Draw a stem-and-leaf diagram for the following list of times taken, in seconds, to complete the test:

7.2, 8.9, 5.8, 7.4, 10.3, 6.1, 8.3, 6.2, 7.4, 6.7, 7.2, 9.1, 7.4, 8.3, 9.4, 5.7

b What is **i** the mean
ii the median
iii the modal time?

c What is the range of times?

The second half of the class watched while the first 16 students completed the test. Then the second group were asked to complete the same test. Their results are shown in the following stem-and-leaf diagram

```
 5 | 3
 6 | 0  4  7  8
 7 | 0  0  1  2  4  7
 8 | 3  4
 9 | 1  3
10 | 1
```

Key
5 | 3 means 5.3 seconds

d Work out all three averages and the range for the second set of students. Compare the two groups of students.

e Which average was the least helpful in part **d**? Why?

f Thinking about the two different groups (the first half and the second half), what conclusions can you make?

9 The table below shows the growth in the number of staff at a company over the space of 20 years. (For the purpose of this question, assume no one ever leaves the company.)

Year	Number employed
1	7
4	25
8	40
12	55
16	60
20	65

a Draw a line graph for this time series.

b Estimate how many staff the company employed after 7 years.

c Estimate in which year the 50th person was employed by the company.

10 The pie chart shows how a local government agency allocates a budget to different areas.

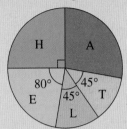

A: Agriculture
E: Education
H: Health
L: Labour
T: Transportation

The agency allocates $30 000 to Education. Calculate:

a the total budget

b the amount allocated to
i Health
ii Labour
iii Agriculture.

c A different region allocates their budget as shown in the pie chart.

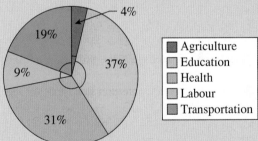

This region spends $6400 of its budget on Agriculture. Copy and complete the table.

Area	Budget allocation ($)
Agriculture	6400
Education	
Health	
Labour	
Transportation	
Total budget	

d Compare the two regions. What conclusions can you make?

e One of the regions is an inner city and one is rural. Which do you think is in the city and why?

Summary

You should know ...

1 How to draw and interpret pie charts.

For example:

The favourite outdoor games of 15 boys are:

Football	7	Basketball	4
Tennis	3	Rugby	1

The 15 boys are represented by
360° so 1 boy is represented by
$\frac{360°}{15} = 24°$

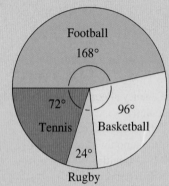

The angles representing favourite games are:

Football: $7 \times 24° = 168°$
Tennis: $3 \times 24° = 72°$
Basketball: $4 \times 24° = 96°$
Rugby: $1 \times 24° = 24°$
(Total angle $= 168° + 72° + 96° + 24° = 360°$)

2 How to draw and interpret stem-and-leaf diagrams.

For example:

Draw a stem and leaf diagram for the prices, in cents, of snacks
sold in a school canteen:

43, 99, 55, 74, 62, 24, 83, 29, 65, 89, 33, 34, 35, 43, 38

Check out

1 The favourite colours of 20 girls are:

Yellow	3
Blue	6
Green	2
Red	5
Orange	4

Draw a pie chart to show this.

2 a Draw a stem-and-leaf diagram for the heights, in centimetres, of these 15 people:

154, 147, 181, 149, 163, 168, 171, 175, 170, 174, 157, 154, 164, 167, 168

b What is the mean height?
c What is the range of heights?
d What is the modal height?

Don't forget to include a key and to write the leaves in numerical order.

```
2 | 4  9
3 | 3  4  5  8
4 | 3  3
5 | 5
6 | 2  5
7 | 4
8 | 3  9
9 | 9
```

Key
2 | 4 means 24 cents

mode = 43
median = 43
range = 75

---

3 How to draw and interpret line graphs for time series.

For example:
The number of tourists visiting a resort were recorded during the first 10 days in August. The total visitors to the resort are recorded on this time series graph. For example, there were 78 tourists on 1 August and 72 tourists on 2 August, so there were 150 tourists in total by 2 August.

During which day in August do you think there was a big tourist market?

Total visitors to resort in August

8 August shows a big jump from the previous day, so it is likely to be the day of the big tourist market.

3 Ruth bought a car in 2007 for $22 000. The estimated value of the car during the following six years is recorded in the table below.

Year	Value ($)
2007	22 000
2008	20 500
2009	17 600
2010	15 500
2011	12 000
2012	8000
2013	4300

Draw a line graph for this time series and comment on what it shows.

4 How to compare data.

For example:

The lists below show the number of books read in a year in a random sample of 15 Class 7 students and 15 Class 11 students.

List A

29, 18, 16, 32, 41, 26, 50, 54, 7, 19, 24, 28, 32, 23, 9

List B

17, 42, 53, 46, 19, 33, 44, 52, 47, 35, 32, 43, 28, 18, 31

Assuming Class 11 do more studying and read more books, which list do you think belongs to Class 11 and why?

List A

In order:	7, 9, 16, 18, 19, 23, 24, 26, 28, 29, 32, 32, 41, 50, 54
Median:	26
Mean:	7 + 9 + 16 + 18 + 19 + 23 + 24 + 26 + 28 + 29 + 32 + 32 + 41 + 50 + 54 = 408
	408 ÷ 15 = 27.2
Mode:	32
Range:	54 − 7 = 47

List B

In order:	17, 18, 19, 28, 31, 32, 33, 35, 42, 43, 44, 46, 47, 52, 53
Median:	35
Mean:	17 + 18 + 19 + 28 + 31 + 32 + 33 + 35 + 42 + 43 + 44 + 46 + 47 + 52 + 53 = 540
	540 ÷ 15 = 36
Mode:	none
Range:	53 − 17 = 36

The best average to compare here is the mean. List B is probably for Class 11, as the mean is higher.

4 These frequency diagrams show the data on the left.

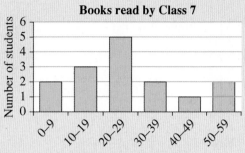

Books read by Class 7

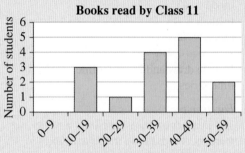

Books read by Class 11

a Why is the mode no good for comparing?

b What does the range tell you?

c Using the frequency diagrams, what is the modal interval for books read for each class?

d Give one advantage of using a stem-and-leaf diagram rather than the diagrams above to represent the data.

5 How to compare proportions in two pie charts that represent different totals.

For example:
The two pie charts below show the drinks sold in a café on different days.

Day A

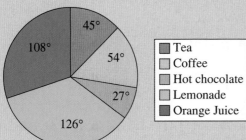

	Tea
	Coffee
	Hot chocolate
	Lemonade
	Orange Juice

Day B

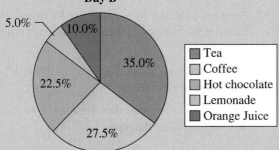

	Tea
	Coffee
	Hot chocolate
	Lemonade
	Orange Juice

On Day A, 36 glasses of orange juice were sold.
On Day B, 28 cups of tea were sold.
Find the total number of drinks sold on each day.

Day A
36 glasses of orange juice are represented by 108°. So one drink is represented by
108° ÷ 36 = 3°
360° ÷ 3° = 120 drinks in total.

Day B
28 cups of tea reprsent 35%. So one drink is represented by 35% ÷ 28 = 1.25%
100% ÷ 1.25% = 80 drinks in total.

5 a Using the pie charts on the left, copy and complete these tables.

Day A

Drink	Number sold
Tea	
Coffee	
Hot chocolate	
Lemonade	
Orange juice	36
Total	120

Day B

Drink	Number sold
Tea	28
Coffee	
Hot chocolate	
Lemonade	
Orange juice	
Total	80

b One of the days was in the summer and the other was in the winter. Which day do you think was in the summer, **A** or **B**? Give a reason for your answer.

Review B

1 Copy and complete:

a $\dfrac{2}{3} = \dfrac{\square}{6}$ **b** $\dfrac{7}{8} = \dfrac{21}{\square}$ **c** $\dfrac{8}{12} = \dfrac{\square}{3}$ **d** $\dfrac{15}{25} = \dfrac{3}{\square}$

2 Solve these equations.

a $2x + 3 = 9$ **b** $x + 4 = 2x - 2$

c $3(x + 1) = 6$ **d** $2x - 3 = x + 6$

3 Find the size of angles a–d.

a
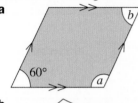

b
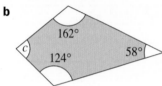

c
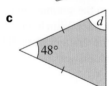

4 The graph below shows the temperature of a cold store after first switching on the cooling unit.

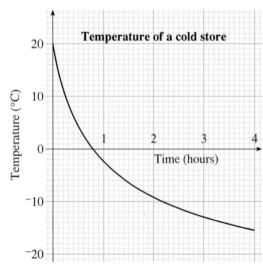

a After how long is the temperature
  i 0°C **ii** ⁻10°C?

b What was the temperature:
  i when the unit was switched on
  ii after 1 hour
  iii after $3\frac{1}{2}$ hours?

5 Work out:

a $\dfrac{7}{10}$ of 6 t **b** $\dfrac{3}{5}$ of 22 kg

c $\dfrac{2}{3}$ of 20 ml **d** $\dfrac{3}{5}$ of 8 m

6 Solve:

a $7m - 8 = 41$ **b** $47 = 5x - 3$
c $6n + 4 = {}^-20$ **d** $3 = 8 - y$
e $60 = 7x + 4$ **f** $39 = 54 - 5r$

7 Sami has $15. She buys a book for $8.79 and two bookmarks costing $2.38 each. How much change does she get?

8 15 students were asked to try to draw a line 4.5 cm long without measuring.

The lengths of the lines they drew are shown in this stem-and-leaf diagram.

2	4	9			
3	1	2	7		
4	0	1	4	4	5	7
5	4	7	9		
6	1				

key
2 | 4 means 2.4 cm

a What was the shortest length drawn?
b What was the longest length drawn?
c What was the **i** mean **ii** median **iii** modal length?

9 Amy has a sweets and Ben has b sweets. Derive an equation for each of these statements:
a Amy and Ben have a total of 50 sweets.
b Amy has 4 times as many sweets as Ben.
c If Amy gave Ben 10 sweets they would both have the same amount of sweets.

10 Find the sizes of angles a–e.

a

b

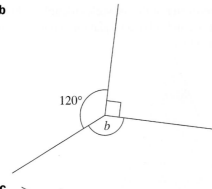

c

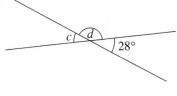

d

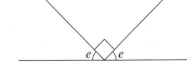

11 Work out:
   a $4 + 0.65 + 3.7$
   b $9 - 5.73$
   c $2.458 + 0.7 + 3.62$
   d $102 - 9.784$

12 Write these fractions in their simplest form.
   a $\frac{6}{8}$ **b** $\frac{12}{16}$ **c** $\frac{16}{20}$ **d** $\frac{20}{24}$

   e $\frac{15}{35}$ **f** $\frac{49}{84}$ **g** $\frac{48}{64}$ **h** $\frac{51}{85}$

13 If $\frac{1}{f} = \frac{1}{u} + \frac{1}{v}$ find f when:
   a $u = 2$ and $v = 4$ **b** $u = 8$ and $v = 12$.

14 A traffic survey counted the number of occupants of 40 cars:

Occupants	Frequency
1	23
2	7
3	4
4	5
5	1

   a Calculate the mean number of occupants per car.
   b On a different day the mean number of occupants per car was 2.3. Which day you think was at the weekend? Why?

15 Find the missing numbers:
   a $3\frac{4}{5} + \square = 6\frac{11}{20}$ **b** $\square - 1\frac{1}{3} = 4\frac{1}{5}$
   c $\square \div 2\frac{2}{7} = 5$ **d** $\square \times 2\frac{3}{4} = 7$

16 Solve:
   a $6(x + 4) = 5(x + 7)$
   b $3(x - 3) = 7(x - 7)$
   c $4(x - 9) = 2(x + 7)$
   d $10(p - 5) = 5(p + 2)$
   e $7(4x - 10) = 9(2x - 5)$

17 Find the midpoint of the line segment AB where
   a A $(6, 8)$ and B $(12, 2)$
   b A $(^-2, 3)$ and B $(14, 7)$
   c A $(^-5, 6)$ and B $(8, ^-4)$

18 Work out:
   a 0.3×8 **b** 6×2.7
   c 12×0.04 **d** 0.08×9
   e 0.006×4

19 Using $v = u + at$, find u when
   a $v = 110$, $a = 6$ and $t = 10$
   b $v = 40$, $a = ^-5$ and $t = 9$

20

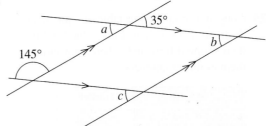

Copy and complete:
   a Angle a is _____° because

   _____.
   b Angle b is _____° because

   _____.
   c Angle c is _____° because

   _____.

21 Work out:
   a $\frac{3}{8} + \frac{4}{8}$ **b** $\frac{7}{12} - \frac{5}{12}$ **c** $\frac{6}{7} + \frac{9}{7}$ **d** $\frac{17}{4} - \frac{9}{4}$

   e $1\frac{3}{5} + 3\frac{2}{5}$ **f** $7\frac{1}{4} - 5\frac{3}{4}$ **g** $\frac{3}{8} + \frac{1}{4}$ **h** $\frac{5}{12} - \frac{1}{6}$

   i $\frac{7}{9} + \frac{2}{3}$ **j** $\frac{11}{15} - \frac{3}{5}$ **k** $\frac{4}{5} + \frac{3}{7}$ **l** $\frac{9}{10} - \frac{6}{7}$

22 Given $a = 2$, $b = ^-1$, $c = 3$, evaluate:
   a $a + 2b$ **b** $ab - c$
   c $ac - 3b$ **d** $ab + bc$
   e $\frac{a}{b} - \frac{c}{b}$ **f** $\frac{abc}{^-6}$

197

23 A cricket ball is thrown up into the air and caught again. The graph shows the height h of the ball at time t.

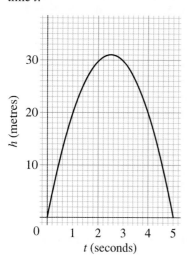

a How long does the ball take to reach its greatest height?
b Estimate the height at $t = 3.5$
c At what times is $h = 15$?

24 Copy and complete this table of equivalent decimals and fractions. Cancel the fractions to their simplest form and, where appropriate, show them as mixed numbers.

Decimal	Fraction
0.8	
	$\frac{1}{4}$
0.3 (or 0.333...)	
	$\frac{3}{10}$
0.05	
1.06	
	$\frac{11}{100}$
	$2\frac{4}{5}$
	$\frac{7}{9}$
8.2	

25 The perimeter of this isosceles triangle is 39 cm. Construct and solve an equation to find missing side length, t.

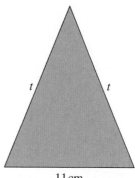

11 cm

26 Using $v = u + at$, find t when
a $v = 40$, $u = 30$ and $a = 5$
b $v = 20$, $u = 48$ and $a = {}^-7$

27 The midpoint of a line segment AB is $(4, 7)$. Write down some possible coordinates for A and B.

28 Work out:
a 2.6×0.5 **b** 4.15×0.4
c 0.3^2 **d** 3.25×0.09
e 6.5×0.07 **f** 3.2×0.004

29 Find:
a $1\frac{3}{4} + 2\frac{4}{5}$ **b** $3\frac{5}{8} - 1\frac{2}{3}$
c $5\frac{1}{6} + 3\frac{5}{9}$ **d** $4\frac{3}{8} - 3\frac{5}{12}$

30 **How Clive spends his week**

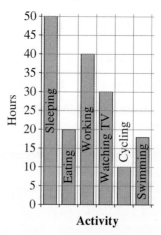

The bar chart shows how Clive spends his week. How many hours are there in a week?
Write down the fraction of each week Clive spends:
a cycling   **b** working
c sleeping   **d** swimming
e awake   **f** exercising.

31 Work out:

a $7 \times \frac{3}{5}$ **b** $\frac{4}{9} \times 5$ **c** $12 \times \frac{5}{6}$

d $8 \times 1\frac{1}{4}$ **e** $2\frac{2}{3} \times 7$ **f** $6 \times 4\frac{1}{4}$

g $4 \div \frac{1}{2}$ **h** $8 \div \frac{2}{3}$ **i** $5 \div \frac{1}{3}$

j $8 \div \frac{3}{4}$ **k** $9 \div 2\frac{1}{4}$

32 The speed of a model airplane at time t seconds is shown in the table.

Time (s)	0	2	4	6	8	10	12
Speed (m/s)	5	4	4	5	7	6	5

Draw a graph from this information and use it to estimate:
a the time at which the speed first reaches 6 m/s.
b the speed after 9 s.

33 These are the lengths, in centimetres, of a catch of 15 fish.
28, 38, 40, 31, 30, 38, 32, 35
29, 30, 32, 37, 38, 38, 39
a Put the numbers in order and find the mean, median, mode and range of the lengths of the catch.
b A different catch of fish had the following data:

      mean   =   24.7
      median =   25
      mode   =   38
      range   =   23

Compare the two catches.
c Do you think the two catches are the same species? Why?

34 Find the value of
a $x^3 + 4x$ when $x = {}^-2$
b $4x^2 - 4$ when $x = 3$
c $5y + y^3$ when $y = {}^-1$
d $3m^2 + 2m - 4$ when $m = {}^-3$
e $3p^3 - p^2 - 7$ when $p = 2$

35 Find the sizes of angles $a-e$.

a

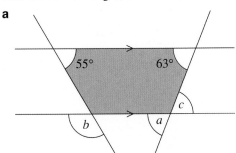

b

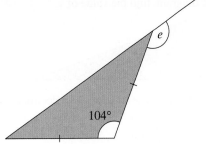

c

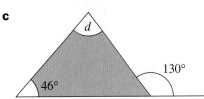

36 Copy and complete:
a $7.35 \div 0.5 = (\square \times 10) \div \square$
    $= 73.5 \div \square = \square$
b $68.4 \div 0.04 = (\square \times 100) \div \square$
    $= \square \div \square = \square$

37 Find the perimeter of this triangle.

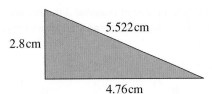

38 The pie chart shows the favourite colours of Class 5.

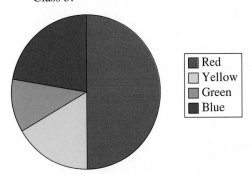

Red
Yellow
Green
Blue

Copy and complete the table.

Colour	Angle	Number of children
Red	180°	9
Yellow	60°	
Green	40°	
Blue	80°	

39 In the diagram, find the value of x.

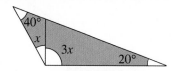

40 Write each of these fractions as a decimal.

a $\frac{7}{10}$ **b** $\frac{13}{100}$ **c** $\frac{429}{1000}$

d $\frac{4}{5}$ **e** $\frac{3}{4}$ **f** $\frac{7}{8}$

g $\frac{17}{20}$ **h** $\frac{23}{50}$ **i** $\frac{49}{50}$

j $\frac{5}{4}$ **k** $\frac{109}{25}$ **l** $\frac{413}{200}$

41 a A tin of paint has a mass of $\frac{2}{3}$ kg. What is the mass of 8 tins?

b Jan buys $2\frac{1}{2}$ kg of seed in one store and $1\frac{3}{5}$ kg in another. How much does he buy altogether?

c Sandra spent $\frac{2}{3}$ of her pocket money on sweets and $\frac{1}{7}$ on a comic. What fraction of her pocket money did she have left?

d Julia bought a piece of material 3 metres wide and $3\frac{1}{4}$ metres long. What was the area of this material?

42 Given $x = 10, y = 25, z = {}^{-}5$, find:

a $\frac{x}{z}$ **b** $\frac{xy}{-z}$ **c** $\frac{x}{z}$ **d** $\frac{10z}{-x}$

43 These are the heights, in centimetres, of 25 children:

140, 155, 157, 142, 167, 153, 163
154, 154, 152, 145, 152, 153, 143
158, 143, 157, 159, 147, 161, 163
147, 159, 161, 170

Draw a stem-and-leaf diagram for this data.

44 If $24 \times \frac{5}{8} = 15$, what is

a $2400 \times \frac{5}{8}$

b $2400 \times \frac{5}{80}$

c $24000 \times \frac{5}{800}$?

45 Sharad is 8 years younger than Gurucharan. How old is each man if the sum of their ages is 52?

46 Write a proof to show that
a the angle sum of a triangle is 180°
b the angle sum of a quadrilateral is 360°
c the exterior angle of a triangle is equal to the sum of the two interior opposite angles.

47 The height, in centimetres, of 30 plants was measured after 8 weeks of growth. The results are shown in the table below.

Height (cm)	Frequency
5-9	3
10-14	8
15-19	11
20-24	7
25-29	1

Draw a histogram for this data.

48 Work out:

a $4.9 \div 0.7$ **b** $4.8 \div 0.04$
c $7.5 \div 0.005$ **d** $1.38 \div 0.06$
e $23.8 \div 0.7$ **f** $3.2 \div 0.004$

49 Using $s = \frac{1}{2}ut + at^2$ find u when $s = 80, t = 5$ and $a = 4$.

50 Find the size of the marked angles.

a

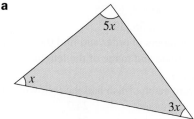

b

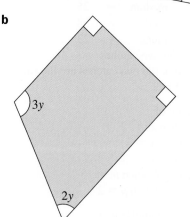

51 Two people start out on the same journey at the same time. One person walks and the other cycles.

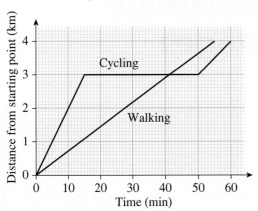

a For how long did the cyclist stop?
b Describe what happened just after 41 minutes.
c After how long did each person reach the destination?
d What speed, in km/h, was the cyclist travelling
  i before they stopped
  ii after they stopped?

52 Derive a formula for the perimeter, P of this rectangle. Use this formula to find y when $P = 96$ and $x = 9$.

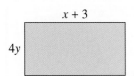

53 Write these in order of size, smallest to largest.
a $0.25, \frac{1}{5}, \frac{21}{100}, 0.03$
b $\frac{3}{4}, 0.8, \frac{87}{100}, \frac{4}{50}, \frac{7}{8}, 0.07$

54 The pie charts below show how two different groups of children travel to school.

How Group A travel to school

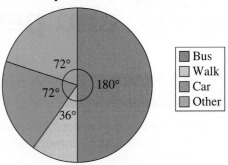

How Group B travel to school

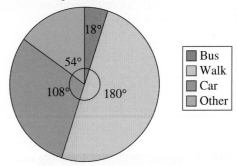

In Group A, 15 students get the bus.
In Group B, 12 students travel by car.
a How many students are there altogether in Group A?
b How many students are there altogether in Group B?
c How many students in each group walk to school?
d Write down some comparisons between the two pie charts.
e One group of students are age 15 and the other group are age 9. Which group do you think was made up of 15-year-olds, A or B? Give a reason for your answer.

13 Fractions, decimals and percentages

Objectives

- Find equivalent fractions, decimals and percentages by converting between them.
- Use equivalent fractions, decimals and percentages to compare different quantities.

- Calculate and solve problems involving percentages of quantities and percentage increases or decreases; express one number as a fraction or percentage of another.
- Recall simple equivalent fractions, decimals and percentages.
- Use known facts and place value to calculate simple fractions and percentages of quantities.

What's the point?

Percentages are everywhere. Your teachers probably give you test results as percentages. Percentages are regularly used in the news. For example, 'Unemployment has fallen by 2%'. Percentages are used in the economy – for example, a percentage of your earnings will be paid in tax.

Before you start

You should know ...

1 How to convert a fraction to a decimal by using equivalent fractions or dividing.
   For example:

   $$\frac{2}{5} = \frac{4}{10} = 4 \div 10 = 0.4$$

   $$\frac{3}{8} = \begin{array}{r} 0.\,3\,7\,5 \\ 8\overline{)3.\,3^{0}6^{0}4^{0}0} \end{array}$$

Check in

1 Write these fractions as decimals.

   a $\frac{3}{10}$ b $\frac{4}{5}$

   c $\frac{5}{8}$ d $\frac{7}{20}$

2 How to convert a decimal to a fraction and cancel the fraction to its lowest terms.
For example:

$$0.6 = \frac{6}{10} = \frac{3}{5}$$

$$0.65 = \frac{65}{100} = \frac{13}{20}$$

3 That a recurring decimal can be written as a fraction.
For example:

$$0.\dot{5} = \frac{5}{9}$$

2 Write these decimals as fractions.

a 0.8 **b** 0.45

c 0.52 **d** 0.04

3 a Write these fractions as decimals.

i $\frac{2}{3}$

ii $\frac{4}{9}$

b Write these recurring decimals as fractions.
i 0.$\dot{7}$

ii 0.$\dot{2}$

13.1 Equivalent fractions, percentages and decimals

You often see percentages in the newspapers.

Daily Journal

3% rise in population since 2000

The ultimate compendium for die-hard automotive enthusiasts and casual onlookers alike, the Automotive Intelligentsia 2009–2010 Sports Car Guide spotlights 54 of the most coveted rides on the road. Full-length profiles place the reader firmly behind the wheel and chronicle each model's heart-pounding performance, advanced technology and storied and specifications. From rough

Experts discuss long-term agriculture strategy

When you are given test results at school, the total mark for tests can differ. For example, one test might be marked out of 10, while another is marked out of 50. Changing your marks from fractions to percentages will help you compare the tests to see how your performance has changed (you will see this later on).

The term 'per cent' comes from the Latin *per centum*. The Latin word *centum* means 100. You will see *cent* in various words, for example a *century* is 100 years and there are 100 *cents* in a dollar.

You need to be able to convert fractions and decimals to percentages.

EXAMPLE 1

Write the fraction $\frac{36}{80}$ as a percentage.

Cancel the fraction using equivalent fractions: $\frac{36}{80} = \frac{9}{20}$

Then multiply by 100: $= \frac{9}{{}_1 \cancel{20}} \times \cancel{100}^5$

$= 45\%$

For harder fractions you can convert the fraction to a decimal using a calculator.

EXAMPLE 2

Using a calculator, write $\frac{19}{32}$ as a percentage rounded to the nearest whole per cent.

First convert the fraction to a decimal using a calculator:

$$19 \div 32 = 0.59375$$

Then multiply by 100 to convert to a percentage:
$0.59375 \times 100 = 59.375\% = 59\%$ to the nearest whole per cent.

You also need to be able to work backwards and convert a percentage into a fraction or decimal.

EXAMPLE 3

Change these percentages into fractions in their lowest terms and decimals:

a 35% **b** 8%

....

a

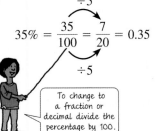

$$35\% = \frac{35}{100} = \frac{7}{20} = 0.35$$

$\div 5$

To change to a fraction or decimal divide the percentage by 100.

b

$\div 4$

$$8\% = \frac{8}{100} = \frac{2}{25} = 0.08$$

$\div 4$

You need to be able to express one quantity as a percentage of another. You use the same ideas, just with a context for the problem.

EXAMPLE 4

450 g of peas contain 18 g of sugar. What percentage of the peas is sugar?

....

Write this as a fraction: $\frac{18}{450}$

Cancel the fraction to its lowest terms:

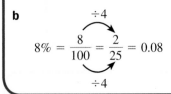

$\div 18$

$$\frac{18}{450} = \frac{1}{25}$$

$\div 18$

Change the fraction to a percentage:

$$\frac{1}{25} \times 100 = 4\%$$

So 4% of the peas is sugar.

Exercise 13A

Do these questions without a calculator unless the question says otherwise.

1 Write these percentages as fractions in their simplest form.

   a 25%   **b** 50%   **c** 75%

   d 16%   **e** 85%   **f** 40%

   g 35%   **h** 3%   **i** 12%

   j 8%

2 Write these fractions as percentages.

   a $\frac{38}{100}$ **b** $\frac{3}{100}$ **c** $\frac{4}{10}$

   d $\frac{7}{50}$ **e** $\frac{17}{25}$ **f** $\frac{9}{20}$

   g $\frac{6}{15}$ **h** $\frac{8}{200}$ **i** $\frac{300}{500}$

   j $1\frac{4}{5}$ **k** $\frac{51}{68}$ **l** $\frac{36}{400}$

3 Copy and complete the table.

Fraction	Equivalent per cent fraction	Percentage
$\frac{3}{5}$	$\frac{60}{100}$	60%
$\frac{1}{4}$	$\frac{25}{100}$	
$\frac{6}{200}$		
$1\frac{3}{4}$		
$\frac{7}{10}$		
$\frac{3}{20}$		
$\frac{40}{250}$		
$\frac{9}{50}$		
$\frac{1}{8}$		
$\frac{96}{120}$		
$\frac{54}{90}$		
$\frac{84}{175}$		

4 You may use a calculator for this question.

Write these fractions as percentages rounded to the nearest whole per cent.

a $\dfrac{17}{32}$ **b** $\dfrac{49}{64}$

c $\dfrac{17}{18}$ **d** $\dfrac{62}{70}$

5 Change each of these test scores to
    i a fraction in its lowest terms
    ii a percentage.
   a 31 out of 50
   b 135 out of 150
   c 60 out of 75
   d 141 out of 200

6 Jenny has to cycle 8 km to school. She has cycled 5 km already. What **a** fraction **b** percentage of her journey has she travelled?

7 Change these decimals to percentages.
   a 0.13   **b** 0.07   **c** 0.2
   d 0.325   **e** 1.25   **f** 2.5

8 48 g out of 200 g in a recipe is sugar. What **a** fraction **b** percentage of sugar is there in the recipe?

9 Change these percentages to decimals.
   a 71%   **b** 5%   **c** 80%
   d 60.5%   **e** 148%   **f** 324%

10 15 cents in each dollar is paid in tax.
   a What fraction is paid in tax?
   b What percentage is *not* paid in tax?

11 Change these decimals to fractions in their simplest form.
   a 0.6   **b** 0.02   **c** 0.55
   d 0.875   **e** 1.75   **f** 2.4

12 Out of a box of 200 apples 10 of them had gone bad.
   a What fraction had gone bad?
   b What percentage of the apples were still good to eat?

13 The table below shows the results when some students were asked if they would like extra maths homework.

	Yes	No	Total
Boys	11	44	55
Girls	9	36	45
Total	20	80	100

Which of these statements are true?
   a More boys than girls wanted extra homework.
   b $\frac{1}{5}$ of the students did not want extra homework.
   c Of the girls, 9% wanted extra homework.
   d Of the boys, 80% did not want extra homework.
   e $\frac{1}{5}$ of the boys wanted extra homework.
   f 36% of the students were girls who did not want extra homework.
   g The percentage of boys that wanted extra homework was higher than the percentage of girls that wanted extra homework.

13.2 Fractions, decimals and percentages of quantities

Finding the percentage of an amount is useful in everyday life. For example, in a 20% discount sale, it would be useful to know how much you will save.

There are a few methods for finding the percentage of an amount. They are in fact all the same calculation, completed in slightly different ways. Here are just some of them – you have lots of choices and the working can also be set out in different ways (you will see this in the next exercise).

Method 1: using fractions
To work out 75% of 800 kg:

Using fractions, 75% is $\dfrac{3}{4}$

$\dfrac{3}{4} \times 800 = 3 \times \dfrac{1}{4}$ of 800

Since $\dfrac{1}{4}$ of 800 = 200, then $\dfrac{3}{4} \times 800 = 3 \times 200$
$= 600\,\text{kg}$

Method 2: using a ratio method (in this case by finding 10%)
To work out 30% of $400:

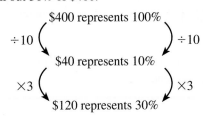

Method 3: using decimals

To work out 15% of 350 m:

15% of 350 m = 0.15 × 350 = 52.5 m

Change 15% to a decimal by dividing by 100
Change the 'of' to '×'

Each method is useful in different circumstances. Methods **1** and **2** are often useful for non-calculator work or working out percentages in your head. Method **3** is useful for working out percentages with a calculator and a good method if you are moving on to percentage increase and decrease questions.

You can use equivalent fractions, decimals and percentages to compare different quantities.

EXAMPLE 5

Mr Green and Mrs Brown both have classes of 40 students.

$\frac{7}{8}$ of Mr Green's class are in school today.

90% of Mrs Brown's class in school today.
Whose class has the most students present today?

....

Mr Green has $\frac{7}{8} = 7 \div 8 = 0.875 = 87.5\%$ students

Since both classes have the same number of students you can just compare the fractions or percentages – there's no need to work out the actual numbers.

Mrs Brown has 90% = 90 ÷ 100 = 0.9 or $\frac{9}{10}$ students

The fractions in their simplest terms ($\frac{7}{8}$ and $\frac{9}{10}$) are hard to compare. It is easier to compare the decimals or percentages.

90% > 87.5% so Mrs Brown had the most students present.

Exercise 13B

1 Sam and Joshua worked out the answer to the question.
'Find 10% of $6.30.'
They both wrote:

$10\% = \dfrac{10}{100}$

10% of $\$6.30 = \dfrac{10}{100} \times \$6.30$

$\qquad\qquad\quad = \dfrac{1}{10} \times \$6.30$

Then Sam wrote:

$= \dfrac{\$6.30}{10}$

$= \$0.63$

And Joshua wrote:

$= \dfrac{\$6.30}{10} = \dfrac{\$6.3\cancel{0}}{1\cancel{0}} = \$6.3$

Which student has made a mistake? What was the mistake?

2 Work these out using a mixture of Methods **1**, **2** and **3**.

  a 5% of $100   **b** 12% of 100 km
  c 25% of 20 ℓ   **d** 75% of $80
  e 37% of $200   **f** 20% of $7.50
  g 8% of 300 g   **h** 15% of 60 children
  i 3% of $1000   **j** 4% of $25

3 Lintang worked out the answer to 35% of $600. She wrote:

$\qquad 600 \div 100 = 6$
$\qquad 6 \times 35 = \$210$

She hasn't quite set this out in the same way as the methods shown above. Out of Methods **1**, **2** and **3**, which do you think she has used? Explain your answer.

4 In each of these sets of quantities, which is the odd one out? Why?

  a $17\%, \dfrac{17}{100}, 0.17, \dfrac{17}{10}$

  b $\dfrac{1}{5}, 20\%, 0.02, \dfrac{10}{50}$

  c $\dfrac{5}{8}, 63\%, \dfrac{15}{24}, 0.625$

  d $300\%, 0.3, \dfrac{30}{10}, 3$

  e $\dfrac{2}{3}, 66.6\%, \dfrac{333}{500}, 0.666$

5 Find:

  a 10% of $9.30   **b** 10% of $11.40
  c 20% of $6.50   **d** 30% of $3.40
  e 15% of $25   **f** 23% of $42

g 3% of $20   **h** 5% of $10.50
i 7% of $95   **j** 4% of $7.50

6 Anya and Sue each have $90.
Anya spends $\frac{5}{8}$ of her money.
Sue spends 60% of her money.
Who has spent the most money?

7 You may use a calculator for this question.
Find:
a 12% of $458   **b** 17.5% of 32 kg
c 18% of 240 ℓ   **d** 1.4% of 5750 km

8

 350g cheese 32% fat

 500g cheese $\frac{1}{4}$ fat

a Which cheese contains the highest percentage of fat?
b Which cheese contains the most grams of fat?
c Are your answers to parts **a** and **b** the same?

9 Which is larger:
a $\frac{3}{5}$ of 400 or 20% of 1250
b 15% of 220 or 0.04 × 800
c 0.25 × 90 or $\frac{1}{3}$ of 60?

10 In an exam 55% of the candidates passed. How many passed if there were
a 200 candidates
b 3000 candidates
c 40 000 candidates?

11 Mr Tang has a 472 km journey to drive. He has driven 45% of this journey already. How many kilometres has he left to drive?

12 Of the 900 children in a school 45% walk to school, 25% go by bus 22% cycle and the rest come by car.
a What percentage come by car?
b Copy and complete this table:

Mode of transport	Walk	Bus	Cycle	Car
Percentage	45%	25%	22%	
Number of children				

13 7.5 tonnes of rubbish were collected. If 42% of the rubbish can be recycled, how many tonnes can be recycled?

14 In each of these sets which is the odd one out? Why?
a 10% of 300, $\frac{2}{5}$ of 75, 0.2 × 15
b 0.3 × 60, 5% of 36, $\frac{3}{4}$ of 24
c $\frac{4}{9}$ of 45, 0.02 × 10, 80% of 25

15 Jamie and Indra share $372. Jamie receives 35% of the money. How many dollars do they each receive?

16 At a garage, 720 cars were given a safety test. If 85% of them passed the safety test, how many cars was this?

13.3 Percentage increase and decrease

A teacher earning $30 000 has a well-earned 5% increase in annual salary. There are two ways of working out their new salary.

The slow way:
Work out 5% of 30 000:
0.05 × 30 000 = 1500
30 000 + 1500 = 31 500

> Change 5% to a decimal, 0.05, and 'of' to 'x'
> Add the increase on to the original

The faster way:
This can be done in a single calculation. An increase of 5% means that you have 105% of the original.

Work out 105% of 30 000:
1.05 × 30 000 = 31 500

> Change 105% to a decimal, 1.05, and 'of' to 'x'

Both methods show that the teacher's new salary is $31 500.

EXAMPLE 6

In a sale, a coat costing $350 was reduced by 20%. What was its sale price?

....

100% − 20% = 80% so the coat now costs 80% of its original price.

80% of $350 = 0.8 × 350 = 280

The sale price is $280

Exercise 13C

1 Match the boxes. The first is done for you.

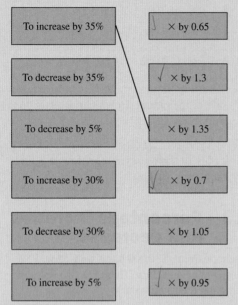

To increase by 35%	× by 0.65
To decrease by 35%	× by 1.3
To decrease by 5%	× by 1.35
To increase by 30%	× by 0.7
To decrease by 30%	× by 1.05
To increase by 5%	× by 0.95

2 Kulwinder and Lee were working out the price of a car originally costing $24 000 after a tax of 8% was added on. Here is their working:

Kulwinder
24 000 × 1.8 = 432 000
New cost $432 000

Lee
24 000 × 0.92 = 22 080
New cost $22 080

Both have made mistakes. Describe their mistakes and write down what the correct working should be.

3 To increase something by 15% the multiplier you should use is 1.15. What multiplier should you use for

a an increase of 12%
b a decrease of 17%
c a decrease of 60%
d an increase of 24%
e a decrease of 4%
f an increase of 8%
g a decrease of 18%
h an increase of 80%
i an increase of 12.5%
j a decrease of 5.5%?

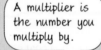

A multiplier is the number you multiply by.

4 A farmer produced 7000 tonnes of crop last year. This year the mass of his crop was down 23%. What was the mass of crop he produced this year?

5 Work out the new values.
a Decrease 200 km by 15%
b Increase $750 by 50%
c Increase 425 ml by 8%
d Decrease 5750 cm by 24%

6 A radio is priced at $120. What is the selling price if 15% value added tax is charged?

7 Factory workers want a 12% increase in pay. How much would a worker get if they currently receive
a $225 a week
b $350 a week?

8 In a restaurant a service charge of 10% is added to the price of the meal. What will be the total bill, when service charge is added, for a meal costing $78?

9 After harvesting, the mass of honey in a row of beehives went down by 12%. If there was originally a mass of 85 kg, how much honey was left in the beehives?

10 Last year at Southfield High School, 150 students passed their maths paper. This year there was a 4% increase in passes. How many students passed their maths paper this year?

11 The insurance premium on Mr Masood's car normally costs $380. With a no-claims discount the premium is reduced by 25%. What is his reduced premium?

12 A runner has a mass of 60 kg at the start of a marathon. During the race his body mass is reduced by 4%. What is his body mass at the end of the race?

13 A machine produces car parts at a rate of 260 per hour. How many car parts per hour does the machine produce if
a it develops a fault and the rate is reduced by 15%
b an improvement is made to the machine and the rate is increased by 20%?

14 Would you rather
   a   have 10% of $5 or 75% of 80 cents?
   b   sit in a queue of traffic for 66% of 1 hour and 10 minutes, or 32% of 2 hours and 5 minutes?
   c   have 80% of 15 sweets or 26% of 50 sweets?

Ask your teacher for a copy of the Percentages game from Teacher Book 2. Play to see who can win.

Using known facts and place value

We can use known facts and place value to make calculations easier.

You know that $4 \times 7 = 28$, so $40 \times 700 = 28\,000$. The same applies to fractions and percentages of quantities.

EXAMPLE 7

a   If $180 \times \frac{2}{5} = 72$, what is 80% of 180?

b   If 60% of 6800 = 4080, what is

   i $6800 \times \frac{3}{10}$ **ii**   15% of 340?

....

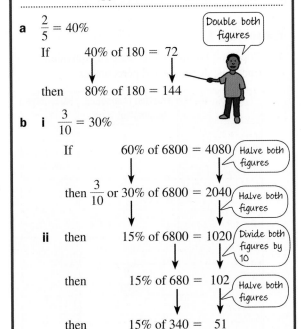

a $\frac{2}{5} = 40\%$

   If   40% of 180 = 72

   then   80% of 180 = 144

   Double both figures

b   **i** $\frac{3}{10} = 30\%$

      If   60% of 6800 = 4080   Halve both figures

      then $\frac{3}{10}$ or 30% of 6800 = 2040   Halve both figures

   ii   then   15% of 6800 = 1020   Divide both figures by 10

      then   15% of 680 = 102   Halve both figures

      then   15% of 340 = 51

Exercise 13D

1   Rewrite the following questions as percentage questions. You do not need to work out the answer. The first one is done for you.
   a $\frac{6}{10}$ of 800 m = 60% of 800 m
   b $0.23 \times \$18$
   c $\frac{3}{8}$ of 20 kg
   d 0.05×700 mm

2   If $6400 \times \frac{7}{10} = 4480$, what is
   a   35% of 6400
   b   35% of 64 000?

3   If $520 \times \frac{2}{5} = 208$, what is
   a   20% of 520
   b   80% of 520
   c   20% of 5200?

4   If 35% of 320 is 112 then $\frac{7}{10}$ of 3200 is 2240.

   Without working out $\frac{7}{10}$ of 3200, and instead using known facts and place value, decide if this statement is true or false. Explain why.

Consolidation

Example 1

a Change 32% to
   i a decimal   **ii** a fraction
b Change $\frac{6}{25}$ to
   i a percentage   **ii** a decimal
c Change 0.05 to
   i a percentage   **ii** a fraction in its lowest terms

·······

a **i** To change 32% to a decimal, divide 32 by 100:
$32 \div 100 = 0.32$

   ii To change 32% to a fraction write 32 over 100 then cancel:
$\frac{32}{100} = \frac{8}{25}$

b **i** To change $\frac{6}{25}$ to a percentage multiply by 100:

$\frac{6}{25} \times 100 = \frac{6}{{}_1\cancel{25}} \times \cancel{100}^4 = 24\%$

   ii To change $\frac{6}{25}$ to a decimal divide the percentage by 100:
$24 \div 100 = 0.24$

c **i** To change 0.05 to a percentage multiply by 100:
$0.05 \times 100 = 5\%$

   ii To change 0.05 to a fraction in its lowest terms write the percentage over 100 then cancel:
$\frac{5}{100} = \frac{1}{20}$

Example 2

Work out:

a 15% of $500

b 3% of 24 km

·······

a 15% of $\$500 = \frac{15}{100} \times \$500$

$= \frac{15}{{}_1\cancel{100}} \times \$\cancel{500}^5$

$= \$\frac{75}{1} = \$75$

b 3% of $24\,km = 0.03 \times 24$

$$\begin{array}{r} 24 \\ \times\ \ 3 \\ \hline 72 \\ 1 \end{array}$$

So $0.03 \times 24 = 0.72\,km$

Example 3

A newspaper headline reads:

5% wage increase for teachers

For a teacher earning $2500 a month, his new pay can be found.

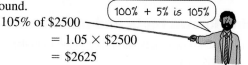

$100\% + 5\%$ is 105%

105% of $2500
$= 1.05 \times \$2500$
$= \$2625$

Example 4

An apple tree had 475 apples on it last year. This year there were 24% fewer apples on the tree. How many apples were on the tree this year?

·······

To decrease 475 by 24%, find 76% of 475:
$0.76 \times 475 = 361$

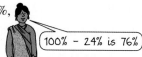

$100\% - 24\%$ is 76%

There were 361 apples on the tree this year.

Exercise 13

1 Copy and complete this table of equivalent fractions, decimals and percentages.

Decimal	Fraction (cancelled to simplest form)	Percentage
0.2		
		25%
	$\frac{7}{8}$	
0.$\dot{3}$ (or 0.333...)		
		12.5%
	$\frac{19}{100}$	
0.75		
		8%
	$\frac{3}{5}$	
		70%

2 Mr Shah has to drive 20 km to work. He has driven 18 km already. What percentage of his journey has he travelled?

3 David said that 3% is 0.3 as a decimal. Explain why he is wrong.

4 The table below records the masses of 2000 babies born during 3 months in a city hospital.

	Less than 4 kg	4 kg or more	Total
Boys	620	280	900
Girls	880	220	1100
Total	1500	500	2000

a What percentage of the total number of babies were
  i boys with a mass of less than 4 kg
  ii girls with a mass of 4 kg or more?
b What percentage of the girls had a mass of less than 4 kg?

5 Write the first amount as
  i a fraction
  ii a percentage of the second.
a 120 cm out of 480 cm
b 140 g out of 2000 g
c 108 ℓ out of 450 ℓ
d 0.51 m out of 3.4 m

6 Find:
a 20% of $460   **b** 17.5% of 8800 kg
c 4% of 575 ℓ   **d** 72% of 2250 km

7 At St Joseph's Academy there are 800 students. 65% of the students are boys. How many students are girls?

8 Out of 160 boxes of bananas, 15% were rejected. How many boxes were accepted?

9 To decrease something by 32%, the multiplier you should use is 0.68. What multiplier should you use for
a an increase of 2%
b a decrease of 27%
c a decrease of 53%
d an increase of 35%
e an increase of 40%?

10 A sales tax of 3% is placed on goods. What is the selling price of a bicycle priced at $950 before tax?

11 A manufacturer produced 9000 cars last month. This month production was down by 12% from last month. How many cars were produced this month?

12 Marlene's meal at the Top Class restaurant cost $95. How much did she pay altogether if there was a 10% service charge and an 8% government tax on all bills?

13 If $3200 \times \frac{7}{20} = 1120$, what is
a 70% of 3200
b 35% of 32 000?

14 Would you rather have $\frac{1}{50}$ of $3500 or 95% of $80?

Summary

You should know ...

1 To convert fractions and decimals to percentages multiply by 100.
To convert percentages to fractions and decimals divide by 100.
For example:
$$\frac{3}{20} \times 100 = 15\%$$
$$0.025 \times 100 = 2.5\%$$
$$35\% = 35 \div 100 = 0.35 = \frac{35}{100} = \frac{7}{20}$$

Check out

1 Copy and complete this table of equivalent fractions, decimals and percentages.

Decimal	Fraction (cancelled to simplest form)	Percentage
0.4		
		65%
	$\frac{3}{8}$	
0.03		
		6%
	$\frac{17}{100}$	

2 How to find percentages of amounts.
For example:

$$6\% \text{ of } \$200 = \frac{6}{100} \times \$200$$

$$= \frac{\$12}{1} = \$12$$

or $0.06 \times \$200 = \$12$

2 Work out:

a 10% of $50   **b** 5% of $10
c 3% of $6   **d** 12% of $25

---

3 How to express a given number as a fraction or percentage of another number.
For example:
Write 169 g as
a a fraction **b** a percentage of 260 g.

a $\dfrac{169}{260} = \dfrac{13}{20}$ using equivalent fractions

b $\dfrac{13}{20} = \dfrac{65}{100}$ using equivalent fractions

$$\frac{65}{100} = 65\%$$

Or (if you are allowed your calculator)
$169 \div 260 = 0.65 = 65\%$

3 Write the first amount as
i a fraction
ii a percentage of the second:

a 437 mm out of 1900 mm
b 4 t out of 50 t
c 70 ml out of 200 ml
d 42 minutes out of an hour

---

4 How to increase and decrease by a given percentage.
For example:

To increase 325 by 28%
find 128% of 325:
$1.28 \times 325 = 416$

100% + 28% is 128%

To decrease 3800 by 15%
find 85% of 3800:
$0.85 \times 3800 = 3230$

100% − 15% is 85%

4 Work out the new values.
a Increase $80 by 5%
b Decrease 800 km by 45%
c Increase 625 g by 16%
d Decrease 32 ℓ by 12.5%

14 Sequences, functions and graphs

Objectives

- Generate terms of a linear sequence using term-to-term and position-to-term rules; find term-to-term and position-to-term rules of sequences, including spatial patterns.

- Use a linear expression to describe the nth term of a simple arithmetic sequence, justifying its form by referring to the activity or practical context from which it was generated.

- Express simple functions algebraically and represent them in mappings.

- Construct tables of values and use all four quadrants to plot the graphs of linear functions, where y is given explicitly in terms of x; recognise that equations of the form $y = mx + c$ correspond to straight-line graphs.

What's the point?

A famous mathematician called Leonardo Fibonacci was well known for writing a book called *Liber Abaci* ('Book of Calculation') early in the thirteenth century. In this book he wrote about a special number sequence now called the Fibonnaci sequence. The sequence is 1, 1, 2, 3, 5, 8, . . . , in which the next number is found from adding the previous two. The Fibonnaci sequence is commonly seen in nature, for example in the arrangement of leaves on the stem of a plant. In computing the Fibonacci search technique is faster than a binary search.

Before you start

You should know ...

1 How to work with negative numbers.
   For example:
   $$2 - 7 = {}^-5$$
   $${}^-3 + 5 = 2$$

Check in

1 Work out:
   a $6 - 10$
   b ${}^-3 + 12$
   c ${}^-8 - 5$
   d ${}^-12 + 5$

2 How to find the value of an expression.
For example:

What is the value of $5n - 3$ if $n = 4$?
When $n = 4$,
$$5n - 3 = 5 \times 4 - 3$$
$$= 20 - 3$$
$$= 17$$

3 How to complete simple sequences by looking at the common difference.
For example:

The next two terms of 4, 6, 8, 10, . . . are 12 and 14 (add 2 each time).

2 Find the value of these expressions when $n = 4$.
 a $2n + 7$
 b $5n - 8$
 c $7n - 10$
 d $3n + 15$

3 Write down the next two terms in these sequences.
 a 15, 20, 25, 30, 35, . . .
 b 30, 40, 50, 60, . . .
 c 4, 7, 10, 13, 16, . . .

14.1 Rules of sequences

Simple investigations

Mathematical investigations are often open-ended questions or problems.

For example:
• What two numbers sum to 5?
• What are the dimensions of a rectangle with area 15cm^2?
• Which shapes have two lines of symmetry?

There is more than one possible answer to such questions. Further, such questions are often just the starting point of an investigation, as other follow-up questions spring to mind.

For example:
• What two numbers sum to:
 a 6 **b** 7 **c** 13?
• What are the dimensions of a triangle with area 15cm^2?
• Which shapes have three lines of symmetry? Four lines of symmetry?

Exercise 14A

Find at least five solutions to each of these open-ended questions.

1 Which two numbers sum to 8?

2 Which two numbers have a product of 36?

3 You can write the number 7 as the sum of two consecutive whole numbers: $3 + 4 = 7$.

Which other numbers can be expressed as the sum of consecutive whole numbers?

4 Which numbers can be written as the difference between two square numbers?

5 The number 141 reads the same when written backwards as it does when written forwards. It is also a multiple of 3. Which other such numbers are there?

6 Which numbers have exactly four factors?

7 You have an unlimited supply of 3 cm and 5 cm rods. What lengths can you make with them?

8 Which prime numbers can be written as the sum of two squares?

Looking for patterns

In many open-ended questions it is quite easy to see patterns.

EXAMPLE 1

Which pairs of numbers sum to 7?

$0 + 7$
$1 + 6$
$2 + 5$
$3 + 4$
$4 + 3$
$5 + 2$
$6 + 1$
$7 + 0$

The numbers on the left-hand side increase by one each time while those on the right decrease by one.

Searching and finding patterns is basic to mathematics. When you find a pattern the next step is to see if you can write down a related rule for the pattern.

EXAMPLE 2

Write down the next three numbers of the sequence:

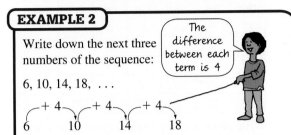
The difference between each term is 4

6, 10, 14, 18, . . .

+ 4   + 4   + 4
6   10   14   18

The difference between 10 and 6 is 4, the difference between 14 and 10 is 4, etc., so the rule is: add 4.

The next three numbers will be 22, 26 and 30.

A **term-to-term** rule describes how to get from one term to the next. For the sequence in Example 2 the term-to-term rule is *add 4*. By continuing the pattern we can find terms further along in the sequence. We can find the 8th term by continuing to add 4 until we have 8 numbers:

6, 10, 14, 18, 22, 26, 30, 32

So the 8th term is 32.

A sequence can be found if you know the term-to-term rule and one number in the sequence.

EXAMPLE 3

The first term of a sequence is 5. The term-to-term rule is: *multiply by 2 then subtract 4*.

1st term: 5
2nd term: $5 \times 2 - 4 = 10 - 4 = 6$
3rd term: $6 \times 2 - 4 = 12 - 4 = 8$
4th term: $8 \times 2 - 4 = 16 - 4 = 12$
5th term: $12 \times 2 - 4 = 24 - 4 = 20$

An **arithmetic sequence** or **linear sequence** is one in which the difference between the terms is the same each time. This is called the **common difference**. In Example 2 the difference was 4 each time so 6, 10, 14, 18, . . . is an arithmetic sequence. In Example 3 the sequence is 5, 6, 8, 12, 20, ... Notice that the differences are 1, 2, 4 and 8. While Example 3 is still a sequence it is not an arithmetic sequence, because the common difference is not the same each time.

A **position-to-term** rule describes how to calculate the term from its position in the sequence. This is often more useful than the term-to-term rule, particularly when you want to find terms that are a long way into the sequence (e.g. the 100th term) without having to work out the sequence from the beginning. In the next exercise you will be using position-to-term rules. You will learn how to work them out later.

EXAMPLE 4

Find the first five terms of the sequences with the position-to-term rules

a multiply by 6

b multiply by 3 then add 2

a

Position	Sequence
1	$1 \times 6 = 6$
2	$2 \times 6 = 12$
3	$3 \times 6 = 18$
4	$4 \times 6 = 24$
5	$5 \times 6 = 30$

$\times 6$

b

Position	Sequence
1	$1 \times 3 + 2 = 5$
2	$2 \times 3 + 2 = 8$
3	$3 \times 3 + 2 = 11$
4	$4 \times 3 + 2 = 14$
5	$5 \times 3 + 2 = 17$

$\times 3 + 2$

Exercise 14B

Write the next three numbers in these sequences.

1 **a** 3, 8, 13, 18, . . .
  b 2, 3, 5, 8, 12, . . .
  c 1, 2, 4, 8, 16, . . .
  d 600, 300, 150, 75, . . .
  e 12, 26, 40, 54, . . .
  f 89, 86, 83, 80, . . .
  g 243, 81, 27, 9, . . .

2 For each of the sequences in Question **1**, write down the term-to-term rule.

3 For each of the sequences in Question **1**, write down the 10th term.

4 What are the missing numbers in these sequences?

  a 23, 29, 35, □, 47, 53, □, 65, . . .

  b □, 3, 8, 13, 18, □, 28, . . .

  c ⁻2, □, 2, □, □, 8, 10, 12, 14, 16, . . .

  d 5, □, □, ⁻1, ⁻3, ⁻5, ⁻7, □, . . .

5 Write down the first five terms of these sequences.

  a The first term is 100, the term-to-term rule is *subtract 5*

  b The first term is 1, the term-to-term rule is *multiply by 3*

  c The fourth term is 11, the term-to-term rule is *add 3*

  d The third term is 90, the term-to-term rule is *subtract 10*

  e The second term is 6, the term-to-term rule is *multiply by 2*

  f The fifth term is 1, the term-to-term rule is *divide by 3*

  g The first term is 3, the term-to-term rule is *multiply by 3 then subtract 5*

  h The first term is 4, the term-to-term rule is *multiply by 5 then add 2*

  i The first term is 116, the term-to-term rule is *divide by 2 then add 2*

6 For each of the sequences in Question **5** say whether or not they are an arithmetic sequence.

7 Write down the first five terms of these sequences.

  a The position-to-term rule is *subtract 2*

  b The position-to-term rule is *add 7*

  c The position-to-term rule is *multiply by 3 then add 4*

  d The position-to-term rule is *multiply by 10 then subtract 1*

  e The position-to-term rule is *multiply by 5 then subtract 3*

  f The position-to-term rule is *multiply by 4 then add 2*

8 You should have found that the sequences in Question **7** are all arithmetic sequences. For each sequence write down the term-to-term rule. Can you see any connection between the term-to-term rule and the position-to-term rule?

9 Here is a page from a calendar:

August						
Mon	**Tue**	**Wed**	**Thu**	**Fri**	**Sat**	**Sun**
			1	2	3	4
5	6	7	8	9	10	11
12	13	14	15	16	17	18
19	20	21	22	23	24	25
26	27	28	29	30	31	

Write down as many linear sequences as you can find. Do they have anything in common?

10 a Write down three different arithmetic sequences with a third term of 10.

  b Write down three different arithmetic sequences with a fifth term of 100.

  c Try to find position-to-term rules for the sequences you have made in parts **a** and **b**.

11 The 10th, 12th and 15th terms of an arithmetic sequence are 35, 43 and 55. Write down the first five terms of this sequence and the position-to-term rule.

14.2 The *n*th term

When we use algebra to describe terms using a position-to-term rule, this is known as finding the **nth term**.

If you wish to predict the 10th or the 52nd number in a sequence you will need to write the rule algebraically. This is can be a more challenging task.

You need to identify the position of each term of the sequence.

EXAMPLE 5

What is the 75th term in the sequence
  2, 4, 6, 8, . . . ?
What is the *n*th term?

....

Position	Term
1	2
2	4
3	6
4	8

The table clearly shows that each term is double the position, so the 75th term will be

$$75 \times 2 = 150$$

The position-to-term rule is $\times 2$. Using algebra, the *n*th term will be $2n$.

Finding the nth term or the algebraic formula for the sequence is not always as easy.

EXAMPLE 6

What is the nth term in the sequence
2, 5, 8, 11, ...?

Look at the differences:

$$+3 \quad +3 \quad +3$$
$$2 \quad 5 \quad 8 \quad 11$$

The common difference is 3 so the sequence will be related to the 3 times table.

Put the terms in a table:

Position	Term
1	2
2	5
3	8
4	11

Compare the sequence with the 3-times table:

2	5	8	11
3	6	9	12

You can see that each term in the sequence is just one less than the 3 times table so the position-to-term rule is $\times$ 3 then -1

Using algebra, the nth term is $3n - 1$.

There is another way to do Example 6 that doesn't involve comparing with the 3-times table. This is the 'position zero' method. Put an extra row at the start of your table with position 0 in it. Continue the sequence backwards by subtracting 3: you get $^-1$ for position zero. The term in position zero is what you need to add or subtract to the $3n$ part.

Position	Term
0	$^-1$
1	2
2	5
3	8
4	11

-3 / $+3$ nth term is $3n - 1$

EXAMPLE 7

What is the nth term in this sequence?
28, 24, 20, 16, 12, ...

Look at the differences:

This is what is written in front of the n

$$-4 \quad -4 \quad -4 \quad -4$$
$$28, \quad 24, \quad 20, \quad 16, \quad 12,$$

Notice you subtract 4 each time.

Work backwards by adding 4 to the first term to get to the term in position zero. This is written after the ^-4n

Position	Term
0	32
1	28
2	24
3	20
4	16
5	12

$+4$ / -4 nth term is $^-4n + 32$

This can also be written with the positive number first. So the nth term is $32 - 4n$.

Exercise 14C

Find the nth term and hence the 43rd term in each of these arithmetic sequences.

1 3, 5, 7, 9, ...

2 207, 205, 203, 201, ...

3 7, 10, 13, 16, ...

4 3, 8, 13, 18, ...

5 34, 31, 28, 25, ...

6 5, 9, 13, 17, ...

7 511, 498, 485, 472, ...

8 Find the nth term for the multiples of 19.

9 Write down some arithmetic sequences of your own and find the nth term for each one.

10 Write down the first five terms of the sequence with nth term $7n + 3$.

11 Match the cards. The first is done for you.

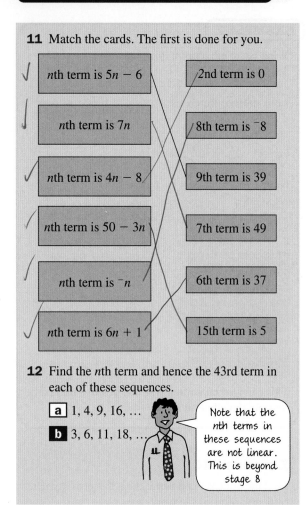

√ *n*th term is $5n - 6$	2nd term is 0
*n*th term is $7n$	8th term is ⁻8
*n*th term is $4n - 8$	9th term is 39
*n*th term is $50 - 3n$	7th term is 49
*n*th term is ⁻*n*	6th term is 37
*n*th term is $6n + 1$	15th term is 5

12 Find the *n*th term and hence the 43rd term in each of these sequences.

a 1, 4, 9, 16, …

b 3, 6, 11, 18, …

Note that the *n*th terms in these sequences are not linear. This is beyond stage 8

Many sequences are formed naturally or grow from simple patterns. In each case the mathematician's task is to find the underlying structure. This structure is usually expressed as an algebraic relationship or a formula. The relationship or formula can then be used to predict terms.

EXAMPLE 8

Look at the sequence generated by the shapes:

$\triangle$ $\triangle\!\!\triangledown$

a How many sticks are needed for the 4th and 5th shapes?

b What is the rule for generating these shapes? Use it to find the number of sticks required for the 76th shape.

a The 4th shape is: $\triangle\!\!\triangle\!\!\triangledown$

It has 9 sticks.

The 5th shape is: $\triangle\!\!\triangle\!\!\triangle$

It has 11 sticks.

b Put the terms in a table:

Shape	Number of sticks
1	3
2	5
3	7
4	9
5	11

The sequence is:

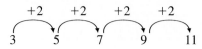

$$\begin{array}{ccccc} +2 & +2 & +2 & +2 \\ 3 & 5 & 7 & 9 & 11 \end{array}$$

The common difference is 2 so the terms are related to the 2 times table:

2	4	6	8	10
3	5	7	9	11

In fact the terms in the sequence are 1 more than the two times table.

The *n*th shape will have $2n + 1$ sticks.

Hence, the 76th shape will have $2 \times 76 + 1 = 153$ sticks.

It is important to be able to understand and justify where the *n*th term comes from using the picture.

You will see from the diagram that there is one blue stick to start the pattern. This is the $+1$ part of the formula. Then when each pattern is formed two extra sticks are added – shown in red, green, and black – which is where the $2n$ comes from.

Exercise 14D

1 Look at the sequences generated by these shapes.

a

b

In each case find the number of sticks required for:

i the 4th and 5th shapes

ii the nth shape

iii the 58th shape.

iv Explain where the nth term comes from using the picture.

2 The table below shows a sequence of shapes made from equilateral triangles with sides of one unit.

Shape	Area of shape (in triangles)	Perimeter of shape
△	1	3
◹◺	2	4
◹◺◹	3	5

a Draw the next three shapes that continue the sequence.

b Copy and continue the table for the next three shapes, indicating their area and perimeter.

c A shape has area 9 triangles. What is its perimeter?

d If the perimeter of a shape is 61 triangles what is its area?

e What would be the perimeter if the area was n triangles?

f Explain your answer to part **c** using the picture to justify where the nth term comes from.

g Write a formula for the perimeter, p, of the shape when the area is n.

3 The table below shows a sequence of octagons made from line segments.

Shape	Number of line segments
⬡	8
⬡⬡	15
⬡⬡⬡	22

a Draw the next two shapes that continue the sequence.

b Continue the table for the next two shapes.

c How many line segments are needed to make the 8th shape in this sequence?

d Write down a formula to show the number of line segments, ℓ, needed to make the nth shape in the sequence.

e Use your formula to find the position in the sequence of the shape with 351 line segments.

4 Look at the tiling pattern shown in the table below.

Tile pattern	Number of tiles
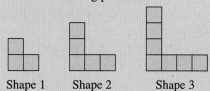	4
	7
	10

a Copy and continue the table for the next three rows.

b How many tiles will be in the 10th tile pattern?

c How many tiles are needed for the nth tile pattern?

d In which position in the sequence is the tile pattern that has 334 tiles?

e Explain where the nth term comes from using the picture.

5 Look at this tiling pattern.

Shape 1   Shape 2   Shape 3

a Draw the next two shapes.

b How many tiles will be needed for the 10th shape?

c How many tiles will be needed for the nth shape?

d Explain where the nth term comes from using the picture.

e Which shape contains 121 tiles?

6 Look at the pattern of blue circles in these diagrams.

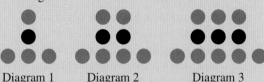

Diagram 1   Diagram 2   Diagram 3

   a Write down the nth term for the number of blue circles in each diagram.

   b How many blue circles will there be in the 100th pattern?

   c If a diagram in the sequence contains 82 blue circles, which diagram is it?

7 Look at the pattern of blue circles in these diagrams.

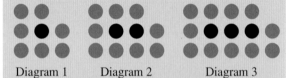

Diagram 1   Diagram 2   Diagram 3

   a Write down the nth term for the number of blue circles in each diagram.

   b How many blue circles will there be in the 50th pattern?

   c If a diagram in the sequence contains 65 blue circles, which diagram is it?

8 Draw some patterns of your own with blue and black counters. (Make sure there is one black counter in the first diagram, two in the second diagram and so on.) Find the nth term for the number of blue counters in your diagrams.

9 The table below shows a sequence of shapes made from dots.

Shape	Number of dots
•	1
•.•:	3
•.•:.•:	6

   a Draw the next two shapes that continue the sequence.

   b Continue the table for the next two shapes.

   c How many dots has the 10th shape in the sequence?

   d A shape is made up of 78 dots. Where does this shape come in the sequence?

   e Copy and complete:

Number of dots in nth

shape $= \frac{1}{2}n \times (........)$

14.3 Functions

This is a function machine:

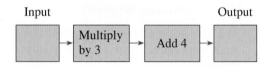

For example:
If 7 is input, then $7 \times 3 + 4 = 25$ is output.

A number entered into the function machine is called the **input**. When a function machine uses the **function** (in this case multiply by 3 then add 4) it produces an **output**. Here, if the input is 2 then the output is 10 because $2 \times 3 + 4 = 10$. If the input is 10 the output is 34, and so on. If the input is n the output is $3n + 4$. We can use a **mapping diagram** like this one to show what outputs go with particular inputs:

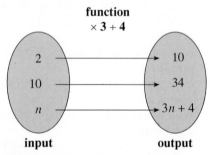

Or it can be shown in a table:

Input	Output
2	10
10	34
n	$3n + 4$

EXAMPLE 9

Complete the table for the function machine.

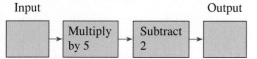

Input   Output

Input	Output
1	
2	
3	
4	
n	

Input	Output
1	3
2	8
3	13
4	18
n	$5n - 2$

Notice when the input is n the output $5n - 2$ is the nth term of the sequence 3, 8, 13, 18, … .

Exercise 14E

1 Use the function machine to fill in the table below.

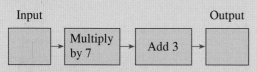

Input   Output

Input	Output
1	
2	
3	
4	
n	

2 Copy and complete this mapping diagram using the function shown.

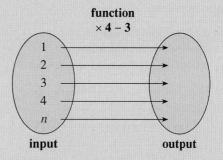

3 For each mapping diagram,
  i draw a function machine
  ii write the output when the input is n.

a

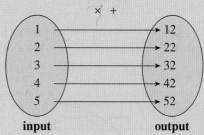

b   function
             × −

input	output
1	2
2	8
3	14
4	20
5	26

c

input	output
1	9
2	14
3	19
4	24
5	29

d

input	output
1	13
2	28
3	43
4	58
5	73

Mapping diagrams don't have to have their inputs in the sequence 1, 2, 3, 4, … . Be careful with these next questions.

4 For these mapping diagrams
  i draw a function machine
  ii write the output when the input is n.
  a

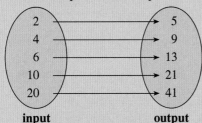

  b

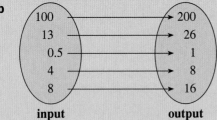

5

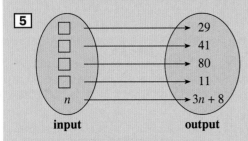

Using the mapping diagram,
Leila wrote the equation $\quad 3n + 8 = 29$
Then she solved it: $\qquad\qquad 3n = 21$
$\qquad\qquad\qquad\qquad\qquad n = 7$

Then Leila wrote:
'The input is 7 for the output of 29.'

Write and solve equations like this to find the missing inputs.

6 Copy and complete this mapping diagram.
  a

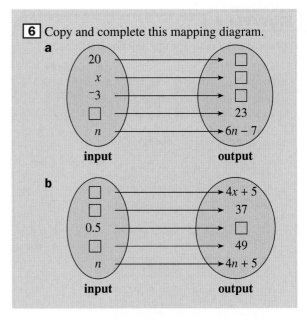

  b

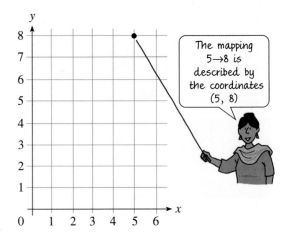

14.4 Linear graphs

Coordinates

On a graph the horizontal axis is often called the x-axis and the vertical axis the y-axis. The **ordered pairs** that describe a mapping on rectangular axes are called **coordinates**.

In the graph below, the black point shows that 5 maps to 8. The coordinates of the point are (5, 8).

> The mapping 5→8 is described by the coordinates (5, 8)

5 is the x-coordinate.

8 is the y-coordinate.

The x-coordinate is *always* written first.

* The coordinates (a, b) of a point on a graph represent the x- and y-coordinates respectively.

Exercise 14F

The graphs show mappings on rectangular axes.
a For each graph list the set of coordinates shown.
b What rule has been used for the mapping?
c Draw a function machine which will produce this mapping.

1

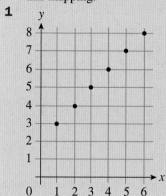

2

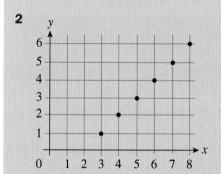

3

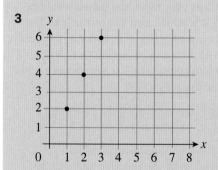

4

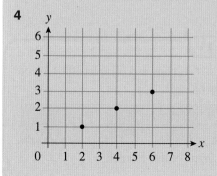

Example 10 illustrates how the rule for a mapping can be shown on a graph.

EXAMPLE 10

Draw a graph of the mapping
$$x \to x + 4$$

The rule is *add 4* so

$$1 \to 5, 2 \to 6, 3 \to 7, 4 \to 8, 5 \to 9$$

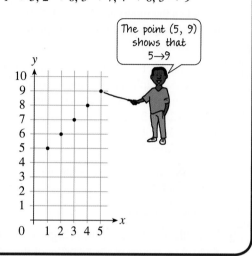

Exercise 14G

1 For the inputs 1, 2, 3, 4 and 5:
   a Draw a graph of the mapping $x \to x + 2$.
   b Draw a mapping diagram of the mapping $x \to x + 2$.
   c List the ordered pairs of the mapping $x \to x + 2$.
   d Which of the three ways of showing the mapping $x \to x + 2$ do you prefer? Why?

2 For the inputs 1, 2, 3, 4 and 5:
   The x-coordinate is mapped to the y-coordinate by the rule $x \to 2x$, so $1 \to 2, 2 \to 4, 3 \to 6$, and so on.
   a List the set of coordinates of the mapping.
   b Draw a graph of the mapping.

3 For the inputs 1, 2, 3, 4 and 5, draw a graph of the mapping of the x-coordinate to the y-coordinate using:
   a $x \to 3x - 2$
   b $x \to 2x + 3$
   c $x \to x$

223

4 The function machine produces coordinates (input, output).

If 3 is the input, the output is 10. The coordinates are (3, 10).

Input 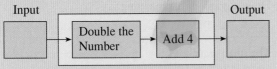 Output

a List the set of coordinates produced when the inputs are 1, 2, 3, 4, 5, 6

b Draw a graph to show the mapping. Label the axes.

5 Using the inputs ⁻2, ⁻1, 0, 1, 2, draw, on the same axes, graphs of the mapping of the *x*-coordinate to the *y*-coordinate using

a $x \rightarrow 2x + 3$

b $x \rightarrow 2x + 1$

c $x \rightarrow 2x - 3$

In each case, join the five points with a straight line.

What do you notice?

6 Using the inputs ⁻2, ⁻1, 0, 1, 2, draw, on the same axes, graphs of the mapping of the *x*-coordinate to the *y*-coordinate using

a $x \rightarrow 2x + 2$

b $x \rightarrow x + 2$

c $x \rightarrow 3x + 2$

In each case, join the five points with a straight line.

What do you notice?

A very simple relation is given by the rule *add 3*. This rule is shown in the mapping diagram.

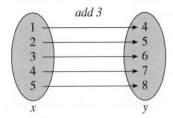

As a number machine this is:

Input → +3 → Output
  x   *y*

or as a mapping:

$x \rightarrow x + 3$

Alternatively, the mapping $x \rightarrow x + 3$ can be shown as ordered pairs on a graph:

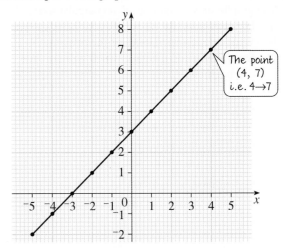

The point (4, 7) i.e. 4→7

Instead of drawing a mapping diagram it is more usual to complete a table for the mapping:

x	1	2	3	4	5
y	4	5	6	7	8

The rule for the mapping $x \rightarrow x + 3$ is rewritten as the equation $y = x + 3$.

The points on the graph can be joined with a straight line.

This is because all of the points in between $x = 1, 2, 3$, 4 and 5 also follow the same rule. For example, if you had $x = 1.5$ as your input, the output would be $y = 1.5 + 3 = 4.5$, which also lies on the line. The same applies to inputs before 1 and after 5: the line has been extended back to $x = {}^-5$, and in fact carries on forever in both directions. The line represents all the mappings of the linear function.

We say the line is represented by the equation
$$y = x + 3$$

The equation $y = x + 3$ means that to find *y*-coordinates you apply the function *add 3* to the *x*-coordinates.

EXAMPLE 11

a If $y = 2x + 1$, copy and complete the table for the mapping.

x	⁻1	0	1	2	3	4	5
y							

b Make a graph of this relation.

a The equation $y = 2x + 1$ can be written as the mapping

$$x \rightarrow 2x + 1$$

or as the function machine

$$x \rightarrow \boxed{\times 2} \rightarrow \boxed{+1} \rightarrow y$$

so when $x = 5$, $y = 2 \times 5 + 1 = 11$

when $x = 4$, $y = 2 \times 4 + 1 = 9$

so the table becomes:

x	$^-1$	0	1	2	3	4	5
y	$^-1$	1	3	5	7	9	11

$(^-1, ^-1)$ $(1, 3)$ $(3, 7)$ $(5, 11)$
$(0, 1)$ $(2, 5)$ $(4, 9)$

b These ordered pairs can be plotted on a graph:

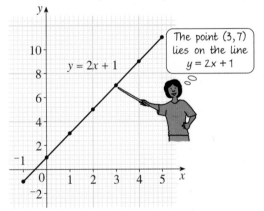

The point $(3, 7)$ lies on the line $y = 2x + 1$

$y = 2x + 1$

Exercise 14H

1 a Copy and complete the mapping diagram for the mapping $x \rightarrow 2x$.

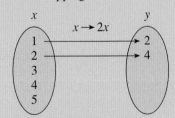

b Write down the ordered pairs of the mapping.
c Show the mapping on a coordinate graph.
d Join the points on the graph. What is the equation of the line?

2 Repeat Question **1** for each of these mappings.

a $x \rightarrow 3x$

1 → 3
2 → 6
3
4
5

b $x \rightarrow x + 2$

1 → 3
2 → 4
3
4
5

c $x \rightarrow x - 3$

1 → $^-2$
2 → $^-1$
3
4
5

d $x \rightarrow 2x + 2$

1 → 4
2 → 6
3
4
5

3 Copy and complete the sentences using the cards (cards can be used more than once).

multiply	add	2	$^-1$	subtract	halve

a When $y = 5x$, to find the y-coordinates you .... the x-coordinates by 5.
b When $y = x - 7$, to find the y-coordinates you .... 7 from the x-coordinates.
c When $y = 2x - 3$, to find the y-coordinates you .... the x-coordinates by ..... then you .... 3.
d When $y = \frac{1}{2}x + 8$, to find the y-coordinates you .... the x-coordinates then you .... 8.
e When $y = 6 - x$, to find the y-coordinates you multiply the x-coordinates by .... then .... 6 or you can .... the x-coordinate from 6.

4 The function machine for the equation
$y = 2x + 2$ is

$x \rightarrow \boxed{\times 2} \rightarrow \boxed{+2} \rightarrow 2x + 2 = y$

Write down function machines for the
following equations:

a $y = 3x$ **b** $y = x + 4$

c $y = x - 6$ **d** $y = 2x + 4$

e $y = 3x - 9$ **f** $y = \dfrac{x}{4} + 7$

g $y = \dfrac{x + 2}{5}$ **h** $y = \dfrac{3x + 2}{4}$

5 To draw a straight line you only need to plot
2 points. Why is it a good idea to actually
work out 3 or more points?

6 a If $y = 2x + 3$, copy and complete the
table for the mapping.

x	$^-2$	$^-1$	0	1	2	3
y				5		9

b Plot the points on a graph and join them
with a straight line.

7 For each equation, copy and complete the
table and then draw its graph.

x	$^-2$	$^-1$	0	1	2	3
y						

a $y = x + 4$ **b** $y = 3x + 4$

c $y = 3x - 2$ **d** $y = 2x - 5$

e $y = 5x - 6$ **f** $y = \frac{1}{2}x + 2$

8 For each equation, copy and complete
the table.

x	$^-5$	$^-4$	$^-3$	$^-2$	$^-1$	0	1	2	3	4	5
y											

a $y = 5x + 5$ **b** $y = 3x + 7$

c $y = 2x + 3$

9 Copy and complete these coordinate pairs for
the equation $y = 10x + 7$ by calculating the
missing y-coordinate using the given
x-coordinate.

a $(7, \square)$ **b** $(5, \square)$

c $(^-2, \square)$ **d** $(^-1, \square)$

10 One of these points does not lie on the line
$y = 2x + 7$. Which one is it?

$(1, 9)$ $(4, 15)$ $(3, 12)$ $(0, 7)$

11 Which of these points lie on the line
$y = 7 - x$?

$(3, 4)$ $(0, 7)$ $(^-1, 6)$ $(^-2, 9)$

12

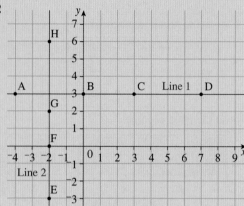

a Complete the coordinate pairs for Line 1
and Line 2 drawn on this graph.

Line 1:   A $(^-4, \square)$ B $(0, \square)$
           C $(3, \square)$ D $(7, \square)$

Line 2:   E $(\square, ^-3)$ F $(\square, 0)$
           G $(\square, 2)$ H $(\square, 6)$

b What are the equations of these lines?

13

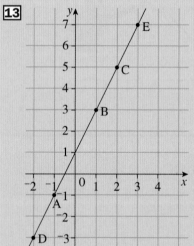

a Use the graph to complete the coordinate
pairs.

A $(^-1, \square)$ B $(1, \square)$ C $(2, \square)$

D $(^-2, \square)$ E $(3, \square)$

b How is the y-coordinate connected to the
x-coordinate?

c What is the equation of the line?

14 a On centimetre-square paper, draw a pair of rectangular axes. On the x-axis, show all the numbers from $^-5$ to 5. On the y-axis show all the numbers from $^-20$ to 30.

b Using these axes, show each set of ordered pairs (x, y) from your tables in Question **8**. Use a different colour for each set of points.

c Join the points in each set with a straight line.

d Write down the coordinate of the point where each line crosses the y-axis (intercept).

e Compare your answer to part **d** with the equations. What do you notice?

f Which line is the steepest?

15 a For each equation, copy and complete the table.

x	$^-2$	$^-1$	0	1	2	3
y						

i $y = x$
ii $y = 2x$
iii $y = 3x$
iv $y = 4x$

b On suitable axes, plot each of these four graphs.

c Which graph is the steepest?

d Repeat parts **a**, **b** and **c** for
i $y = {}^-x$
ii $y = {}^-2x$
iii $y = {}^-3x$
iv $y = {}^-4x$

e What do you notice?

TECHNOLOGY

Review how to plot graphs of equations by visiting
www.purplemath.com/modules
(look for 'graphing linear equations')

Equations in the form $y = mx + c$

There are many different equations which make many different graphs. Some produce straight lines, some produce curves. It is important to be able to tell just from looking at an equation whether or not it will produce a straight line.

In Book 1 you learned that the equation $y = 4$ means that the y-coordinate is always 4, no matter what the x-coordinate is. Any equations of the form $y = $ 'a number' are horizontal lines. Vertical lines are described by $x = $ 'a number'. For example, $x = 2$ means that the x-coordinate is always 2, no matter what the y-coordinate is.

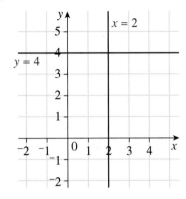

Exercise 14H involved mostly diagonal lines. Diagonal lines contain an x and a y in the equation. There are no powers of x other than x^1 in the equation (for example there is no x^2, $\frac{1}{x}$, x^3, etc.). We usually write the equation for a diagonal line in the form $y = mx + c$, where m and c stand for numbers. If the value of c is 0 the line will still be diagonal. If the value of m is 0 then the line will be horizontal rather than diagonal. The values m and c can be positive or negative. Diagonal straight lines can be described in other ways – you will learn about this in Book 3. You will also learn more about the significance of the values of m and c in stage 9.

Exercise 14I

1 Kanika says that in the line $y = 8 - 3x$, $m = 8$ and $c = {}^-3$. She has made a mistake. What mistake has she made?

2 For these equations of straight lines in the form $y = mx + c$, write down the value of
    i m **ii** c
a $y = 5x + 8$
b $y = {}^-x + 2$
c $y = 3x$
d $y = 7 - x$
e $y = x + 6$
f $y = 9 - 2x$
g $y = 4 + x$
h $y = {}^-3x - 7$

3 Sort these equations of lines into one of the three straight line categories shown.

$y = x$ $y = 3$ $y = {}^-3x$ $x = 7$

$y = 3x - 4$ $y = {}^-x + 3$ $y = 2x$

Categories:

Horizontal line   Vertical line   Diagonal line

4 Which of these are straight lines?

$y = 4x - 7$ $y = x^2$ $x = 7$ $y = \dfrac{1}{x}$

$y = 8 - x$ $y = x^2 - 3x + 4$ $y = 7 - x^3$

$y = {}^-3$

5 Sort these lines into the two groups shown below.

$y = x$ $y = {}^-4x$ $y = 6x - 2$ $y = 7x$

$y = {}^-x + 9$ $y = 5x + 5$ $y = 11 - 6x$

$y = {}^-x$

Groups:

Line sloping   Line sloping
upwards   downwards

6 For each of these straight lines write down five points on the line and work out the equation of the line.

a

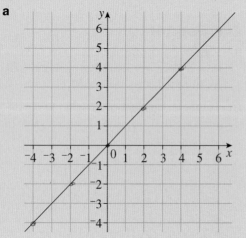

b

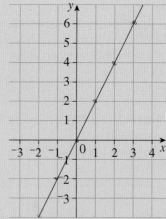

c

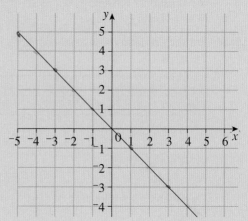

d

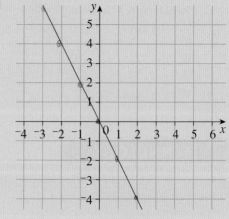

Consolidation

Example 1

Find the first five terms of these sequences.

a The first term is 5, the term-to-term rule is *add 3*.

b The position-to-term rule is *multiply by 2 then add 5*.

a
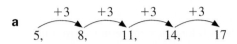

b

Position	1	2	3	4	5
Sequence	1×2 $+ 5$ $= 7$	2×2 $+ 5$ $= 9$	3×2 $+ 5$ $= 11$	4×2 $+ 5$ $= 13$	5×2 $+ 5$ $= 15$

$\times 2 + 5$

So the first five terms of the sequence are 7, 9, 11, 13, 15

Example 2

a What is the sequence of the areas made by the shapes below?

b What is the *n*th term of the areas made by the shapes below?

c Justify the *n*th term by relating it back to the diagram.

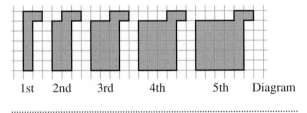

1st   2nd   3rd   4th   5th   Diagram

a The sequence in the areas of the shapes is 7, 12, 17, 22, 27, …

b To find the *n*th term look at the differences:

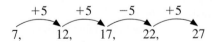

The difference is *add 5* each time so it is a linear relationship to do with multiples of 5.
Compare the terms of the sequence with the sequence 5*n*:

7,   12,   17,   22,   27   Sequence
5,   10,   15,   20,   25   5*n* (or multiples of 5)

The terms in the sequence are each 2 more than the corresponding multiple of 5. The *n*th term is therefore $5n + 2$.

c Each time you change diagrams you have 5 extra squares added. This is where the 5*n* comes from. The $+ 2$ comes from the extra two squares at the top of the final column of squares.

Example 3

a Complete this mapping diagram using the function shown.

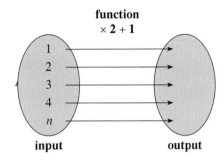

b If the inputs are the *x*-coordinates and the outputs are the *y*-coordinates write the equation of the line.

c Copy and complete the table below using the equation you found in part **b**.

x	⁻2	⁻1	0	1	2
y					

d Write down the coordinates of the five points you found from the table.

e Draw the line of the equation you found in part **b**.

a
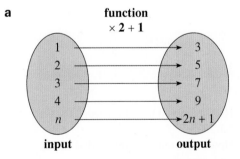

b Using $x \rightarrow 2x + 1$ gives us the equation $y = 2x + 1$. Notice this equation is of the form $y = mx + c$ so it will be a straight line graph.

c

x	⁻2	⁻1	0	1	2
y	⁻3	⁻1	1	3	5

d The table gives us the coordinates:
$(^-2, ^-3)$ $(^-1, ^-1)$ $(0, 1)$ $(1, 3)$ $(2, 5)$

e Plotting these points and joining with a straight line gives:

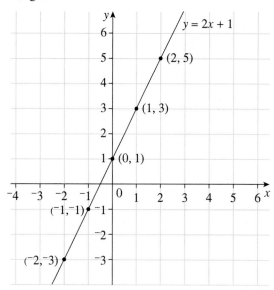

Exercise 14

1 Write down the next three terms in each arithmetic sequence.
  a 7, 15, 23, 31, ...
  b 104, 93, 82, 71, ...
  c 10, 20, 30, 40, ...
  d 5, 11, 17, 23, ...
  e 17, 13, 9, 5, ...

2 Find the nth term of each of the sequences in Question **1**.

3 Look at the tiling pattern:

Shape 1   Shape 2   Shape 3

  a How many red tiles are needed for the 5th shape?
  b How many red tiles are needed for the 15th shape?
  c How many red tiles are required for the nth shape?
  d If the mth shape has 100 red tiles, what is the value of m?
  e Justify your nth term by using the diagrams to explain where it comes from.

4 Write down the first five terms of these sequences.
  a The first term is 6, the term-to-term rule is *add 4*
  b The first term is 72, the term-to-term rule is *subtract 10*
  c The first term is 64, the term-to-term rule is *divide by 2*
  d The first term is 3, the term-to-term rule is *multiply by 3*
  e The first term is 3, the term-to-term rule is *multiply by 4 then add 2*

5 For each of the sequences in Question **4**, say whether it is an arithmetic sequence or not.

6 Write down the first five terms of these sequences.
  a The position-to-term rule is *subtract 5*
  b The position-to-term rule is *add 6*
  c The position-to-term rule is *multiply by 4 then add 7*
  d The position-to-term rule is *multiply by 8 then subtract 3*

7 **a** Copy and complete this mapping diagram using the function shown.

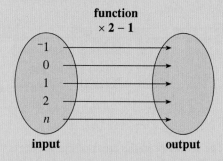

  b If the inputs are the x-coordinates and the outputs are the y-coordinates write the equation of the line.
  c Copy and complete the table below using the equation you found in part **b**.

x	$^-1$	0	1	2
y				

  d Write down the coordinates of the four points you found from the table.
  e Draw the line of the equation you found in part **b**.

8 What can you say about your answers to Question **7**, parts **a**, **c** and **d**?

9 a Using the inputs $^-2$, $^-1$, 0, 1, 2, draw mapping diagrams for the functions
    i $\times 2 + 4$ **ii** $\times 3 - 5$

  b Using your answers to part **a**, draw the lines
    i $y = 2x + 4$ **ii** $y = 3x - 5$

10 What are the missing numbers in these sequences?
  a 32, 38, 44, ☐, 56, 62, ☐, 74, …
  b ☐, 46, 42, 38, 34, ☐, 26, …

11 Find the nth term for the multiples of 8.

12 One of these points does not lie on the line $y = 8x - 3$. Which is it?
  $(1, 5)$ $(4, 29)$ $(0, ^-3)$ $(7, 59)$

13 Copy and complete the table for the equation $y = 2x - 3$.
Use it to draw the graph of $y = 2x - 3$

x	$^-2$	$^-1$	0	1	2
y					

14 Write down the first five terms of the sequence with nth term $11n + 4$.

15 Using the inputs $^-2$, $^-1$, 0, 1, 2, draw graphs of the mapping of the x-coordinate to the y-coordinate using:
  a $x \rightarrow x - 1$ **b** $x \rightarrow 3x + 1$
  c $x \rightarrow 2x + 4$ **d** $x \rightarrow 5 - 2x$

16 Using the equation $y = 9x + 10$, copy and complete these coordinate pairs by calculating the missing y-coordinate using the given x-coordinate.
  a $(6, ☐)$ **b** $(3, ☐)$
  c $(^-4, ☐)$ **d** $(^-5, ☐)$

17 Write the equation that goes with each sentence.
  a To find the y-coordinate multiply the x-coordinate by 4 then subtract 3.
  b To find the y-coordinate subtract the x-coordinate from 2.
  c To find the y-coordinate halve the x-coordinate then add 8.
  d To find the y-coordinate add 3 to the x-coordinate then halve that result.

18 Which of these points lie on the line $y = 8 - 2x$?
  $(3, 4)$ $(0, 8)$ $(^-1, 10)$ $(^-2, 4)$

Summary

You should know …

1 How to find terms in a sequence.
For example:
  a Find the next four terms of the sequence with first term 81 and term-to-term rule *subtract 4*
  b Find the first five terms of the sequence with position-to-term rule *multiply by 10 then add 5*

a
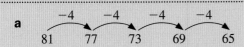

So the next four terms are 81, 77, 73, 69, 65

b

Position	1	2	3	4	5
Sequence	1×10 $+ 5$ $= 15$	2×10 $+ 5$ $= 25$	3×10 $+ 5$ $= 35$	4×10 $+ 5$ $= 45$	5×10 $+ 5$ $= 55$

$\times 10 + 5$

So the first five terms are 15, 25, 35, 45, 55

Check out

1 a Find the next three terms in these sequences.
   i 17, 22, 27, 32, …
   ii 3, 6, 11, 18, …

  b Find the first five terms of the sequence with first term 7 and term-to-term rule *add 11*

  c Find the first five terms of the sequence with position-to-term rule *multiply by 6 then subtract 10*

2 How to find the rule for a sequence.
For example:
6, 13, 20, 27, . . .

+7   +7   +7
6   13   20   27

7 is added to the previous term.
The sequence is related to the 7-times table:
7   14   21   27
Each term is one less than the 7-times table,
so the nth term is $7n - 1$.

2 Find the:
    i   term-to-term rule
    ii nth term
of these sequences.
a   3, 5, 7, 9, . . .
b   19, 22, 25, 28, . . .
c   88, 81, 74, 67, . . .

---

3 Graphs with equations in the form $y = mx + c$ are straight lines.
For example:
$y = 12x - 4$ and $y = 8 - x$ are straight lines.
For $y = 12x - 4$, $m = 12$ and $c = {}^-4$
For $y = 8 - x$, $m = {}^-1$ and $c = 8$

3 For these lines with equations in the form $y = mx + c$ write down the value of
    i m
    ii c
a $y = 2x + 7$
b $y = x - 4$
c $y = 8x$
d $y = 10 - x$

---

4 The function $x \to x - 3$ represents the equation $y = x - 3$. To find the y-coordinates you need to subtract 3 from the x-coordinates.

Completing a table gives you

x	$^-1$	0	1	2	3
y	$^-4$	$^-3$	$^-2$	$^-1$	0

The graph of $y = x - 3$ is

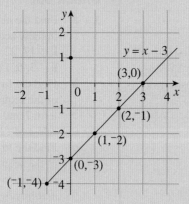

4 For each equation, complete the table and draw the graph.

x	$^-2$	$^-1$	0	1	2
y					

a $y = 2x$
b $y = x + 3$
c $y = 2x - 3$
d $y = 4 - x$

Transformations

Objectives

- Transform 2D shapes by rotation, reflection and translation, and by simple combinations of these transformations.

- Understand and use the language and notation associated with enlargement; enlarge 2D shapes, given a centre of enlargement and a positive integer scale factor.

- Interpret and make simple scale drawings.

What's the point?

A suitable enlargement of a photograph of yourself will result in a life-size image. In the same way the transformation of enlargement is vital for using and making maps and scale drawings.

Before you start

You should know ...

1 About rotational symmetry.
For example:

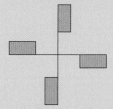

This shape has rotational symmetry of order 4 because it fits back on itself four times when turned through one complete turn.

Check in

1 What is the order of rotational symmetry of these shapes?

a

b

c

d

2 About lines of symmetry.
For example:

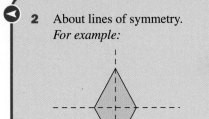

This rhombus has two lines of symmetry.

2 How many lines of symmetry do these shapes have? Copy each shape and draw its lines of symmetry.

a   **b**

c   **d**

15.1 Reflection

You will need tracing paper, squared paper and a small mirror.

- An object reflected in a mirror line produces an **image**.

The object and image are identical, but point in opposite directions. The reflection forms an **opposite image**.

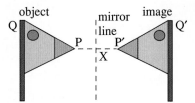

The image of P is P′. We say P maps to P′. In the same way Q maps to Q′.

A line from P to P′ intersects the mirror line at X. The distance PX = P′X.

- A reflection is a 'flipping over' movement.

Exercise 15A

1 In the diagram, PQ is reflected in a mirror line *m* to produce P′Q′.
PP′ meets *m* at X, QQ′ meets *m* at Y.
a What is the image of P′Q′?
b What is the image of PXYQ?
c What is the image of angle PXY?

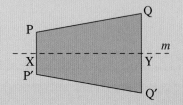

d Without measuring, what can you conclude about the size of angle PXY?
e Do you agree that the line joining an object and its image must intersect the mirror line at right angles?

2 Copy these shapes onto squared paper. Draw their reflections in the dashed line.

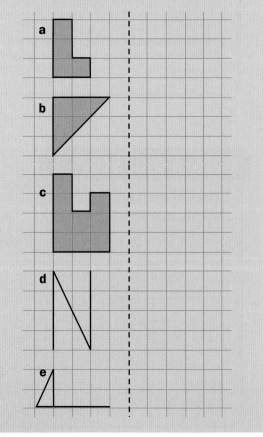

3 Trace each shape and the dashed line. Draw the reflection of the shape in the dashed line.

a

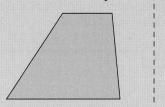

b

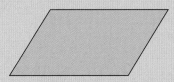

c

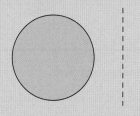

4 Copy each diagram onto squared paper. Reflect the shape in Line y.

a

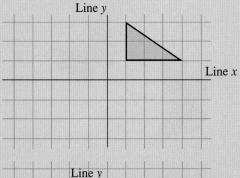

b

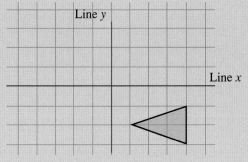

c

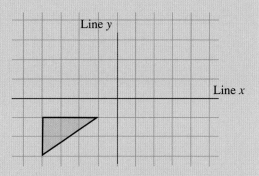

5 On the same diagrams that you drew for Question **4** draw the images formed by reflection in Line x.

6 Copy each diagram onto squared paper. Draw the reflection of the shape in the dashed line.

a

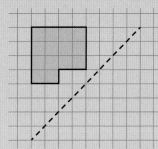

b

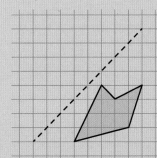

c

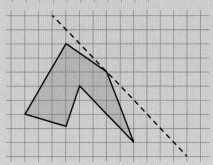

d

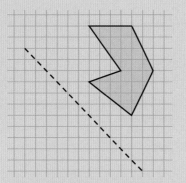

7 a Copy this drawing of a screw, and draw its reflection in the dashed line.

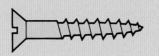

b Would a nut that fitted a bolt also fit its reflection?

8 Each diagram below shows both the object and the image after reflection. Copy the diagrams and draw in the mirror lines.

a

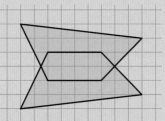

b

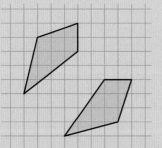

c

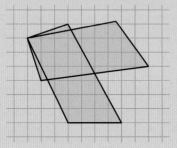

15.2 Rotation

Angles of turn are usually measured in an **anticlockwise** direction. So, in the exercises that follow, rotation is anticlockwise unless stated otherwise.

EXAMPLE 1

Describe these rotations.

S is mapped to S' by the rotation centre C, angle 130°.

S' is mapped to S by the rotation centre C, angle 230°

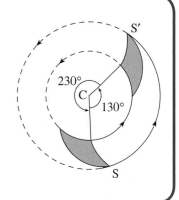

• A rotation is a turning movement.

Exercise 15B

1 Describe the rotation of the butterfly.

2 Copy each of these shapes onto squared paper. Draw the image formed by rotating the shape through the angle shown about the point O. You may use tracing paper to help you.

a

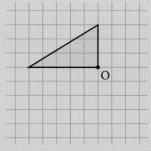

i 90°
ii 180°

b

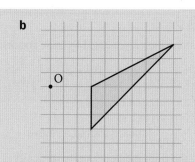

  i   90°
  ii 180°

c

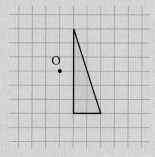

  i   270°
  ii 180°

d

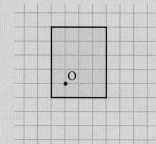

  i   180°
  ii 270°

3 Copy the diagram. You may use tracing paper to help you.

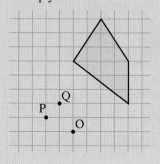

Draw the image formed by rotating quadrilateral ABCD
a   90° about the point O
b   180° about the point P
c   270° about the point Q

4 Use tracing paper to help you describe these rotations.

a

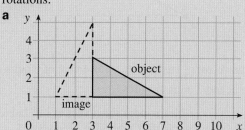

b

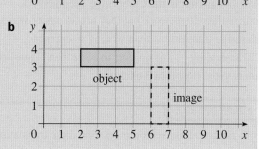

15.3 Translation

You will need tracing paper, squared paper and a ruler.

A reflection and a rotation are two ways of mapping an object onto its image.
Another way is a **translation**.

- **A translation is a sliding movement.**

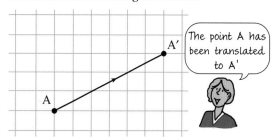

The point A has been translated to A'

Look at the diagram above. The point A has been moved or translated to the point A'.

237

This change of position can be described using a column vector:

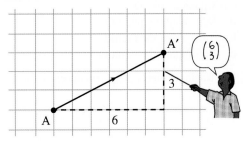

The vector $\binom{6}{3}$ represents a movement of 6 units to the right and 3 units upwards.

Note you will learn more about vectors in Chapter 19. Vectors are not in the Cambridge Secondary 1 curriculum framework but they are an easier way to describe translations than writing a sentence.
'Translate by $\binom{6}{3}$' is quicker to write than 'Translate 6 units to the right then 3 units up.'

EXAMPLE 2

Write a column vector to describe these translations:

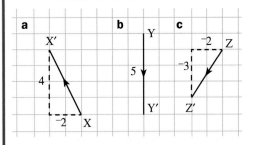

a $X \to X'$ is a movement of 2 units to the left ($^-2$) and 4 units upwards (4),

so $\binom{-2}{4}$

b $Y \to Y'$ is a movement of 0 units to the right (0) and 5 units downwards ($^-5$),

so $\binom{0}{-5}$

c $Z \to Z'$ is a movement of 2 units to the left ($^-2$) and 3 units downwards ($^-3$),

so $\binom{-2}{-3}$

Given the column vector you can easily draw the translation.

EXAMPLE 3

Draw the transformation represented by these vectors.

a $A \to A' = \binom{2}{-5}$ b $B \to B' = \binom{-3}{-4}$

a $A \to A' = \binom{2}{-5}$ means start at A, move 2 units to the right then go 5 units vertically downwards to A'.

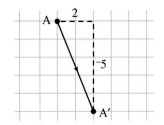

b $B \to B' = \binom{-3}{-4}$ means start at B, move 3 units to the left then go 4 units vertically downwards to B'.

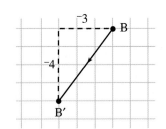

Notice the arrow on the line – this is important as it shows the direction.

Translating shapes

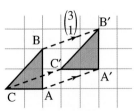

In the diagram above triangle ABC is translated to triangle A'B'C' by the vector $\binom{3}{1}$.

The vertices of triangle ABC are translated to the vertices of A'B'C'.

Notice that the two triangles ABC and A'B'C' are identical in every respect: they both have the same size and shape. In other words triangle ABC is **congruent** to triangle A'B'C'.

Exercise 15C

1 Write column vectors to describe each of these translations.

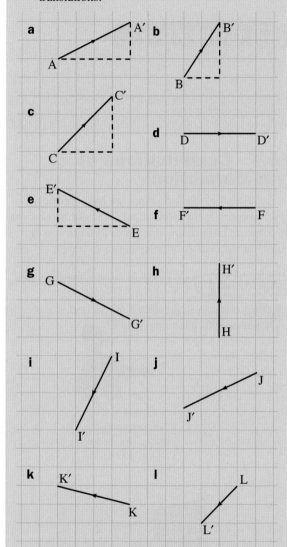

2 On squared paper, draw the translation represented by the column vector:

a $A \to A' = \begin{pmatrix} 2 \\ 3 \end{pmatrix}$ b $B \to B' = \begin{pmatrix} 3 \\ 2 \end{pmatrix}$

c $C \to C' = \begin{pmatrix} 4 \\ 0 \end{pmatrix}$ d $D \to D' = \begin{pmatrix} -3 \\ 2 \end{pmatrix}$

e $E \to E' = \begin{pmatrix} -4 \\ 3 \end{pmatrix}$ f $F \to F' = \begin{pmatrix} 2 \\ -3 \end{pmatrix}$

g $G \to G' = \begin{pmatrix} 5 \\ -1 \end{pmatrix}$ h $H \to H' = \begin{pmatrix} -5 \\ -2 \end{pmatrix}$

i $I \to I' = \begin{pmatrix} -3 \\ 0 \end{pmatrix}$ j $J \to J' = \begin{pmatrix} -4 \\ -5 \end{pmatrix}$

3 Write a column vector to describe the translation of each triangle A to its image B.

a
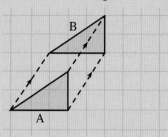

b
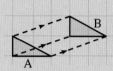

c
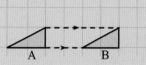

d
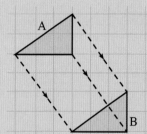

e
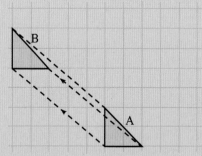

4 Copy this triangle onto squared paper.

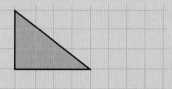

Translate the vertices using the vector $\begin{pmatrix} 5 \\ 1 \end{pmatrix}$. Draw the image. Is it congruent to the first triangle?

5 Repeat Question **4** but this time translate the triangle using the vectors:

a $\begin{pmatrix} 5 \\ 2 \end{pmatrix}$ **b** $\begin{pmatrix} 2 \\ 3 \end{pmatrix}$ **c** $\begin{pmatrix} {}^{-}1 \\ 3 \end{pmatrix}$ **d** $\begin{pmatrix} {}^{-}4 \\ {}^{-}2 \end{pmatrix}$

6 Write a column vector for a single translation from S to U, shown in the diagram.

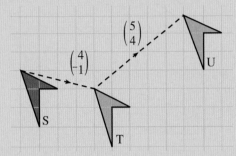

7 a From this diagram write **a**, **b** and **c** as column vectors.

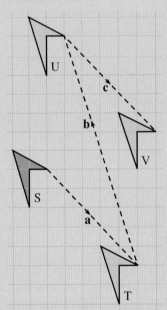

b What can you say about **a** and **c**?
c Write a single column vector for the translation of S to U.
d What single column vector gives the translation of S to V?
e Complete the statements:

$\begin{pmatrix} 5 \\ {}^{-}5 \end{pmatrix} + \begin{pmatrix} {}^{-}4 \\ 12 \end{pmatrix} =$

> You may want to look at Chapter 19 for help.

$\begin{pmatrix} 5 \\ {}^{-}5 \end{pmatrix} + \begin{pmatrix} {}^{-}4 \\ 12 \end{pmatrix} + \begin{pmatrix} 5 \\ {}^{-}5 \end{pmatrix} =$

f What column vectors give the translations of U to S and of V to S?

Transformations such as reflection, rotation and translation leave the size and shape of an object unchanged. The object and image under such transformations are **congruent** to each other.

TECHNOLOGY

Learn more about congruent triangles by visiting the website

www.onlinemathlearning.com/congruent-triangles.html

15.4 Combinations of transformations

So far all of the transformations we have looked at have been single transformations. Transformations can be combined so that an object goes through two or more transformations to map it onto an image.

EXAMPLE 4

Reflect object A in the dotted mirror line. Then rotate 90° about the point O

> Remember this is anticlockwise unless the question says otherwise.

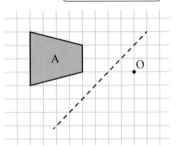

Reflect object A in the dotted mirror line. Label it A′.

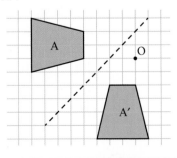

Then rotate image A′ 90° about the point O. Label it A″.

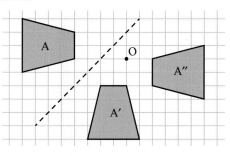

The notation A′ means the image of A after one transformation. A″ is the image after two transformations. If three transformations are used, write A‴.

Exercise 15D

1   In each part of this question, shape A goes through two transformations. The first transformation maps A onto A′. The second transformation maps A′ onto A″. Copy each diagram onto squared paper. Draw each of the images A′ and A″.

a

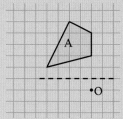

Reflect object A in the dotted mirror line. Then rotate 90° about the point O.

b

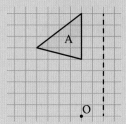

Rotate object A 270° about the point O. Then reflect in the dotted mirror line.

c

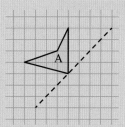

Reflect object A in the dotted mirror line.

Then translate by the vector $\begin{pmatrix} 4 \\ 3 \end{pmatrix}$.

2   Rotate object A 270° about centre O. Label the image A′. Then reflect A′ in the mirror line m_1. Label the image A″.

Finally, translate A″ by the vector $\begin{pmatrix} ^-5 \\ 0 \end{pmatrix}$. Label the image A‴.

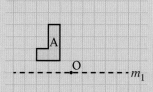

3

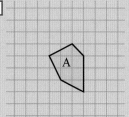

a   Translate A using the vector $\begin{pmatrix} ^-3 \\ ^-5 \end{pmatrix}$. Label the image A′.

Then translate A′ through $\begin{pmatrix} 3 \\ 1 \end{pmatrix}$. Label the image A″.

b   If you translate A″ through $\begin{pmatrix} a \\ b \end{pmatrix}$ you end up back at shape A.

What are the values of a and b?

15.5 Enlargement

Here are two enlargements of a photograph:

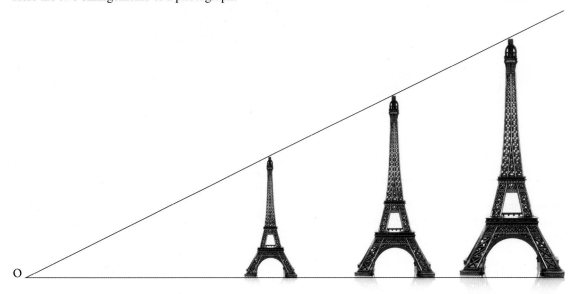

- An **enlargement** is a transformation that changes the size of an object. The point O is called the **centre of enlargement**.

You will need squared paper.

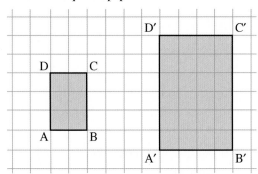

Look at the two rectangles above.

A′B′ = 2 × AB
B′C′ = 2 × BC

We say that rectangle ABCD has been enlarged with scale factor 2.

To define an enlargement you must be given:
- the scale factor of the enlargement
- the centre of the enlargement.

For example, triangle ABC is enlarged by a scale factor of 2 where O is the centre of enlargement:

A maps to A′
B maps to B′
C maps to C′

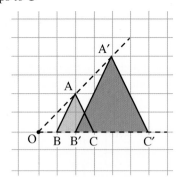

Notice:
From O to A you move 2 squares right then 2 up.
From O to A′ it is 4 squares right and 4 up.
From O to B you move 1 square right.
From O to B′ it is 2 squares right.
From O to C you move 3 squares right.
From O to C′ it is 6 squares right.

Distances OA', OB' and OC' are double OA, OB and OC as the scale factor is 2.

You can use this to help you draw enlargements.

Sometimes the enlargement is not on squared paper so you can't count squares to enlarge the shape.

EXAMPLE 5

Draw the image of triangle ABC after an enlargement with centre O and scale factor 2.

....

Measure the distances OA, OB and OC with a ruler:

OA = 3 cm, OB = 2 cm, OC = 4 cm.

Since the scale factor is 2,

$$OA' = 2 \times 3\,cm = 6\,cm$$
$$OB' = 2 \times 2\,cm = 4\,cm$$
$$OC' = 2 \times 4\,cm = 8\,cm$$

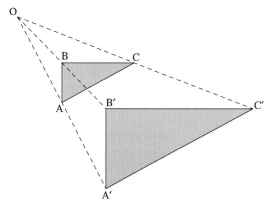

(**Note:** diagram not drawn to scale.)

Exercise 15E

1   Use a ruler to measure the edges of these shapes then write down the scale factor for each enlargement.

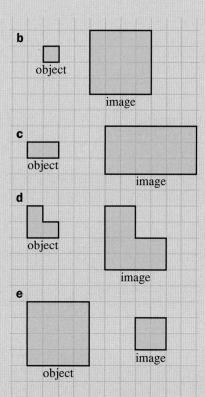

2   Copy these shapes onto squared paper. Using the centre O, enlarge them by the scale factor shown.

a   Scale factor 3

b Scale factor 4

c Scale factor 2

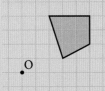

d Scale factor 3

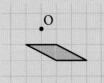

3 Trace each shape and draw the image of the shape after the given enlargement.

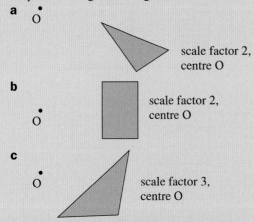

a scale factor 2, centre O

b scale factor 2, centre O

c scale factor 3, centre O

4 The centre of enlargement can be inside the shape. The following diagram shows an enlargement with centre of enlargement O.

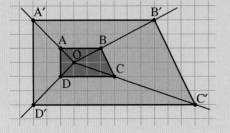

What is the scale factor of the enlargement that maps ABCD onto A′B′C′D′?

5 Copy these shapes and, using the centre O, enlarge them by the scale factor shown.

a Scale factor 2

b Scale factor 3

c Scale factor 4

d Scale factor 2

15.6 Scale drawing

This is a scale drawing of a football field which in real life is 80 m long and 55 m wide. It has a scale of 1 cm to 10 m.

Football field

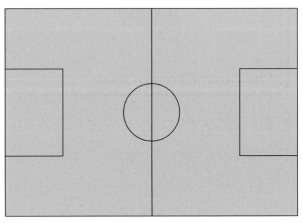

Scale: 1 cm represents 10 m

The scale drawing has a length of 8 cm and a width of 5.5 cm.

EXAMPLE 6

A plan of a house is drawn with a scale of 1 cm to 2 m.

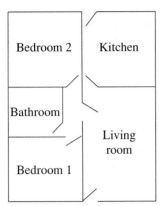

Scale: 1 cm represents 2 m

a What are the dimensions of the living room?
b A 20 m long wall is to be built at the back of the house. What would be the wall's length on the scale drawing?

---

a The living room on the plan is 3 cm long and 2 cm wide.

1 cm represents 2 m
so 3 cm represents 3 × 2 m = 6 m
and 2 cm represents 2 × 2 m = 4 m

The room is 6 m long and 4 m wide

b 2 m is represented by 1 cm
so 20 m is represented by $\frac{20}{2}$ cm = 10 cm

The wall would be shown by a 10 cm line on the plan.

Exercise 15F

1 These lines are scale drawings.
  i _____   **ii** _____
  iii _____   **iv** _____
  a Measure the lengths of the lines in cm.
  b What are the actual lengths of the lines if
    A 1 cm represents 4 cm
    B 1 cm represents 10 cm
    C 1 cm represents 5 m?

2 Copy and complete this table.

Length on drawing	Scale	Actual length
a 7 cm	1 cm represents 5 m	
b 3.5 cm	1 cm represents 10 m	
c 8.2 cm		16.4 m
d 13 cm	1 cm represents 20 km	
e	1 cm represents 10 m	70 m
f	1 cm represents 25 km	170 km
g 12 cm		42 m

3 A rectangular room is 7 m long and 5 m wide.
Make a scale drawing of the room using a scale in which:
  a 1 cm represents 1 m
  b 1 cm represents 2 m

4 Use a scale of 1 cm to 1 m to make a scale drawing of a room measuring 12 m by 9 m. Find the distance between opposite corners of the room.

5 The plan shows the grounds of a school.

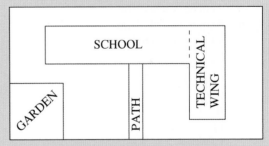

Scale: 1 cm represents 10 m

  a On the plan, measure with your ruler:
    i the length of the school
    ii the length of the technical wing
    iii the width of the school
    iv the width of the technical wing
    v the length of the garden
    vi the length of the path.
  b What are these distances in real life?

6 Measure the dimensions of your classroom. Use a scale where 1 cm represents 1 m to make a plan of your classroom.

7 Make a scale drawing of your bedroom. What scale did you use?

Using ratios

Scales can be written as ratios.

EXAMPLE 7

a Write the scale in which 1 cm represents 20 m in ratio form.

b A plan is drawn using a scale of 1 : 500. Find the actual measurement of a line of length 4 cm on the plan.

....

a 20 m = 20 × 100 cm
     = 2000 cm
So 1 cm represents 2000 cm.
The scale is 1 to 2000,
which is written as 1 : 2000

b 1 : 500 means 1 cm represents 500 cm
So 4 cm represents 2000 cm = 20 m

Exercise 15G

1 Write these scales in ratio form:
  a 1 cm represents 1 m
  b 1 cm represents 5 m
  c 1 cm represents 10 m
  d 1 cm represents 1 km
  e 1 mm represents 10 m

2 Find the actual length and width of these rectangles if they have been drawn to a scale of
    i 1 : 100
    ii 1 : 50
    iii 1 : 2000

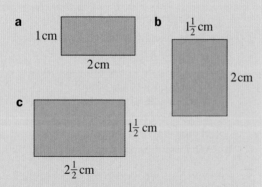

3 A map is drawn showing some buildings in a town. The scale used is 1 : 50 000.
  a Find the actual distance between
    i the town hall and the park when the distance on the map is 3.5 cm
    ii the hospital and the market when the distance on the map is 7.3 cm.

b Find the distance on the map between
    i the high school and the primary school when the actual distance is 1.75 km
    ii the bakery and the supermarket when the actual distance is 0.35 km.

4 The diagram shows the plan of a town.

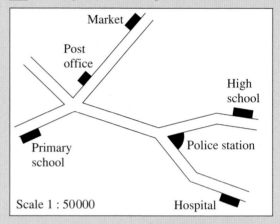

Using the scale of 1 : 50 000, find the shortest possible walking distance between
  a the post office and the market
  b the primary school and the high school
  c the hospital and the police station
  d the market and the hospital
  e the post office and the primary school.

5 Copy and complete this table.

Length on drawing	Scale 1 : n	Actual length
0.85 cm	1 : 1000	
	1 : 50 000	2.1 km
5.2 cm		5.2 km
3.5 cm	1 : 250	
7.2 cm		360 m
	1 : 1 000 000	21 km

Consolidation

Example 1

Find the image of triangle ABC after reflection in the dashed mirror line.

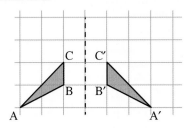

When reflected in the mirror line the triangle is A′B′C′.

Example 2

What is the image of triangle ABC after the translation $\begin{pmatrix} 4 \\ 1 \end{pmatrix}$?

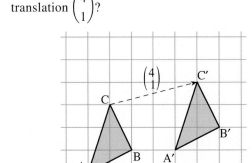

The translation $\begin{pmatrix} 4 \\ 1 \end{pmatrix}$ shifts the triangle 4 units to the right and 1 unit upwards.

The image of ABC is A′B′C′.

Example 3

Find the image of triangle ABC after an enlargement with scale factor 2 and centre O.

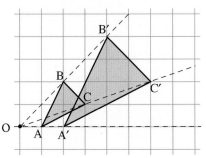

For an enlargement scale factor 2,

OA′ = 2 × OA

OB′ = 2 × OB

OC′ = 2 × OC

Where O is the centre of enlargement.

The image of ABC is A′B′C′.

Example 4

Find the image after triangle ABC is rotated 90° anticlockwise about the point O, then translated through $\begin{pmatrix} 0 \\ -4 \end{pmatrix}$.

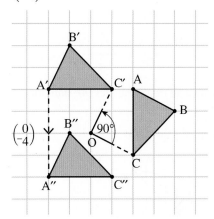

Triangle A′B′C′ is the image after rotating triangle ABC 90° anticlockwise about the point O. Then A″B″C″ is the image after translating A′B′C′ through no units to the right and four units down.

Example 5

A map has a scale of 1 : 50000.
a What is the actual distance represented by 2.7 cm on the map?
b How is a distance of 6 km represented on the map?

a   Actual distance = 2.7 × 50 000 cm
$\qquad\qquad\qquad$ = 135000 cm
$\qquad\qquad\qquad$ = 1350 m (or 1.35 km)
b   Map distance = 6 km ÷ 50 000
$\qquad\qquad\qquad$ = 6000 m ÷ 50 000
$\qquad\qquad\qquad$ = 0.12 m
$\qquad\qquad\qquad$ = 12 cm

Exercise 15

1 Copy the diagram.

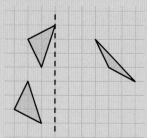

Reflect all three triangles in the dashed mirror line.

2 Copy the diagram.

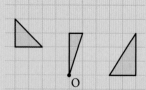

Rotate all three triangles 90° anticlockwise about O.

3 Copy the diagram.

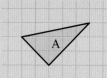

Drawing all the images on the same diagram, draw the image of triangle A after translation through

a $\begin{pmatrix} 0 \\ 3 \end{pmatrix}$. Label this A′.

b $\begin{pmatrix} 4 \\ 0 \end{pmatrix}$. Label this B.

c $\begin{pmatrix} {}^-2 \\ {}^-3 \end{pmatrix}$. Label this C.

d $\begin{pmatrix} 5 \\ 3 \end{pmatrix}$. Label this D.

e $\begin{pmatrix} 8 \\ {}^-1 \end{pmatrix}$. Label this E.

f $\begin{pmatrix} 3 \\ {}^-3 \end{pmatrix}$. Label this F.

4 Repeat Question **2** but this time rotate the triangles
  a 90° clockwise
  b 180°

5 Copy these shapes and, using the centre O, enlarge them by the scale factor shown.
  a Scale factor 4

  b Scale factor 2

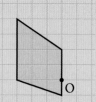

  c Scale factor 2

  d Scale factor 3

6 Quadrilateral X is mapped onto quadrilateral X′ by a reflection. Out of lines A, B and C, which is the correct mirror line?

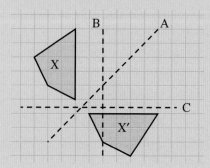

7 a

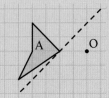

Reflect object A in the dashed mirror line. Then rotate 90° clockwise about the point O.

b

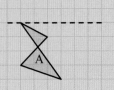

Reflect object A in the dashed mirror line.

Then translate through $\begin{pmatrix} 4 \\ -1 \end{pmatrix}$.

Summary

You should know ...

1 A reflection produces an opposite image of an object.
For example:

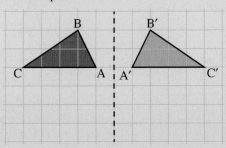

Triangle ABC is reflected in the mirror line to produce its image, triangle A′B′C′.

2 A rotation is described by its centre, angle of turn and direction of turn.
For example:

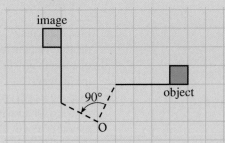

Centre of rotation is O, angle of turn is 90° anticlockwise.
The direction is usually anticlockwise unless stated otherwise.

Check out

1 Copy the diagram.

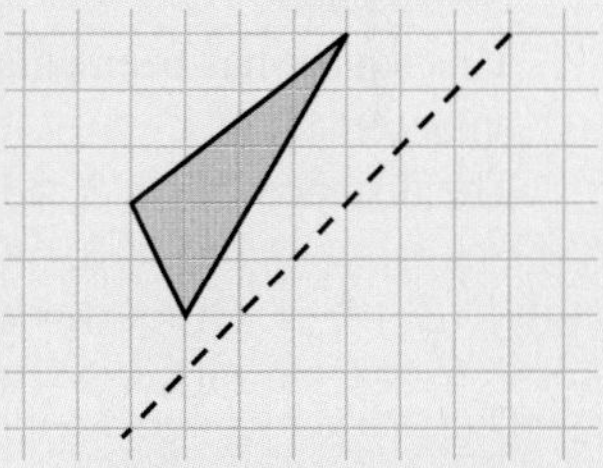

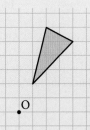

Reflect the triangle in the dashed line.

2 Copy the diagram.

Find the image of this triangle after a rotation about the centre O of
a 90°
b 90° clockwise
c 180°

3 A translation is a sliding movement, with no turning.
You can describe a translation with a column vector.
For example:

$$\begin{pmatrix} 4 \\ ^-3 \end{pmatrix}$$

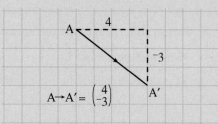

$$A \rightarrow A' = \begin{pmatrix} 4 \\ ^-3 \end{pmatrix}$$

The 4 tells you move *4 units to the right*.
The ¯3 tells you move *3 units down*.

3 Copy the diagram.

Draw the image of this triangle after the translation:

a $\begin{pmatrix} 4 \\ ^-1 \end{pmatrix}$ **b** $\begin{pmatrix} ^-1 \\ 4 \end{pmatrix}$

c $\begin{pmatrix} ^-1 \\ ^-4 \end{pmatrix}$

---

4 An enlargement needs a scale factor and a centre of enlargement.
For example:
This square has been enlarged by a scale factor of 3, with centre of enlargement O.
The distance OA × 3 = OA′
OB × 3 = OB′
OC × 3 = OC′
OD × 3 = OD′

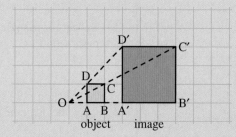

object   image

4 Copy the diagram.

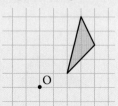

Draw the image of this triangle after an enlargement with centre O and scale factor 2.

---

5 A scale drawing can be used to represent a larger object.
For example:
A distance of 8 m is represented by 4 cm on a plan.
What is the scale used?

4 cm represents 8 m (800 cm)
So   1 cm represents 200 cm
Or, using ratio: 1 : 200.

5 A drawing has a scale of 1 : 500.
a Find the actual measurement in metres of a length of 7.5 cm on the drawing.
b Find the length on the drawing for an actual measurement of 135 m.

16 Ratio and proportion

Objectives

- Recall relationships between units of measurement.
- Simplify ratios, including those expressed in different units; divide a quantity into more than two parts in a given ratio.
- Use the unitary method to solve simple problems involving ratio and direct proportion.
- Solve simple word problems, including direct proportion problems.

What's the point?

People use ratio and proportion every day when they are cooking. For example, to bake a cake you might need 2 eggs and 125 g each of sugar, flour and butter. To make a cake twice as big all the ingredients need to be doubled so that they stay in the same proportion and the recipe will still work.

Before you start

You should know ...

The metric abbreviations:
- mm – millimetres
- cm – centimetres
- m – metres
- km – kilometres
- g – grams
- kg – kilograms
- t – tonnes
- ml – millilitres
- ℓ – litres

Check in

From the list on the left, which are measurements of
- **a** length
- **b** mass
- **c** capacity?

16.1 Units of measurement.

You need to be able to recall relationships between units of measurement.

- Measurements of length:
  - $10\,\text{mm} = 1\,\text{cm}$
  - $100\,\text{cm} = 1\,\text{m}$
  - $1000\,\text{m} = 1\,\text{km}$

- Measurements of mass:
  - $1000\,\text{g} = 1\,\text{kg}$
  - $1000\,\text{kg} = 1\,\text{t}$

- Measurements of capacity:
  - $1000\,\text{ml} = 1\,\ell$

- Measurements of time:
  - 60 seconds in a minute
  - 60 minutes in an hour
  - 24 hours in a day
  - 7 days in a week
  - 365 days in a year (366 in a leap year)

- Measurements of area:
  - $100\,\text{mm}^2 = 1\,\text{cm}^2$
  - $10\,000\,\text{cm}^2 = 1\,\text{m}^2$
  - $1\,000\,000\,\text{m}^2 = 1\,\text{km}^2$

- Measurements of volume:
  - $1000\,\text{mm}^3 = 1\,\text{cm}^3$
  - $1\,000\,000\,\text{cm}^3 = 1\,\text{m}^3$

Exercise 16A

1 Copy and complete:
  - **a** $0.31\,\text{km} = \square\,\text{m}$
  - **b** $48\,\text{hours} = \square\,\text{days}$
  - **c** $68\,000\,\text{ml} = \square\,\ell$
  - **d** $4300\,\text{m} = \square\,\text{km}$
  - **e** $0.7\,\text{t} = \square\,\text{kg}$
  - **f** $300\,\text{minutes} = \square\,\text{hours}$
  - **g** $500\,\text{mm} = \square\,\text{cm}$
  - **h** $16\,\ell = \square\,\text{ml}$
  - **i** $0.8\,\text{kg} = \square\,\text{g}$

2 Copy and complete this crossnumber puzzle using the clues given.

Across

1 Akanni has a mass of 56 000 g. What is his mass in kg?
2 A piece of metal 5.1 m long is to be divided into three equal pieces. How many centimetres is each equal part?
3 4 tins of tomatoes have a total mass of 1.5 kg. What is the mass, in grams, of 1 tin?

Down

1 A piece of material is 6 m long. A 33 cm length is cut from this material. How many centimetres of material are left?
2 Anna is 1.37 m tall. What is her height in centimetres?
3 8 bottles contain a total of 2.6 ℓ of water. How many millilitres of water are there each bottle?

3 Copy and complete:
  - **a** $3\,\text{m}^2 = \square\,\text{cm}^2$
  - **b** $1.7\,\text{km}^2 = \square\,\text{m}^2$
  - **c** $2.6\,\text{m}^3 = \square\,\text{cm}^3$
  - **d** $990\,000\,\text{cm}^2 = \square\,\text{m}^2$
  - **e** $6\,000\,000\,\text{mm}^2 = \square\,\text{m}^2$

16.2 Ratio

A ratio compares the size of two quantities.

In this diagram the ratio of triangles to squares is 3 : 5.

The ratio of squares to triangles is 5 : 3.

We can also write these fractions:

$\frac{3}{8}$ of the shapes are triangles.

$\frac{5}{8}$ of the shapes are squares.

Simplifying ratios

Ratios are **equivalent ratios** if they are in the same proportion.

The ratio of triangles to squares in this diagram is 4 : 6. This means that for every 4 triangles there are 6 squares. However, you can see from the red and blue shapes that for every 2 triangles there are 3 squares, so the ratio 4 : 6 can be simplified to 2 : 3.

To write a ratio in its simplest form you divide by the HCF (highest common factor) of the numbers in the ratio.

EXAMPLE 1

Simplify: **a** 15 : 25   **b** 4 : 8 : 12

....

a
$$\begin{array}{c} 15 : 25 \\ \div 5 \searrow \quad \swarrow \div 5 \\ 3 : 5 \end{array}$$

b
$$\begin{array}{c} 4 : 16 : 12 \\ \div 4 \downarrow \quad \div 4 \downarrow \quad \div 4 \\ 1 : 4 : 3 \end{array}$$

Exercise 16B

1 Write each ratio in its simplest form.

    a 30 : 40   **b** 35 : 25

    c 10 : 40 : 25   **d** 18 : 36

    e 52 : 13   **f** 24 : 84 : 108

    g 17 : 51   **h** 140 : 200 : 80

2 In what ratio are the side lengths of these triangles? Write the smallest number first and largest last. Don't forget to simplify the ratio.

    a 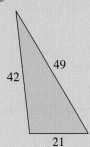   **b**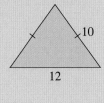

    c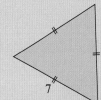

3 Which pairs of ratios are equivalent?

    18 : 27, 2 : 3

    22 : 77, 200 : 700

    8 : 16, 2 : 3

    34 : 51, 4 : 6

    95 : 38, 15 : 6

    32 : 56, 7 : 4

4 Construct two different triangles with sides in the ratio 3 : 4 : 5. What sort of triangles are they?

5 Ratios in their simplest whole number form will not contain decimals. For example, to find 0.8 : 6 as a ratio in its simplest whole number form:

$$0.8 \ : \ 6$$
$$\times 10 \Big\downarrow \qquad \Big\downarrow \times 10$$
$$8 \ : \ 60$$

> Multiply by 10 to clear the decimal.

which then simplifies to 2 : 15.

Write each of these as a ratio in its simplest whole number form.

a 0.5 : 7
b 4.9 : 1.4
c 5 : 6.5 : 3.5
d 14 : 8.2

6 Write each of these as a ratio in its simplest whole number form:

a $\frac{1}{4} : 5$ **b** $2.2 : \frac{1}{5}$

c $36\% : 0.6 : \frac{3}{25}$ **d** $\frac{2}{5} : \frac{3}{4} : \frac{2}{3}$

⇒ INVESTIGATION

Use the digits 0, 1, 2, 3, 4, 5, 6, 7, 8 and 9.
a Choose any three digits and add them up. For example, 3 + 6 + 7 = 16
b Make all possible two-digit numbers using those three digits. Do not use repeated digits, for example 33, 66 and 77 are not allowed. Add up these two-digit numbers. For our example, this is 36 + 37 + 63 + 67 + 73 + 76 = 352
c Write the ratio 'answer to **a** : answer to **b**' For our example, this is 16 : 352
d Simplify the ratio. For our example, 1 : 22
e Repeat steps **a** to **d** for another set of three numbers. Do this a few times. What do you notice?

If you wish to extend this investigation further there are various questions you can consider:
• What happens if you choose four or more digits in part **a**?
• Can you see a pattern between the number of digits chosen and the ratio?
• What happens if you do allow repeated digits (so for our three-digit example 33, 66 and 77 would be added too)?
• Can you explain why this works?

Using the same units

• When writing a ratio, both quantities must be in the **same units**.

EXAMPLE 2

Length of line P is 2 cm: _____ P
Length of line Q is 13 mm: _____ Q
The ratio comparing the length of P to the length of Q is:

$$2 \,\text{cm} : 13 \,\text{mm}$$
$$or \ 20 \,\text{mm} : 13 \,\text{mm}$$
$$or \ \mathbf{20 : 13}$$

The ratio itself has **no units**.

EXAMPLE 3

It takes James 2 hours 14 minutes to walk from Lake Pupuke to Auckland zoo.
It takes him 50 minutes to cycle there.
The ratio of walking time to cycling time is:

$$2 \text{ hours } 14 \text{ minutes} : 50 \text{ minutes}$$
$$or \ 134 \text{ minutes} : 50 \text{ minutes}$$
$$or \ \mathbf{134 : 50}$$
$$or \ \mathbf{67 : 25}$$

Exercise 16C

1 a Which bag contains the most sugar?
b Is it correct to use the ratio 100 : 5 to compare the two quantities of sugar? Why?
c Express 5 kilograms in grams.
d Now write a ratio to compare the quantities of sugar in Ⓧ and Ⓨ.

2 Write a ratio to compare the mass of the first object with the mass of the second.
a

b

3 Use a ratio to compare these quantities.
   a 1 m 10 cm; 57 cm   **b** 100 mm; 1 cm
   c 1.3 cm; 18 mm   **d** 100 mm; 1 m
   e 1.2 kg; 311 g   **f** 5 min; 40 s
   g 350 ml; 1.1 litres   **h** 5; three dozen
   (**Hint:** a 'dozen' means 12)

4 Use a ratio to compare these quantities.
   a 1 hour; 13 min   **b** 1 week; 4 days
   c 0.8 cm; 15 mm   **d** 980 kg; 1 t
   e 1.4 t; 700 kg

5 This was Farhan's homework on simplifying ratios:

	Question	Working	Answer
a	3 g : 3 kg	divide by 3	1 : 1
b	17 cm : 85 cm	divide by 17	5 : 1
c	10 m : 70 cm	divide by 10	1 : 7
d	150 ml : 50 ml	divide by 10	15 : 5

Farhan has made some mistakes. Mark and correct his homework.

6 Use a ratio to compare these quantities.
   a 5 cm^2, 250 mm^2
   b 0.4 m^2, 16 000 cm^2
   c 1.6 cm^3, 20 000 mm^3
   d 0.25 m^3, 75 000 cm^3

Dividing a quantity in a given ratio

You can divide a quantity into two amounts using ratio to decide how much is in each amount.

EXAMPLE 4

$250 is divided between Lola and Yemi in the ratio 3 : 7. How much do they each get?

For every 3 parts Lola gets, Yemi gets 7 parts.
3 + 7 = 10 parts in total

Divide $250 by 10 to find the value of 1 part.
1 part is worth $25

Lola gets 3 × $25 = $75

Yemi gets 7 × $25 = $175

It is a good idea to check these answers by adding:
$75 + $175 = $250

You can also use ratios to divide quantities into more than two amounts.

EXAMPLE 5

Share 50 elephants among Jo, Minny and Sam in the proportion 1 : 2 : 7.

1 + 2 + 7 = 10 parts
There are 50 elephants. So there are 5 elephants in each part.

Jo gets **5 elephants**
Minny gets 5 × 2 = **10 elephants**
Sam gets 5 × 7 = **35 elephants**

Exercise 16D

1 A cake is shared between Annabel, Ria and Faith in the ratio 1 : 2 : 3.

   a If Annabel gets 1 part, how many parts should Ria get? How many parts should Faith get?
   b Into how many equal slices should you cut the cake?

2 Divide:
   a 450 kg in the ratio 3 : 2
   b 135 m in the ratio 7 : 8
   c 700 ml in the ratio 2 : 5
   d $204 in the ratio 5 : 7

3 Sometimes you may want to change units before dividing in a ratio to make calculations easier. For example, dividing 1 hour in the ratio 5 : 7 is easier if you change the hour into 60 minutes – then the answer is 25 : 35. After changing to appropriate units, divide:
   a 0.2 kg in the ratio 3 : 7
   b 0.6 m in the ratio 2 : 3
   c 0.112 ℓ in the ratio 3 : 4
   d 1 hour in the ratio 3 : 7 : 10

4 Share the following among A, B and C.
a 60 nuts, in the ratio $4:1:7$
b 100 ml, in the ratio $2:5:3$
c 75 m, in the ratio $8:2:5$
d 1.8 m, in the ratio $1:7:1$
e $25, in the ratio $2:13:5$
f $54, in the ratio $8:5:14$
g $100, in the ratio $2:4:19$
h 1000 kg, in the ratio $117:62:21$

5 **a** Draw a line 12 cm long. Divide it in the ratio $2:1:3$.
b Draw a line 72 mm long. Divide it in the ratio $2:4:3$.

6 Share:
a 0.9 kg, in the ratio $3:17:10$
b $24.50, in the ratio $1:1:12$
c 1.5 kg, in the ratio $13:2:5$

7 Milly was completing her ratio homework. Below is her work for the question 'Share $56 between Jamil and Harry in the ratio $3:5$'.

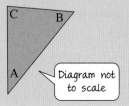

Mark Milly's homework. Write the correct working if she has made a mistake.

8 The angles A, B and C in this triangle are in the ratio $3:5:7$. Work out the size of each angle.

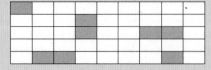

Diagram not to scale

9 Look at the diagram below.

How many more rectangles need to be shaded so that the ratio of shaded to unshaded is $2:7$?

10 Find two numbers which are in the ratio $7:4$ and have the sum 165.

11 The angles of a quadrilateral are in the ratio $1:6:3:5$. What are the angles?

16.3 Proportion

The ratio questions you have looked at so far have involved sharing a total amount in a given ratio. Sometimes you are instead given one of the proportions and asked to find the total or the other proportion.

EXAMPLE 6

The ratio of staff to students in a school is $2:25$. There are 425 students. How many staff are there?

For every 25 students there are 2 staff.

$425 \div 25 = 17$

$17 \times 2 = 34$

Find out how many lots of 25 students there are.

So there are 34 staff.

When one quantity increases and another quantity increases at the same rate this is known as **direct proportion**.

An example of direct proportion is a telephone call costing 8 cents per minute. If you double the number of minutes, you double the cost. If you have 10 minutes you have 10 times the cost (80 cents) of 1 minute (8 cents).

Sometimes the proportions are not as easy as this to work with because they do not involve the number 1. The next two examples show the **unitary method**, which is when you find the value of a single unit then scale it up or down, or change the ratio to the form $1:n$.

EXAMPLE 7

A pink paint mixes white and red paint in the ratio $5:3$.

If 32 litres of white paint are used, how many litres of red paint must be used to make the pink paint?

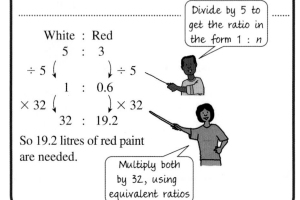

Divide by 5 to get the ratio in the form $1:n$

So 19.2 litres of red paint are needed.

Multiply both by 32, using equivalent ratios

EXAMPLE 8

If ribbon costs $0.80 for 2.5 m, how much does it cost for 26 m?

....

The cost for 1 m of ribbon is
$0.80 ÷ 2.5 = $0.32

So the cost for 26 m is
$0.32 × 26 = $8.32

> Find the value of a single metre then scale it up.

Exercise 16E

1 Solve these direct proportion questions using the unitary method.

a If 15 tickets cost $36, what do 7 tickets cost?

b If 8 boxes of washing powder have a mass of 40 kg what is the mass of 20 boxes?

c If 3 chocolate bars cost $1.59, what do 8 chocolate bars cost?

d If 3 bags of potatoes have a mass of 36 kg, what is the mass of 11 bags?

e A 5-litre tin of paint covers 65 m². What area would 8 litres of paint cover?

2 I change $60 into €45.

a How many euros would I get for $50?

b How many dollars would I get for €36?

3 Pens and pencils in a box are in the ratio 12 : 18

a How many pencils are there if there are 20 pens?

b How many pens are there if there are 33 pencils?

4 Zainab uses 3 tomatoes for every 1.5 litres of sauce he makes.

a How much sauce can she make from 14 tomatoes?

b How many tomatoes does she need for 2 litres of sauce?

5 In a kitchen drawer the ratio of spoons to knives to forks is 7 : 3 : 5

a How many knives are there if there are 21 spoons?

b How many forks are there if there are 21 knives?

c How many spoons are there if there are 32 knives and forks?

6 A market stall sells goods at these prices:

- Apples $1.20 for 4 kg
- Mangoes $2.80 for 5 kg
- Bananas $1.08 for 3 kg

Calculate the price of

a 6 kg of apples

b 3 kg of mangoes

c A bag of fruit with 2 kg of bananas, 2 kg of mangoes and 3 kg of apples.

7 A grey paint is a mix of white and black paint in the ratio 14 : 5.
If 24 litres of black paint are used, how many litres of white paint are needed to make the grey paint?

8 A road 3.2 cm long on a map is 8 km long in real life. A river is 4.1 cm long on the same map. How long is the river in real life?

9 Neema had this question in her homework:
'If Sandra shares some sweets with James in the ratio 4 : 3 and Sandra gets 24, how many does James get?'
Neema wrote:
24 ÷ 3 = 8
Sandra gets 4 × 8 = 32
Is this working correct? If not, correct it.

10 Look at this recipe for English pancakes.

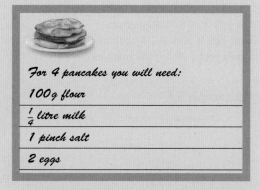

For 4 pancakes you will need:

100 g flour

¼ litre milk

1 pinch salt

2 eggs

a How much flour do you need to make 10 pancakes?

b How many pancakes can you make with a litre of milk (assuming you have plenty of the other ingredients)?

c How much milk do you need for 18 pancakes?

11 A photograph is enlarged to make a poster. The photograph is 10 cm wide and 15 cm high. The poster is 35 cm wide. How high is the poster?

12 It takes Jen half an hour to run 5.4 km.

a At the same speed, how far can Jen run in 40 minutes?

b At the same speed, how long does it take to run 6.3 km?

Consolidation

Example 1

Write as a ratio:

a 2 m 15 cm : 85 cm
    = 215 cm : 85 cm
    = 215 : 85

b 2 hours : 50 minutes
    = 120 minutes : 50 minutes
    = 120 : 50

Example 2

Write 5 : 25 : 45 as a ratio in its simplest form.

Divide by 5:

$$5 : 25 \qquad : 45$$
$$\div 5 \; (\qquad) \div 5 \qquad) \div 5$$
$$1 : 5 \qquad : 9$$

Example 3

Share 80 marbles among Alan, Kamil and Yasmin in the ratio 2 : 3 : 5.

There are 2 + 3 + 5 = 10 parts
Each part = 80 marbles ÷ 10 = 8 marbles
Alan gets 2 × 8 = 16 marbles
Kamil gets 3 × 8 = 24 marbles
Yasmin gets 5 × 8 = 40 marbles

Exercise 16

1 Compare these quantities as ratios.

  a $\frac{1}{2}$ hour; 15 minutes

  b 2 kg; 300 g

  c 250 m; 3 km

  d $6\frac{1}{2}$ hours; 100 minutes

  e 25 mm; 2 m

  f 18 cm; 4 km

2 Write these ratios in their simplest form.

  a 2 : 20 : 200   **b** 36 : 40
  c 85 : 17 : 51   **d** 25 : 625 : 75
  e 184 : 56   **f** 91 : 65 : 117

3 Share the following between Anton and Dannisha:

  a 860 marbles in the ratio 3 : 2
  b 36 pens in the ratio 5 : 7
  c 90 kg in the ratio 1 : 8
  d $216 in the ratio 7 : 5
  e 800 ml in the ratio 7 : 13

4 Divide:

  a 0.3 m in the ratio 3 : 2
  b 0.42 kg in the ratio 6 : 5 : 3
  c 2.7 l in the ratio 2 : 4 : 3
  d 2 hours in the ratio 3 : 7 : 5

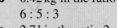 Hint: you may want to change units first.

5 I change $80 into €64.

  a How many euros would I get for $130?
  b How many dollars would I get for €50?

6 If 5 bags of sweets contain 140 sweets in total, calculate how many sweets 8 bags will contain.

7 A grey paint is a mix of white and black paint in the ratio 12 : 5.
If 32 litres of black paint are used, how many litres of white paint must be used to make the grey paint?

8 8 calculators cost $55.60.

  a How much will 3 calculators cost?
  b How many calculators can you buy for $88?

9 In a chemistry lab, acid and water are mixed in the ratio 1 : 5.
A bottle contains 228 ml of the mixture.
How much acid and how much water were needed to make this amount of the mixture?

10 To make 3 glasses of orange squash you need 825 ml of water and 75 ml of orange cordial.

  a How much water do you need to make 10 glasses of orange squash?
  b How many glasses of squash can you make with 0.2 ℓ of orange cordial?

11 Which is the better value for buying a cola drink:
2-litre bottles on offer 3 for $5.10
or 24 cans, each 330ml, for $5.94?

Summary

You should know ...

1 How to compare one thing to another using a ratio.

For example:

The ratio of triangles to squares is 5 : 2.

The ratio of circles to squares to triangles is 1 : 2 : 5

2 To compare two quantities using a ratio, the quantities must be in the same units.

For example:
$$2 \text{ hours} : 15 \text{ minutes}$$
$$= 120 \text{ min} : 15 \text{ min}$$
$$= 120 : 15$$
$$= 8 : 1$$
A ratio itself has no units.

3 How to simplify a ratio by dividing the numbers in the ratio by the HCF of the numbers.

For example:
The ratio 8 : 12 : 20 = 2 : 3 : 5 (divide all numbers by 4)

4 How to share something in a given ratio.

For example:
Share $30 in the ratio 2 : 3 : 1
There are 2 + 3 + 1 = 6 parts
Each part is worth $30 ÷ 6 = $5
so 2 parts are worth 2 × $5 = $10
and 3 parts are worth 3 × $5 = $15
so $15 to $10 to $5

5 How to use the unitary method to solve problems involving ratio and direct proportion.

For example:
The cost of 6 cakes is $8.10. Find the cost of 8 cakes.
The cost of 1 cake is 8.10 ÷ 6 = $1.35
So 8 cakes cost 1.35 × 8 = $10.80

Check out

1

What is the ratio of
a squares to triangles
b triangles to circles
c circles to squares to triangles?

2 Use a ratio to compare these quantities.
a 1 hour; 45 minutes
b 25 mm; 2 cm
c 3 kg; 200 g

3 Write each ratio in its simplest form.
a 6 : 12
b 28 : 70
c 24 : 9 : 33
d 40 : 15 : 35
e $2\frac{1}{4}$ hours : 30 minutes

4 Share $80 in the ratio
a 4 : 1 : 5
b 5 : 3
c 1 : 9
d 8 : 7 : 5
e 1 : 1 : 3
f 3 : 5 : 12

5 Solve these word problems.
a A holiday costs $1047 for 3 adults. How much does it cost for 5 adults?
b Mr Walton earns $388.50 for 30 hours of work. How much does he earn in an 8-hour day?
c 4 books have a mass of 0.92 kg. What is the mass of 7 of these books?

17 Area, perimeter and volume

Objectives

- Know the definition of a circle and the names of its parts; know and use formulae for the circumference and area of a circle.

- Derive and use formulae for the area of a triangle, parallelogram and trapezium; calculate areas of compound 2D shapes, and lengths, surface areas and volumes of cuboids.

- Use simple nets of solids to work out their surface areas.

What's the point?

Estate agents deal with the sale of properties. The key criteria for fixing the price of a piece of land are its location and its area. The larger the area, the more valuable the land. Precise measurements of land area are therefore very important!

Before you start

You should know ...

1 Area can be measured in mm^2, cm^2, m^2 and km^2.

□ 1 mm
1 mm

1 cm
1 cm

Check in

1 Estimate the area of:
   a a postage stamp
   b your thumbnail
   c a page of your exercise book.

2 You can find the area of a shape by counting squares.
For example:

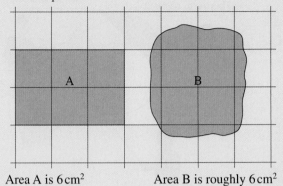

Area A is 6 cm² Area B is roughly 6 cm²

2 Find the areas of these shapes. Each square is 1 cm².

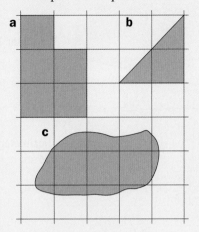

3 How to draw nets.
For example:
The net of this triangular prism

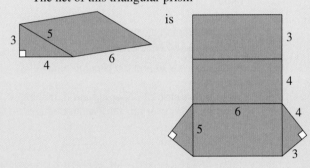

is

3 Draw the nets of these shapes.

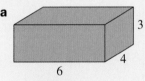

17.1 Perimeter

• The distance around the edge of a shape is called its **perimeter**.

EXAMPLE 1

Find the perimeter of the shape below.

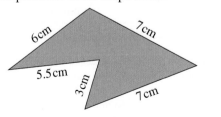

The perimeter is
 6 cm + 7 cm + 7 cm + 3 cm + 5.5 cm
 = 28.5 cm

Exercise 17A

1 Find the perimeter of these shapes:

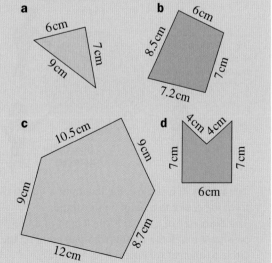

2 Find the lengths of the unknown sides in these shapes.

a

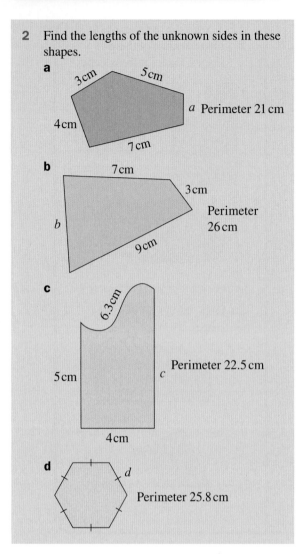

3 cm 5 cm

4 cm

7 cm

a Perimeter 21 cm

b

7 cm

3 cm

b

9 cm

Perimeter 26 cm

c

6.3 cm

5 cm

c Perimeter 22.5 cm

4 cm

d

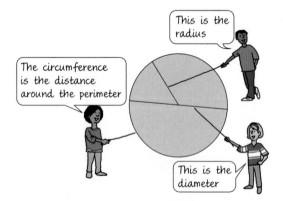

d

Perimeter 25.8 cm

Perimeter of circles – circumference

This is the radius

The circumference is the distance around the perimeter

This is the diameter

- The **radius** of a circle is the length of a line drawn from the centre to a point on the circle.
- The **diameter** of a circle is the length of a line passing through the centre from one point on the circle to another (diameter = 2 × radius).

- The **circumference** of a circle is the distance around it.

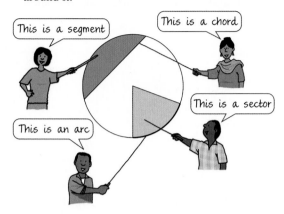

This is a segment

This is a chord

This is a sector

This is an arc

- A **chord** is a straight line from one point on the diameter to another point on the diameter.
- A chord splits a circle into two **segments**.
- An **arc** is part of the circumference.
- A **sector** is formed by joining a radius to an arc to a radius.

▷ ACTIVITY

Here is a good way to find the circumference of a cylindrical tin.

a Wrap a thin strip of paper around it, making sure the ends overlap. Stick a pin through the overlap.

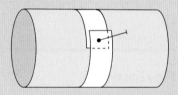

b Unwrap the paper and lay it flat on the table.

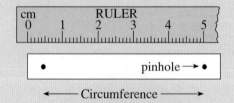

cm   RULER
0   1   2   3   4   5

• pinhole → •

◀— Circumference —▶

c Measure the distance between the two pinholes. This gives the circumference of the tin.
d Use this method to find the circumference, *C*, of five cylindrical objects.
e With your ruler carefully measure the diameter, *D*, of each object.

f Copy and complete the table below.

Object	Circumfernce C cm	Diameter D cm	C ÷ D
tin	25.1 cm	8 cm	
plate			
cup			

What do you notice about the values of C ÷ D in your table?

If you were careful you should have found that

C ÷ D is roughly 3.1

That is:

The circumference of a circle is just over three times the diameter of the circle.

The circumference of a circle is given more accurately by the relation

$C = \pi \times D$

The Greek letter π, or pi (pronounced 'pie'), cannot be found exactly. It is about 3.14 or $3\frac{1}{7}$.

Since

$D = 2 \times$ radius (r)

You can also write

$C = 2\pi r$

• The circumference of a circle is approximately 3.14 × diameter or 2 × 3.14 × radius.

EXAMPLE 2

Find the circumference of a circle with diameter 8 cm. Take $\pi = 3.14$

Circumference = $\pi \times$ diameter
= 3.14 × 8 cm
= 25.12 cm

Exercise 17B

Use 3.14 for π in this exercise.

1 Calculate the circumference of a circle with a diameter of:
  a 2 cm   **b** 10 cm
  c 12 cm   **d** 21 cm

2 Calculate the circumference of a circle with a radius of:
  a 5 cm   **b** 8 cm
  c 13 cm   **d** 39 cm

3 The centre circle on a playing field has a radius of 7.5 m. Find its circumference.

4 A bicycle wheel has a diameter of 70 cm. What is the circumference of the wheel?

5 Each year the Earth goes around the sun in a nearly circular path.

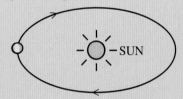

The Earth is about 150 000 000 km from the sun. How far does the Earth travel each year?

6 The distance from the tip of the minute hand to the centre of a clock is 6 cm.
  a How far will the tip of the minute hand move in one hour?
  b How far will it move each minute?

7 A circular toy railway has a radius of 1.4 m. Calculate the time that a toy train will take to travel once round the track at a constant speed of 22 cm/s. Give your answer to the nearest second.

8 The diameter of the Earth is about 12 750 km. Find the distance around the equator.

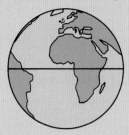

9 Find the perimeters of these shapes.

a

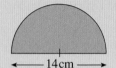

14 cm

b

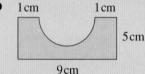

1 cm   1 cm

5 cm

9 cm

c

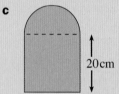

20 cm

21 cm

d

5 cm

4 cm

3 cm

e

8 cm

10 An athletics track consists of two equal semicircles joined by 100-metre straights.

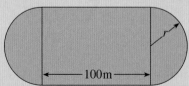

100 m

The radius of each semicircle is 35 m. Find the perimeter of the track.

11 The distance around the inner circle of a rubber tyre is 220 cm.
   a Find the inner radius.
   b If the thickness of the tyre is 14 cm, find the distance around the outside of the tyre.

17.2 Areas of rectangles and triangles

A w

l

- The area of a rectangle, A, is:
    $A = l \times w$
  where l is the length and w the width of the rectangle.

EXAMPLE 3

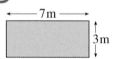

7 m

3 m

Find the area of a rectangular yard 7 m long and 3 m wide.

. . . .

Area of yard = length × width
             = 7 m × 3 m
             = 21 m²

Exercise 17C

1 Find the area of these rectangles.

a   **b**

2 cm   2 cm

3 cm   2 cm

Check your answers by counting squares.

2 Calculate the area of these rectangles.

a   **b**

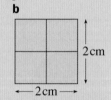

4 m   3 cm

$6\frac{1}{2}$ m   8.4 cm

3 Copy and complete this table for rectangles.

	Length (cm)	Width (cm)	Area (cm²)
a	13	3	
b	25	5	
c		7	126
d	31		279
e	4.5	3.2	

4 Ahmed has a rectangular-shaped lawn with width 5.3 m and length 8.7 m.
Calculate the area of the lawn.

5 Find the missing side lengths.

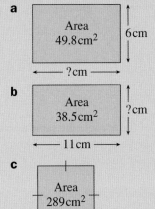

a
Area
49.8 cm²   6 cm
?cm

b
Area
38.5 cm²   ?cm
11 cm

c
Area
289 cm²
?cm

Areas of triangles

EXAMPLE 4

Find the area of the triangle below, drawn on centimetre squares.

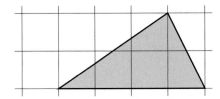

A simple way to find the area is to complete a rectangle surrounding the triangle:

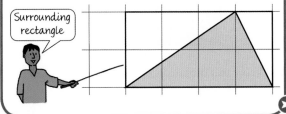

Surrounding rectangle

Then divide the triangle into two triangles, A and B:

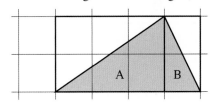

Area of triangle A $= \frac{1}{2}$ of 6 cm²
$= 3$ cm²

Area of triangle B $= \frac{1}{2}$ of 2 cm²
$= 1$ cm²

Area shaded triangle $= 3$ cm² $+ 1$ cm²
$= 4$ cm²

Notice area of surrounding rectangle
$= 8$ cm²

Exercise 17D

1 Using the method from Example 4, find the area of these shaded triangles.

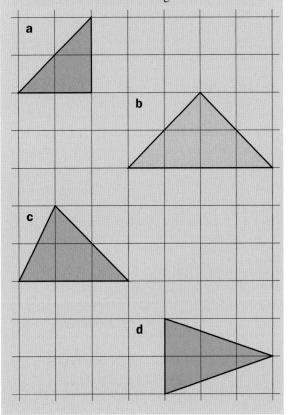

a

b

c

d

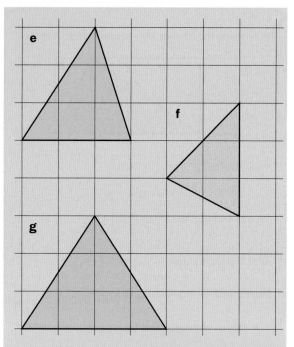

2 **a** Using your answers to Question **1**, copy and complete the table:

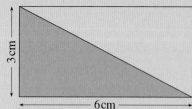

Area of surrounding rectangle	Area of triangle

b What do you notice about the two columns?
c Copy and complete.
Area of triangle = □ × area of surrounding rectangle?

3 **a** Find the area of this rectangle.

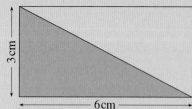

b What fraction of the rectangle is the shaded triangle?
c What is the area of the shaded triangle?

4 Find the area of the shaded triangles.

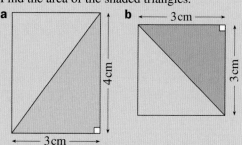

5 Look at this triangle.

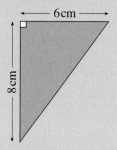

a Is the area of the triangle the same as half of a rectangle with sides 6 cm and 8 cm?
b Find the area of the triangle.

In Exercise 17D you should have found:

• The area of a triangle is half the area of the rectangle that surrounds it.

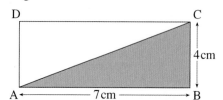

Area of shaded triangle $= \dfrac{1}{2}$ area of rectangle ABCD

$$= \dfrac{1}{2}\,(7\,\text{cm} \times 4\,\text{cm})$$

$$= 14\,\text{cm}^2$$

You need to be able to derive the formula for the area of a triangle.
This rectangle has length b and width h.

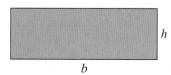

You know that the area is $b \times h$.

If you draw in the diagonal of this rectangle you split it into two equal triangles.

The area of each of these two triangles is half the area of the rectangle

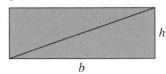

So the area of the triangle is $\dfrac{b \times h}{2}$

If you start with a triangle with base b and height h,

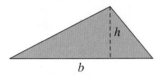

then you can draw a rectangle around it which has twice the area of the triangle:

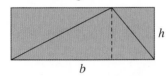

• Generally, the area of a triangle, A, with base length b and vertical height h is

$$A = \frac{b \times h}{2} \quad \text{or} \quad \frac{1}{2}b \times h$$

EXAMPLE 5

Find the area of these triangles.

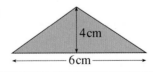

a Area of triangle $= \frac{1}{2}b \times h = \frac{1}{2} \times 4\,\text{cm} \times 3\,\text{cm}$

$= 6\,\text{cm}^2$

b Area of triangle $= \frac{1}{2}b \times h$

$= \frac{1}{2} \times 6\,\text{cm} \times 4\,\text{cm}$

$= 12\,\text{cm}^2$

Exercise 17E

1 Work out the area of these triangles.

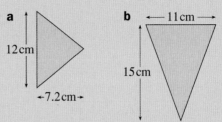

2 Measure the base of each triangle and its vertical height. Calculate their areas.

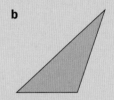

3 Copy and complete the table for triangles.

	Base (cm)	Vertical height (cm)	Area (cm²)
a	10	4	
b	16		96
c	4.5	12	
d		2.4	14.4
e	6.3	9.2	

4 Find the missing side lengths.

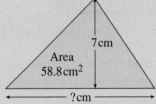

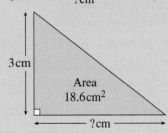

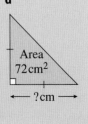

5 What is the area of the shaded part?

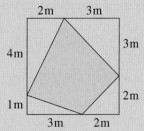

6 Find the area of this shape.

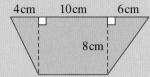

7 Find the area of the quadrilateral.

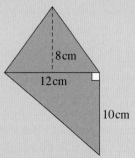

💻 TECHNOLOGY

Watch the video on finding the area of a triangle in the Geometry section of the website

www.mathplayground.com/mathvideos.html

⇒ INVESTIGATION

Look at these triangles.

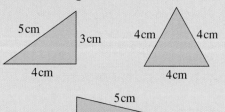

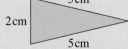

They each have perimeter 12 cm.
a Draw five other triangles with perimeter 12 cm.
b Find the areas of the triangles you drew.
c Which triangle has the largest area?

17.3 Area of a circle

You will need a ruler, a pair of compasses, a protractor, scissors, cm-squared paper and a calculator.

A square has been drawn around a circle of radius r. What is the area of the square?

> Area of this square
> $= r \times r$
> $= r^2$

The area of the square is $4r^2$.

This time the shaded square has been drawn *inside* the circle of radius r. What is the area of the shaded square?

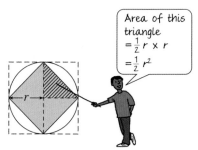

> Area of this triangle
> $= \frac{1}{2} r \times r$
> $= \frac{1}{2} r^2$

The area of the square is $2r^2$.

So the area of the circle of radius r lies between the areas of the two squares; that is, between $4r^2$ and $2r^2$. The area of the circle is about $3r^2$.

Exercise 17F

1 Draw a circle of radius 4 cm on cm-squared paper. Find the area of the circle by counting squares.

2 Repeat Question **1** for a circle of radius
  a 5 cm   **b** 3 cm   **c** 6 cm

3 a Use your answers to Questions **1** and **2** to help you copy and complete the table below.

Radius of circle r (cm)	3	4	5	6
r^2		16		
$3 \times r^2$		48		
Area of circle (cm²)				

b Do you agree that the areas of the circles you found in Questions **1** and **2** are slightly more than $3 \times r^2$?

4 Estimate the area of a circle of radius.
a 8 cm   **b** 3 cm   **c** 6 cm

▷ ACTIVITY

Draw a circle and divide it into 16 equal parts.

Circumference
= $2\pi r$

$\frac{1}{2}$ circumference = πr

Cut out each sector and fit them together. Cut the last sector in half and place one half at each end of your shape.

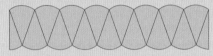

Your new shape is almost a rectangle.
What is its height? (*r*)
What is its base length? (πr)
What is its area? ($\pi r \times r = \pi r^2$)

From Exercise 17F and the last activity you should see that:

* The area *A* of a circle with radius *r* is given by the formula
  $$A = \pi \times r^2$$
  where π is about 3.14.

EXAMPLE 6

Find the area of a circle with radius 4 cm.

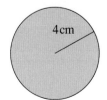
4 cm

Area of circle $= \pi \times r^2$
$\qquad = 3.14 \times 4^2$
$\qquad = 3.14 \times 16$
$\qquad = 50.24 \text{ cm}^2$

Exercise 17G

Use 3.14 for π in this exercise.

1 Find the area of a circle with a radius of:
a 2 cm   **b** 7 cm
c 10 cm   **d** 14 cm

2 Find the area of each circle.

a 7.1 cm   **b** 16.2 cm

3 Write down how to find the area of a circle when you know its diameter.

4 Find the area of a circle with a diameter of:
a 20 cm   **b** 24.6 cm   **c** 102 cm

5 Find the area of the shaded part of each circle:

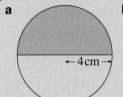
a 4 cm

b 21 cm

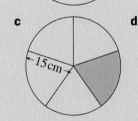
c 15 cm

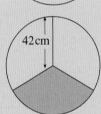
d 42 cm

6 Find the area of these shapes.

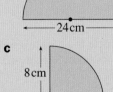
a 24 cm

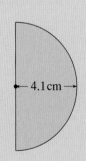
b 4.1 cm

c 8 cm

7 Find the area of the shaded part of each shape.

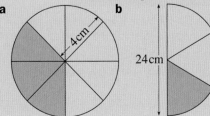
a 4 cm   **b** 24 cm

269

8 Find the area of the shaded part of each shape.

a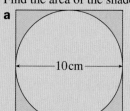

b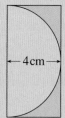

9 Find the area of the shaded part of these shapes.

a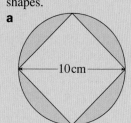

b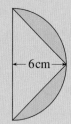

10 Find the radius of a circle with an area of:
a 314 cm² **b** 12.56 cm² **c** 100 cm²

11 Find the area of the shaded parts of these shapes.

a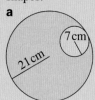

b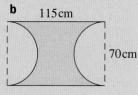

c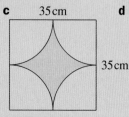

d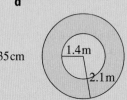

12 The largest possible circle is cut from a square sheet of paper of side 14 cm. What area of paper is left?

TECHNOLOGY

Review what you have learnt about the area of simple shapes by visiting the Geometry section at

www.mathsisfun.com

Then do the review test in the Shape and space section at

www.bbc.co.uk/schools/ks3bitesize/maths/

17.4 Areas of parallelograms and trapeziums

Area of a parallelogram

▷ ACTIVITY

A 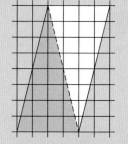 B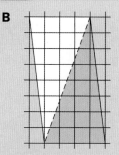

a Make a tracing of the shaded triangle in the first parallelogram above.
b Does your tracing fit exactly on to the white triangle in the first parallelogram?
c What can you say about the white triangle and the shaded triangle?
d Repeat parts **a** to **c** for drawing **B**.
e What is the connection between the area of each parallelogram above and the area of its shaded triangle?
f Can you suggest a quick way to find the area of parallelogram?

A parallelogram is made up of two identical triangles. The area of each triangle is $\frac{1}{2}(b \times h)$. The area of the parallelogram is twice this.

• The area of a parallelogram is $A = b \times h$
  (b is base length, h is vertical height).

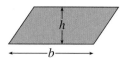

The area of a parallelogram can be shown another way.

Cut a triangle off the end of the parallelogram:

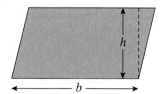

Put the triangle you cut off on the other end of the parallelogram:

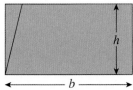

You can see it makes a rectangle.

So a parallelogram has the same area as a rectangle with the same base and height.

EXAMPLE 7

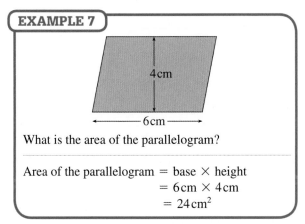

What is the area of the parallelogram?

Area of the parallelogram = base × height
$$= 6\,\text{cm} \times 4\,\text{cm}$$
$$= 24\,\text{cm}^2$$

Exercise 17H

1   Find the area of each parallelogram.

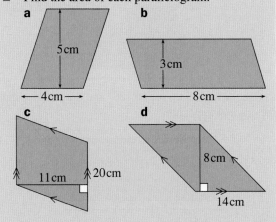

2   Find the area of these parallelograms.

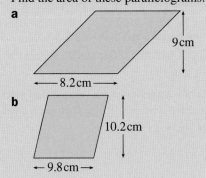

3   By measuring carefully, find the area of this parallelogram.

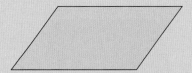

4   Copy and complete the table for parallelograms.

	Base (cm)	Height (cm)	Area (cm²)
a	3.5	8	
b		16	144
c	6.5		52
d	2.3	3.2	
e		7.1	26.27

5   Find the missing side lengths.

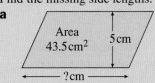

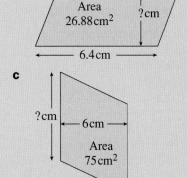

Area of a trapezium

This shape has only one pair of parallel sides. It is called a **trapezium**.

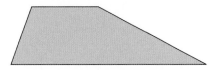

*   A trapezium is a quadrilateral with one pair of sides parallel.

The area of a trapezium can be found by dividing it into two triangles.

271

EXAMPLE 8

Find the area of this trapezium.

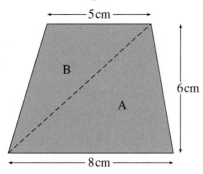

Divide the trapezium into two triangles, A and B.

Area of triangle A $= \frac{1}{2} \times 8\,\text{cm} \times 6\,\text{cm}$
$= 24\,\text{cm}^2$

Area of triangle B $= \frac{1}{2} \times 5\,\text{cm} \times 6\,\text{cm}$
$= 15\,\text{cm}^2$

Area of trapezium $= A + B$
$= 24\,\text{cm}^2 + 15\,\text{cm}^2$
$= 39\,\text{cm}^2$

Exercise 17I

1 For each diagram, find the area of:
   i   the shaded triangle
   ii the unshaded triangle
   iii the trapezium.

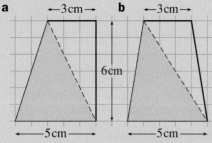

2 Draw a set of trapeziums each with a base of
   5 cm, a height of 6 cm and a top edge of 3 cm.
   Do they all have the same area?

3 Find the area of the trapezium.

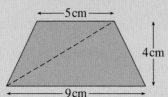

4 Work out the area of this parallelogram made
   up of two identical trapeziums.

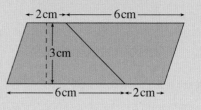

Look at this trapezium.

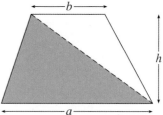

The area of the shaded triangle $= \frac{1}{2}a \times h$

The area of the white triangle $= \frac{1}{2}b \times h$

The area of the trapezium $= \frac{1}{2}a \times h + \frac{1}{2}b \times h$
$= \frac{1}{2}(a + b) \times h$

The formula for the area of a trapezium can be derived
another way.

Look at this trapezium.

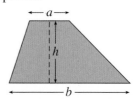

Rotate the trapezium round 180° and draw it touching
the original trapezium:

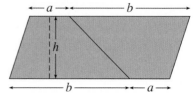

The new shape is a parallelogram with base $a + b$ and
height h.

The area of the parallelogram is $(a + b) \times h$

The trapezium is half this shape and so the area is
   $\frac{1}{2}(a + b) \times h$

• The area of a trapezium with two parallel sides of
  length a and b a perpendicular distance h apart is
      $A = \frac{1}{2}(a + b) \times h$

 **TECHNOLOGY**

Need another look at this?
Watch the video on finding the area of a trapezium
in the Geometry section of

www.mathplayground.com

EXAMPLE 9

Find the area of the trapezium.

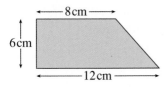

$$\text{Area of trapezium} = \frac{1}{2}(a + b) \times h$$
$$= \frac{1}{2}(12 + 8) \times 6$$
$$= \frac{1}{2} \times 20 \times 6$$
$$= 60\,cm^2$$

Exercise 17J

1 Find the area of each trapezium.

a 3cm, 6.5cm, 7cm

b 3cm, 5cm, 4.6cm

c 8cm, 7cm, 4cm

d 5cm, 7.3cm, 9cm

e 12.5cm, 4.2cm, 8.5cm

f 16m, 8m, 10m

2 By measuring, find the area of the trapezium.

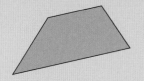

3 Using the formula $A = \frac{1}{2}(a + b) \times h$ find the area of a trapezium when:

i $a = 7$cm, $b = 11$cm and $h = 10$cm
ii $a = 4$cm, $b = 3$cm and $h = 8$cm
iii $a = 2.6$cm, $b = 7.4$cm and $h = 5.6$cm
iv $a = 9.3$m, $b = 6.3$m and $h = 12.2$m

4 Find the area of a trapezium with parallel sides of length 24 cm and 16 cm and perpendicular height of 18 cm.

5 Copy and complete the table for trapeziums.

	Length of parallel sides		Perpendicular distance between	Area of trapezium
	PQ	RS	PQ and RS	
a	9 m	15 m	7 m	
b	16.8 m	12.5 m	8.4 m	
c	23 cm	37 cm	40 cm	
d	12.4 m	6.8 m	4.5 m	
e	24 cm	16 cm	15.5 cm	

6 The area of a trapezium is 80 cm². Its parallel sides are 32 cm and 16 cm in length. Find the perpendicular height.

17.5 Areas of compound shapes

A compound shape is what you get when two or more different shapes are combined to make a shape.

For example, this compound shape is made up of a triangle and a parallelogram:

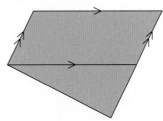

To work out the area of a compound shape, simply work out the area of the separate shapes then add them together to find the total area.

Some exercises you have completed so far have already used the idea of compound shapes. For example, see Exercise 17E, Questions **6** and **7**.

Some areas may be worked out more easily using subtraction, rather than by adding up lots of smaller areas.

273

EXAMPLE 10

Work out the shaded area in these compound shapes.

a

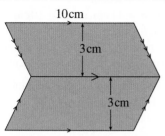

b

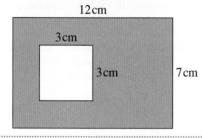

a There are two parallelograms, each with base 10 cm and height 3 cm.

Area of one parallelogram
$= b \times h = 10 \times 3 = 30 \, \text{cm}^2$

Area of both parallelograms
$= 30 \times 2 = 60 \, \text{cm}^2$

b The area of the large rectangle is 12 cm by 7 cm. You need to take from this the area of the 3-cm square.

Area of rectangle $= l \times w = 12 \times 7 = 84 \, \text{cm}^2$

Area of square $= l^2 = 3^2 = 9 \, \text{cm}^2$

Shaded area $= 84 - 9 = 75 \, \text{cm}^2$

Exercise 17K

1 Find the area of each of these shapes.

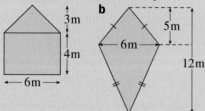

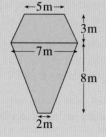

2 Ali's garden is the shaded region.

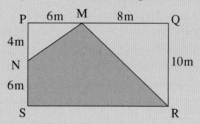

Find the area of Ali's garden.

3 Andy wants to cut the shaded triangle from a piece of cardboard.

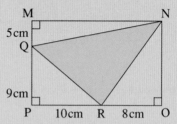

Find the area of the shaded triangle.

4 Find the area of the shaded parts in these figures:

a

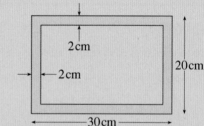

b

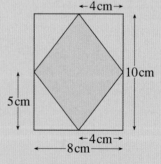

c

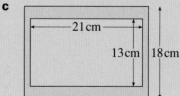

d

e

f

g

5 In her metalwork class, Anja cut a square of tin, of edge 3.7 cm, from a larger square of edge 6.3 cm. What area of tin was left?

6 The diagram shows an antique gramophone record of diameter 29 cm.

The diameter of the label is 7 cm. Find the area of the shaded (green) playing surface of the record.

7 An arrow for a signpost has to be cut from a rectangular metal sheet measuring 30 cm by 40 cm.

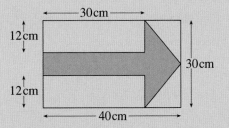

Find the area of the arrow.

8 The diagram shows a piece of metal which is to be used to make the blade of a saw.

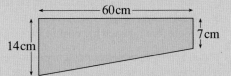

a Find the area of the blade.

b Metal costs $78 per square metre. What is the cost of metal contained in the blade?

9 The diagram shows a piece of wood which has been cut to make the deck of a toy boat. Find its area.

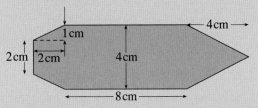

10 Give the base and height of as many triangles with area 24 cm^2 as you can. (Use only whole number lengths).

11 The front of a brick house is 10 m wide and 6 m high. There are four windows, each 2.5 m by 1.5 m, and one door 2 m by 0.8 m. What area of the front is brick?

12 a Calculate the area of a square with side 5 cm. What is the area of a square whose sides are twice as long? How are the two areas connected?

b What happens if the new square has sides three times the original?

13 A rectangular floor measuring 18 m by 15 m is to be covered with carpet, leaving a border 1.6 m wide round the room.

Find the area of the carpet required.

14 Here is the plan of a room.

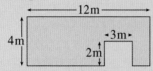

a   What is the area of the room?
b   If the room is to be paved with rectangular tiles 20 cm by 30 cm, what is the area of a tile in   **i** cm²   **ii** m²?
c   How many tiles are needed to pave the room?
d   What would be the cost of paving the whole room if each tile cost $1.30?

15 The inside of an athletics track needs to be re-seeded with grass. The area that needs seeding is sketched below. Find the total amount of seed required if 0.25 kg of seed is needed for 1 m² of ground.

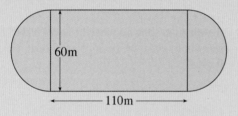

17.6   Volume of a cuboid

Volume is always measured in cubic units: mm³, cm³, m³.

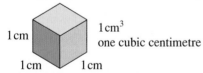

one cubic centimetre

Small objects are measured in mm³:

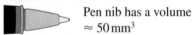

Pen nib has a volume ≈ 50 mm³

Medium-sized objects are measured in cm³:

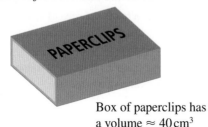

Box of paperclips has a volume ≈ 40 cm³

Larger objects are measured in m³:

A refrigerator has a volume ≈ 1 m³

Exercise 17L

1   Which unit is most suitable to measure the volume of:
a   your classroom
b   a pencil case
c   an orange
d   a car
e   a grain of rice
f   a cricket ball
g   a swimming pool
h   a hen's egg
i   a shoe box
j   an oil drum?

2   Write down five objects which have a volume:
a   more than 1 m³
b   less than 1 cm³.

3   Find the volume of these solids if each cube is 1 cm³.

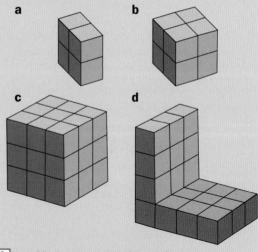

4 a   How many cm make a m?
b   How many cm² make a m²?
c   How many cm³ make a m³?

Cuboids

A cuboid has length l, width w and height h, as shown.

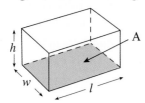

- Its volume is:

  $V = l \times w \times h$ or $V = lwh$

  Since $A = l \times w$ is the area of the shaded face of the cuboid, we can also write $V = Ah$

We can use this idea to help us find missing side lengths when we know two side lengths and the volume of a cuboid.

EXAMPLE 11

Find the missing side length of this cuboid.

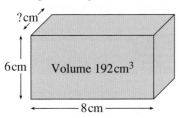

$V = Ah$

Volume $= 192\,\text{cm}^3$

Area of front face $= 6 \times 8 = 48\,\text{cm}^2$

$V = A \times h$

$192 = 48 \times h$

$h = 192 \div 48$

$\quad = 4$

> Solve the equation $192 = 48 \times h$ by dividing both sides by 48.

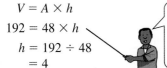

So the missing side length is 4 cm.

Exercise 17M

1 Find the volume of these cuboids:

a   **b**

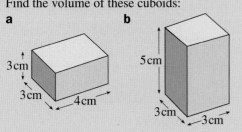

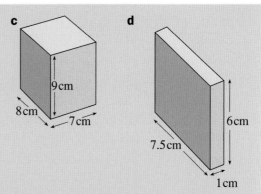

c   **d**

2 Copy and complete the table for cuboids.

	l cm	w cm	h cm	V cm³
a	2	6	12	
b	8		4	64
c		0.5	8.2	82
d	2.4	6.7		48.24

3 Find the area of the shaded face of the cuboid, then find its volume:

a   **b**

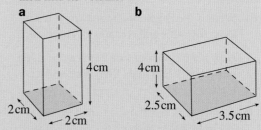

4 Copy and complete the table for cuboids.

	A cm²	h cm	V cm³
a	4		64
b	4	10.4	
c	5.2	2.5	
d		3.55	28.40

5 For each of these cuboids find the missing side lengths:

a

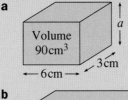

Volume 90 cm³

b

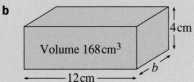

Volume 168 cm³

c

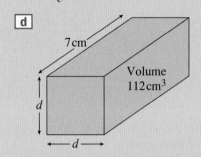

Volume
0.24 m³

2m

0.3m

c

d

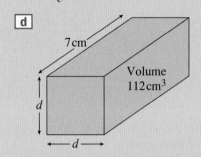

7cm

Volume
112 cm³

d

d

6 What is the height of a room which is 8 m long, 6 m wide and contains 144 m³ of air?

7 How many cubes of side 2 cm can be fitted into a box 12 cm long, 8 cm wide and 4 cm high?

8 A store sells cereal packets that measure 20 cm × 30 cm × 8 cm.
They are delivered to stores in larger boxes measuring 90 cm × 40 cm × 40 cm. Draw a diagram to show how the packets may be packed to fit inside the larger box. How many packets would fill the larger box?

9 What is the capacity in litres of a metal box 20 cm wide, 50 cm long and 30 cm high?
(1 litre = 1000 cm³)

10

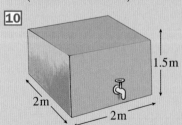

1.5m

2m

2m

a What is the volume of this water tank in
i m³ **ii** cm³?
b How many litres of water does it hold?

11 A water tank holds 3000 litres of water when full. The tank has a rectangular base with area 3 m². Find the height of the tank.

12 The volume of this cuboid is 527.28. Find the value of *y*.

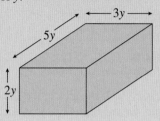

3*y*

5*y*

2*y*

17.7 Surface area

The total area of all faces of a solid is called the surface area. For a cuboid this is the sum of the areas of its six rectangular faces. You can draw the net of the shape to help you work out the surface area. You learned about drawing nets of solids in Chapter 3.

EXAMPLE 12

Find the surface area of this cuboid.

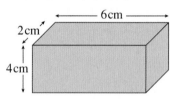

6cm

2cm

4cm

Draw the net and work out the areas of the six rectangular faces.

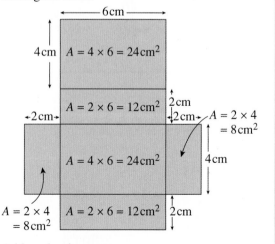

6cm

4cm $A = 4 \times 6 = 24 \text{cm}^2$

$A = 2 \times 6 = 12 \text{cm}^2$ 2cm

2cm 2cm $A = 2 \times 4 = 8 \text{cm}^2$

2cm

$A = 4 \times 6 = 24 \text{cm}^2$ 4cm

$A = 2 \times 4 = 8 \text{cm}^2$ $A = 2 \times 6 = 12 \text{cm}^2$ 2cm

Add up the six areas:
24 + 12 + 24 + 12 + 8 + 8 = 88

So surface area = 88 cm²

Exercise 17N

1 **a** Draw the net of a cube with side 2 cm.
       What is the surface area of a cube with
       side 2 cm?
   b Repeat the questions from **a** but this time
       for a cube with side 3 cm.

2 Draw the nets of these cuboids and work out
   their surface areas.

   a

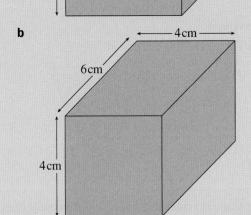

   b

3 Angela said 'You don't need to draw the net to
   work out surface area. Just work out the area
   of the three faces you can see in the diagram,
   add them together and double this answer to
   include the three faces that you can't see.'
   Use Angela's method to work out the surface
   area of these cuboids.

   a

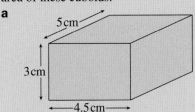

   b

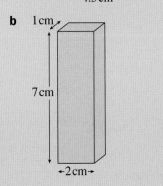

c

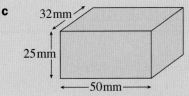

d

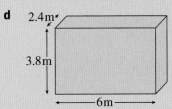

4 Each of the side lengths in Cuboid Y are
   double those in Cuboid X.

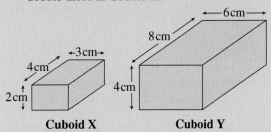

   Cuboid X   **Cuboid Y**

   a Find the surface area of
       i Cuboid X   **ii** Cuboid Y
   b How many times bigger is the surface
       area of Cuboid Y than the surface area of
       Cuboid X?
   c Find the volume of
       i Cuboid X   **ii** Cuboid Y
   d How many times bigger is the volume of
       Cuboid Y than the volume of Cuboid X?

5 The diagrams show a prism and its net. Using
   the net, work out the surface area of the prism.

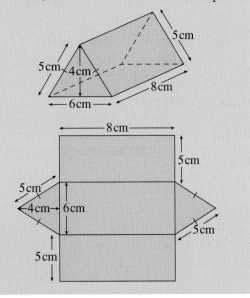

Consolidation

Example 1

Find the perimeter of these shapes.

a

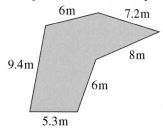

b

....

a Perimeter = distance around shape
      = 6m + 7.2m + 8m + 6m + 5.3m + 9.4m
      = 41.9m

b Perimeter = circumference of circle
      = $2\pi \times$ radius
      = $2 \times 3.14 \times 8$
      = 50.24m

Example 2

Find the area of these shapes.

a   **b**

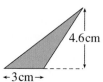

....

a Area of triangle = $\frac{1}{2} \times$ base $\times$ height
      = $\frac{1}{2} \times 3$cm $\times 4.6$cm
      = 6.9cm^2

b Area of circle = πr^2
      = 3.14×9^2cm^2
      = 3.14×81cm^2
      = 254.34cm^2

Example 3

Find the area of the trapezium.

....

Area of trapezium = $\frac{1}{2}(a + b) \times h$

$= \frac{1}{2}(4m + 12m) \times 6m$

$= 48\,m^2$

Example 4

Find the area of this compound shape.

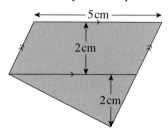

....

This shape is made up of a parallelogram and a triangle.

Area of the parallelogram
$= b \times h = 5$cm $\times 2$cm $= 10$cm^2

Area of the triangle

$= \frac{1}{2}b \times h = \frac{1}{2} \times 5$cm $\times 2$cm $= 5$cm^2

Total area $= 10$cm$^2 + 5$cm$^2 = 15$cm^2

Example 5

Find the missing side length in this cuboid.

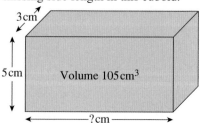

$V = Ah$
Volume = 105 cm^3
Area of side face = 3 cm $\times$ 5 cm = 15 cm^2
$V = A \times h$
105 cm^3 = 15 cm$^2 \times h$
h = 105 cm$^3 \div 15$ cm^2
  = 7 cm

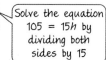

Solve the equation
105 = 15h by
dividing both
sides by 15

So the missing side length is 7 cm.

Example 6

Find the volume and surface area of this cuboid.

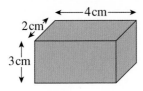

Volume $= l \times w \times h = 4\,\text{cm} \times 2\,\text{cm} \times 3\,\text{cm} = 24\,\text{cm}^3$

To find the surface area of the cuboid you can draw a net to help:

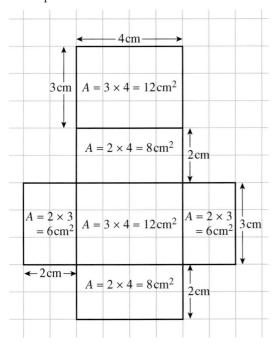

The total surface area is
$12 + 8 + 12 + 8 + 6 + 6 = 52\,\text{cm}^2$

Exercise 17

1 Find the area of these triangles.

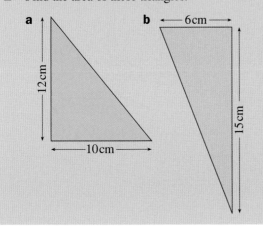

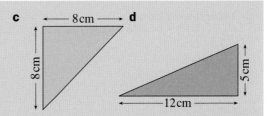

2 a Find the lengths of the unknown sides in these figures.

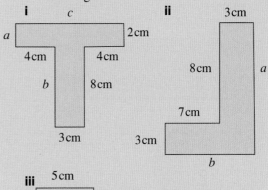

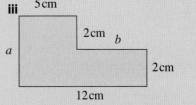

b Find the perimeter of each figure.
c Find the area of each figure.

3 Draw a circle and label on it the following parts:
a radius   b diameter
c circumference

4 Work out
i the circumference   ii the area of these circles.

a

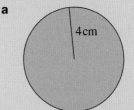

b

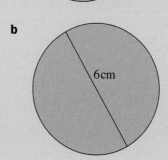

281

5 Find the shaded area.

a

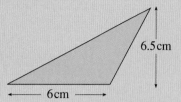

b

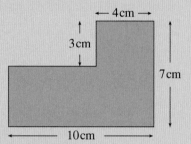

c 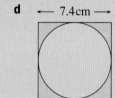 **d**

6 What is the area of this field?

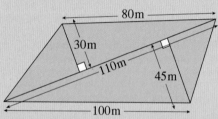

7 Draw a circle and label on it the following parts:

a arc   **b** chord
c sector   **d** segment

8 Find the area of these shapes.

a

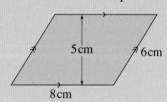

b

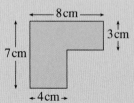

c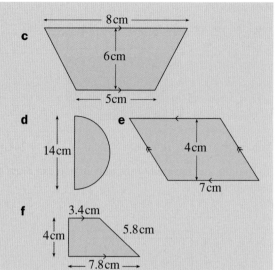

d

e

f

9 Find the **i** volume and **ii** surface area of these cuboids:

a

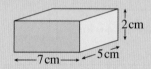

b

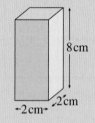

c

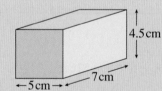

10 Find the missing side lengths.

a

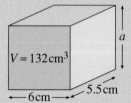

$V = 132 \text{cm}^3$

b

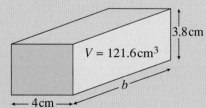

$V = 121.6 \text{cm}^3$

Summary

You should know ...

1 Perimeter is the distance around the edge of a shape.
For example:

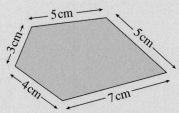

Perimeter = $(3 + 5 + 5 + 7 + 4)$ cm
= 24 cm

2 The circumference of a circle is πD or $2\pi r$, where $\pi = 3.14$
For example:

Circumference = πD
= 3.14×10 cm
= 31.4 cm

3 Area of a triangle = $\frac{1}{2}$ base $\times$ height = $\frac{1}{2}bh$

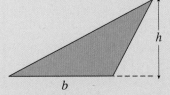

Area of a circle = πr^2
For example:
What is area of a circle
with radius 6 cm?
Area = $\pi r^2 = 3.14 \times 6^2$ cm^2
= 3.14×36 cm^2
= 113.04 cm^2

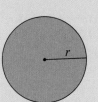

Check out

1 The perimeter of this shape is 30 cm. Find the length of the unknown side.

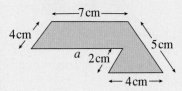

2 Find the circumference of these circles.

a

b

3 Find the area of these shapes.

a

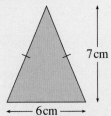

b

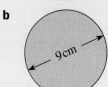

4 The area of a parallelogram, A, is given by the formula

$A = b \times h$

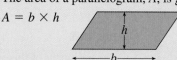

For example:

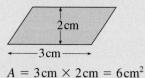

$A = 3\,\text{cm} \times 2\,\text{cm} = 6\,\text{cm}^2$

4 Find the area of these parallelograms.

a

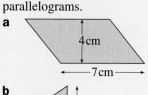

b

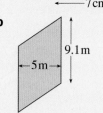

---

5 The area of a trapezium, A, is given by the formula

$A = \frac{1}{2}(a + b) \times h$

For example:

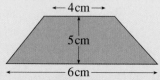

$A = \frac{1}{2}(4\,\text{cm} + 6\,\text{cm}) \times 5\,\text{cm}$

$\quad = 25\,\text{cm}^2$

5 Find the area of these trapeziums.

a

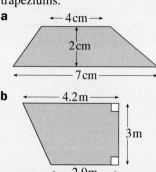

b

---

6 The volume of a cuboid, V, is given by the formula
$V = l \times w \times h$
For example:

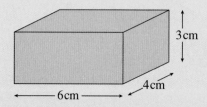

$V = l \times w \times h = 6\,\text{cm} \times 4\,\text{cm} \times 3\,\text{cm} = 72\,\text{cm}^3$
The surface area of a cuboid is the total area of the six rectangular faces. Drawing a net may help here.

For example:
To find the surface area of the cuboid above:
Area of top $= 6\,\text{cm} \times 4\,\text{cm} = 24\,\text{cm}^2$
Area of bottom $= 24\,\text{cm}^2$
Area of front $= 6\,\text{cm} \times 3\,\text{cm} = 18\,\text{cm}^2$
Area of back $= 18\,\text{cm}^2$
Area of right-hand side $= 4\,\text{cm} \times 3\,\text{cm} = 12\,\text{cm}^2$
Area of left-hand side $= 12\,\text{cm}^2$
Surface area $= 24\,\text{cm}^2 + 24\,\text{cm}^2 + 18\,\text{cm}^2 + 18\,\text{cm}^2 +$
$\qquad\qquad 12\,\text{cm}^2 + 12\,\text{cm}^2 = 108\,\text{cm}^2$

6 Find the
i volume
ii surface area of these cuboids.

a

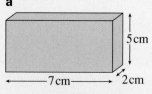

b

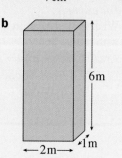

18 Probability

- Compare estimated experimental probabilities with theoretical probabilities, recognising that:
  - when experiments are repeated different outcomes may result
  - increasing the number of times an experiment is repeated generally leads to better estimates of probability.

- Know that if the probability of an event occurring is p, then the probability of it not occurring is $1 - p$.

- Find probabilities based on equally likely outcomes in practical contexts.

- Find and list systematically all possible mutually exclusive outcomes for single events and for two successive events.

What's the point?

How likely is it that your team will win a football match? What are the chances that it will rain today? Probability gives answers to such questions and many more.

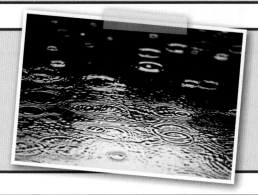

Before you start

You should know ...

1 How to read information from a bar chart.
   For example:
   Here is a bar chart of students' favourite colours:

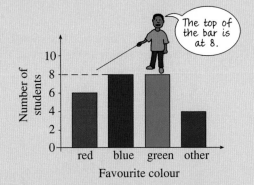

The top of the bar is at 8.

Blue is the favourite colour of 8 students.

Check in

1 From the bar chart in the example on the left, find:
   a the number of students that said red was their favourite colour
   b the number of students that did not choose red, blue or green as their favourite colour.

2 How to simplify a fraction by dividing the numerator and denominator by the highest common factor.
For example:

$$\frac{8}{12} \overset{\div 4}{\underset{\div 4}{=}} \frac{2}{3}$$

4 is the HCF of 8 and 12.

2 Write these fractions in their simplest form.

a $\dfrac{9}{18}$ **b** $\dfrac{4}{16}$

c $\dfrac{12}{30}$ **d** $\dfrac{15}{35}$

3 How to add and subtract decimals.
For example:
$0.2 + 0.35 = 0.55$
$1 - 0.6 = 0.4$

3 Work out:
a $0.4 + 0.2$
b $0.5 + 0.25$
c $1 - 0.8$
d $1 - 0.35$

4 How to add and subtract fractions using a common denominator.
For example:

$\dfrac{1}{2} + \dfrac{1}{4} = \dfrac{2}{4} + \dfrac{1}{4} = \dfrac{3}{4}$ or 0.75

$1 - \dfrac{2}{5} = \dfrac{5}{5} - \dfrac{2}{5} = \dfrac{3}{5}$ or 0.6

4 Work out:

a $\dfrac{1}{4} + \dfrac{2}{5}$ **b** $\dfrac{3}{8} + \dfrac{4}{7}$

c $1 - \dfrac{3}{5}$ **d** $1 - \dfrac{7}{8}$

18.1   The idea of probability

Some events are more likely to happen than others. In mathematics, the word **probability** is used to describe this situation.

There is a **high probability** that you will put on shoes tomorrow.

There is a **low probability** that you will stay awake for 24 hours.

In mathematics, the probability of an event is given as a number from 0 to 1.

The diagram shows the meaning of different probabilities:

```
1   ┬ certain
0.9 ┤ nearly certain
0.8 ┤
0.7 ┤
0.6 ┤
0.5 ┼ just as likely to happen as not (even chance)
0.4 ┤
0.3 ┤
0.2 ┤
0.1 ┤ very unlikely
0   ┴ impossible
```

You can see that the probability of a **certain** event is **1**; the probability of an **impossible** event is **0**; if an

event is **as likely to happen as not**, its probability is 0.5.

Probability can be written as a fraction as well as a decimal. Probability can also be written as percentages. When using percentages the probability scale goes from 0% to 100%. Most of the work in this chapter will be using decimals and fractions.

Never write probability in words or using ratios. For example, '3 : 4' or '2 out of 5' are both commonly used but incorrect ways of writing probability.

Exercise 18A

1 *I will live to be a hundred.*

How likely is this event?

2 Three students answered Question **1**. Their answers are shown on the diagram. Try to write their answers in words.

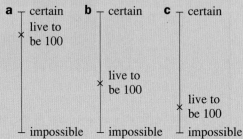

3 Draw a diagram like those in Question **2**.

Mark and label a cross on it to show whether you think each event is nearly certain, nearly impossible, or as likely as not.

a I will get married.

b If I buy a new pen it will write.

c It will rain tomorrow.

d I will get to school on time tomorrow.

e I will take 20 catches in a cricket match.

4 Are some of the events in Question **3** more likely to happen than others? Pick out the most likely event.

5 Write down three events that you think are:

a impossible

b certain

c just as likely to happen as not.

6 Try to decide for yourself how certain or impossible an event is, if its probability is described as:

a 0   **b** 1   **c** $\frac{1}{2}$

d $\frac{9}{10}$ **e** $\frac{2}{10}$

7 Look again at Question **3**. Give each event a probability between 0 and 1.

8 Assign a probability of 0 or 1 to each event.

a One day I will die.

b Tomorrow it will snow.

c I will get 100% in my maths exam.

d If I put ice outside in the sunshine it will melt.

e I shall grow 2 cm taller today.

9 Choose a fraction between 0 and 1 that you think describes the probability of each event:

a The next baby born at the local hospital will be a boy.

b A throw of a dice will give a 5.

c A number chosen from 1 to 10 will begin with F when spelled out in words.

18.2 Experimental probability

You will need a coin, a plastic cup and a dice.

You can define the probability that an event occurs, i.e. a 'successful outcome' P(S), as

• $$P(S) = \frac{\text{number of successful outcomes}}{\text{total number of outcomes}}$$

For example, if you throw a coin 100 times the probability of getting a head is

$$P(\text{Head}) = \frac{\text{number of times heads occurred}}{\text{total number of throws}}$$

$$= \frac{\text{about } 50}{100} \text{ or about } \frac{1}{2} \text{ or } 0.5$$

Many probabilities can be found either by experiment or survey.

EXAMPLE 1

Fifty people were asked their favourite sport.

The results are shown in the table.

Sport	Cricket	Basketball	Netball	Football
No. of people	17	8	20	5

Calculate the probability that:

a a person's favourite sport is football

b a person's favourite sport is *not* netball.

....

a P(favourite sport is football)

$$= \frac{\text{number of people liking football}}{\text{total number of people}}$$

$$= \frac{5}{50} = \frac{1}{10}$$

b P(favourite sport is not netball)

$$= \frac{\text{number of people who like sports other than netball}}{\text{total number of people}}$$

$$= \frac{17 + 8 + 5}{50} = \frac{30}{50} = \frac{3}{5}$$

Exercise 18B

1 Call the face of a one-cent coin with a 1 written on it *face one*. Call the other face *face two*.

Put the coin in the plastic cup, shake it, and tip it on to your desk top. Repeat this 100 times.

Keep a record of which face turns up each time in a table like this:

Face one	Face two
⌿⌿⌿ //	⌿⌿⌿ ⌿⌿⌿

For what fraction of the total number of throws was the outcome:

a face one   **b** face two?

2 Now collect the results from nine students in your class and record them in a table like this:

	Face one	Face two
My result	48	52
Student 1		
2		
⋮		
9		
Total		

a Compare your results with friends.
b For what fraction of the total number of throws was the outcome:
  i face one   **ii** face two?

3 Look at your results for Questions **1** and **2**. When you throw a coin are you:
a more likely to get face one *or*
b more likely to get face two *or*
c equally likely to get either?

4 Throw a dice 30 times.

a Copy and complete the table.

Score	1	2	3	4	5	6
Frequency						

b What is the probability of throwing:
  i a 2   **ii** an even number?

c Now collect the results from four students in your class and record them in a table like this:

	1	2	3	4	5	6
My result						
Student 1						
2						
3						
4						
Total						

d Compare your results with friends.
e Repeat part **b** using the data in part **c**, from 150 throws.

5 Packets of flower seeds are checked for purity, in case any weed seeds have been included. Here is a frequency table showing the results of checking 100 packets:

Number of weed seeds	0	1	2	3	4	5	6	7	8
Number of packets	5	16	26	19	13	12	5	2	2

What is the probability that a packet contains:
a 6 weed seeds   **b** 1 weed seed
c more than 6 weed seeds?

6 A scientist weighs kittens at birth with the following results:

Mass (g)	190	200	210	220	230	240	250
Frequency	13	34	57	50	29	12	5

a How many kittens did the scientist weigh?
b Find the probability that a kitten has a mass of:
  i 250 g   **ii** less than 250 g.

7 A test report on 32 cars gives these maximum speeds:

Maximum speed (km/h)	Number of cars
100-119	3
120-139	9
140-159	5
160-179	5
180-199	5
200-219	3
220-239	1
240-259	1

Find the probability that a car picked at random has a maximum speed of:
a 120–139 km/h
b less than 120 km/h.

8 These are the marks of 50 people who took a skills test. The maximum mark was 240.

Marks	Number of candidates
0-29	2
30-59	5
60-89	9
90-119	16
120-149	8
150-179	7
180-209	2
210 or over	1

What is the probability that a person chosen at random scored:
a 120–149
b less than 30
c 120–179
d 180 or over?

18.3 Theoretical probability

Often it is impractical or too time consuming to carry out an experiment. In these cases you can still calculate the theoretical probability of an event. It is defined in the same way as experimental probability, that is:

$$P(\text{success}) = \frac{\text{number of successful outcomes}}{\text{total number of outcomes}}$$

EXAMPLE 2

One letter is chosen at random from the word STATISTICS. What is the probability that it is
a S **b** a vowel **c** not a vowel?

....

a $P(S) = \dfrac{\text{number of Ss}}{\text{total number of letters}} = \dfrac{3}{10}$

b $P(\text{vowel}) = \dfrac{\text{number of vowels}}{\text{total number of letters}} = \dfrac{3}{10}$

c $P(\text{not a vowel}) = \dfrac{\text{number of consonants}}{\text{total number of letters}} = \dfrac{7}{10}$

You will see from the answers to Example 2, parts **b** and **c** that the probability of getting a vowel and the probability of not getting a vowel add up to 1:

$$\frac{3}{10} + \frac{7}{10} = 1$$

So there is a quick way of working out the answer to part **c**:

$$1 - \frac{3}{10} = \frac{7}{10}$$

- If the probability of an event occurring is p, then the probability of it not occurring is $1 - p$.

Exercise 18C

1 A dice is thrown.
    a How many different ways could it land?
    b What is the probability that a 3 is thrown?
    c What is the probability of not throwing a 3?

2 What is the probability of picking a 6 at random from 10 cards labelled 1 to 10?

3 A bag contains two white beads and three black beads. One bead is chosen at random.

What is the probability that the bead is
    a white   **b** black?

4 A letter is chosen at random from the word MATHEMATICS. What is the probability that the letter is
    a A   **b** M   **c** not A   **d** not I?

5 25 yellow cards numbered 1 to 25 and 25 blue cards numbered 1 to 25 are mixed together.

Here are *some* of the cards:

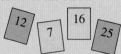

What is the probability that a card picked at random is

a a 3

b blue

c not a 7

d not blue

e a number which is a multiple of 3

f a blue 24?

6 a One month of the year is to be chosen. List the possible outcomes.

b How many of the possible outcomes in part **a** begin with the letter J?

c What is the probability that the chosen month will begin with the letter:
i J **ii** M **iii** D **iv** P?

d What is the probability that the month will not begin with J?

7 A class has 15 boys and 30 girls. The teacher chooses the class monitor as follows:

The students' names are written on slips of paper, which are folded and put in a bag; the bag is shaken and one name drawn out. This person will be the monitor.

What is the probability that the monitor will be

a a boy **b** a girl?

8 Look at your answers for Question **7**.

a Is it more likely that the monitor will be a boy or a girl?

b What can you say about the sum of the two probabilities?

c What does a probability of 1 mean?

9 A bag contains 3 blue beads, 5 green beads and 2 red beads. One bead is picked from the bag. Which colour is:

a most likely to be picked

b least likely to be picked?

10 A box of buttons is made up as follows: 5 blue, 6 green, 8 red, 4 yellow, 7 white. One button is picked from the box.

a How many buttons are there altogther?

b Are these true or false?

i $P(R) = \dfrac{4}{15}$ **ii** $P(B) = \dfrac{1}{5}$

iii $P(G) = 20\%$ **iv** $P(\text{not Y}) = 0.8$

11 A bag contains the following currency notes: ten \$1, twenty \$5, forty \$10, thirty \$20. The notes are shaken up and one is chosen.

a How many notes are there altogether?

b Calculate:

i $P(\$1)$ **ii** $P(\$5)$

iii $P(\$10)$ **iv** $P(\$20)$

12 A box contains 20 pens. Three of them are faulty. One pen is chosen from the box. What is the probability that it is:

a faulty **b** not faulty?

13 A dice is rolled once. Calculate:

a $P(5)$

b $P(6)$

c $P(\text{even number})$

d $P(\text{odd number})$

e $P(\text{number} < 5)$

f $P(\text{number} > 2)$

g $P(\text{multiple of } 3)$

h $P(\text{number not divisible by } 3)$

You should have seen in Exercise 18B, Questions **2** and **4** that when you repeat an experiment different outcomes usually happen. If every student in your class throws a dice 20 times, it is unlikely that everyone will throw the same number of sixes.

Look at Exercise 18B, Question **4**. When you throw a dice 30 times you would expect, in theory, to get these results:

Score	1	2	3	4	5	6
Frequency	5	5	5	5	5	5
Probability	$\dfrac{1}{6}$	$\dfrac{1}{6}$	$\dfrac{1}{6}$	$\dfrac{1}{6}$	$\dfrac{1}{6}$	$\dfrac{1}{6}$

Did you or any of your friends get these results?

You should know that the theoretical probability of getting each of the dice scores is $\frac{1}{6}$ or, as a decimal, 0.17 (to 2 d.p.). Were any of your results close to this? It may be easier to change your fractions to decimals to compare.

Now look at the data when you had 150 results (Exercise 18B, Question **4,** part **c**). Were your probabilities closer to 0.17?

When comparing estimated, experimental probabilities with theoretical probabilities you should find that increasing the number of times an experiment is repeated generally leads to better estimates of probability.

Note that experimental probabilities and theoretical probabilities are often not the same, but the more times you repeat the experiment the closer the experimental probability should get to the theoretical probability, assuming that the experiment is fair.

Exercise 18D

1 A biased coin was thrown 10 times. The results are recorded in this table.

Outcome	Heads	Tails
Frequency	3	7

a What is the experimental probability of this coin landing heads up? Give your answer as a decimal.

The same coin was thrown 100 times. The results are recorded in this table.

Outcome	Heads	Tails
Frequency	38	62

b What is the new experimental probability of this coin landing heads up, using this data? Give your answer as a decimal.

The same coin was thrown 1000 times. The results are recorded in this table.

Outcome	Head	Tail
Frequency	410	590

c What is the new experimental probability of this coin landing heads up, using this data? Give your answer as a decimal.

d Out of the three probabilities in parts **a, b** and **c**, which do you think is closest to the theoretical probability? Why?

2 Throw a drawing pin in the air 50 times. Copy the table below and record whether the drawing pin lands point up or point down.

Outcome	Point up	Point down
Tally		
Number of times		

a What is the experimental probability of the drawing pin landing point up? Give your answer as a decimal.

b Now combine your results with a friend to find the experimental probability of a drawing pin landing point up for 100 throws. Give your answer as a decimal.

c Now combine your results with 8 more friends to find the experimental probability of a drawing pin landing point up for 500 throws. Give your answer as a decimal.

d Out of the three probabilities in parts **a, b** and **c**, which do you think is closest to the theoretical probability? Why?

3 Conduct an experiment of your choice for which you know the theoretical probabilities. Work out the experimental probabilities after 10 trials, then 100 trials. See how close your experimental probabilities are to the theoretical ones.

18.4   Listing outcomes

•   **Mutually exclusive** outcomes means 'outcomes that cannot happen at the same time'.

For example, if you roll a dice, this is an event with lots of possible outcomes.

Two possible outcomes are rolling a 1 and rolling a 6. These two outcomes are mutually exclusive because they can't both happen at the same time.

Two other possible outcomes are rolling a 1 and rolling an odd number. These are not mutually exclusive because if you roll a 1 both events have happened.

In Exercise 18C, Question **6**, part **a** you were asked to list all the possible outcomes. When there is only a single event, listing all possible mutually exclusive outcomes is easy.

When there are two or more events you need to make sure that you follow a logical method when listing outcomes to make sure that you do not miss any out. For example, if you were writing the possible outcomes for throwing two dice, it does not make sense to list them randomly, e.g. 1, 4 or 6, 2 or 2, 5 … . You can see that it would be very easy to miss out some outcomes. The next example shows a more logical way of listing.

EXAMPLE 3

A dice is rolled and a coin is thrown.
a Write down all the possible outcomes.
b What is the probability of throwing a head and a 5?

....

a First, list all the possible outcomes with heads thrown with the coin: H, 1 or H, 2 or H, 3 or H, 4 or H, 5 or H, 6

Then list all the possible outcomes with tails thrown with the coin: T, 1 or T, 2 or T, 3 or T, 4 or T, 5 or T, 6

b You can see there are 12 possible outcomes. Only one outcome is H, 5, so $P(H, 5) = \frac{1}{12}$

Some people prefer to use a two-way table to list outcomes. This is shown in the next example.

EXAMPLE 4

a Two tetrahedral (4-sided) dice are rolled. Each dice is numbered 1 to 4. Draw a two-way table to show all the possible outcomes.

b What is the probability of the score on the second dice being higher than the score on the first dice?

....

a

		Dice 1			
		1	**2**	**3**	**4**
Dice 2	**1**	1, 1	2, 1	3, 1	4, 1
	2	1, 2	2, 2	3, 2	4, 2
	3	1, 3	2, 3	3, 3	4, 3
	4	1, 4	2, 4	3, 4	4, 4

b In the table, those outcomes for which the score on the second dice is higher than the score on the first dice are shown in blue: there are 6. There are 16 possible outcomes, so:

$$P(\text{score on the second dice is higher}) = \frac{6}{16} = \frac{3}{8}$$

Exercise 18E

1 Two coins are thrown. This is Sarah's working:

The possible outcomes are HH, HT, TT so $P(TT) = \frac{1}{3}$

a Sarah has made a mistake. What mistake has she made?
b What should the probability of two tails be?

2 Two dice are rolled. Each dice is numbered 1 to 6.

a Copy and complete this two-way table to show all the possible outcomes.

		Dice 1					
		1	**2**	**3**	**4**	**5**	**6**
Dice 2	**1**		2, 1				
	2						
	3						
	4				5, 4		
	5						
	6						

b What is the probability of rolling
i a 3 on the first dice and a 6 on the second dice
ii the same score on both dice
iii a higher score on the first dice than the second?

3 Two children are born in a family. List all possible outcomes and work out the probability that they are both boys.

4 On a restaurant set menu you can choose a starter and a main course:

Starter	Main course
Mushrooms	Lasagne
Sardines	Trout
Fruit	Vegetarian curry
Parsnip soup	Chicken rice

a Copy and complete this list of possible outcomes (using just the first letter of each dish):
M, L   M, T   etc.

b What is the total number of possible combinations of meals?

c Assuming a starter and main course are chosen at random, what is the probability that at least one of those dishes will contain fish (sardines or trout)?

d If there were 6 starters and 8 main courses, what would the total number of possible combinations of meals be?

5 Alex, Budi, Carl and Dinesh all run a race.

a Write down all the possible outcomes for who wins and who comes second.

b What is the probability Carl wins and Dinesh comes second (assuming that the runners all have roughly the same ability)?

6 Jane is having pizza. She is allowed two toppings on her pizza. (If she really likes something she is allowed two lots of the same topping.)

a If Jane chooses her toppings from pineapple, meat, sweetcorn, mushroom and olives, list all the possible outcomes for her two toppings.

b Assuming Jane chooses her two toppings at random, what is the probability that her pizza is vegetarian (has no meat)?

7 Jamil wants to phone a friend. He knows the phone number contains 6 digits. He knows the first four digits of the phone number are 3027 but he doesn't remember the last two digits.

a Jamil decides to phone all the possible numbers until he gets the right one. Calls cost $0.25 each. How much would it cost if he had to try all the numbers?

b What is the probability of him calling the correct number first time?

8 Danushka has 3 bags full of coloured sweets. Each bag has an equal number of red, yellow, green and purple sweets. Without looking, Danushka picks a sweet from each of the three bags at random.

a List all the possible outcomes for the colours of his three sweets.

b What is the probability that at least two of the sweets are red?

c What is the probability that at most one of the sweets is yellow?

ACTIVITY

This is a game for two people, Player A and Player B.
You will need two dice and some paper.

One player rolls both dice. The score is the sum of the two rolls.

Player A gets a point if the score is 2, 3, 4, 10, 11 or 12.
Player B gets a point if the score is 5, 6, 7, 8 or 9.

Jane said: 'This game is not fair because Player A gets a point if 6 different scores are rolled but Player B only gets a point for 5 different scores.'

Play the game for 10 rolls. Who won?
Play the game for 100 rolls. Who won?

Do you think this game is fair? Why?

Change the scoring system of the game to make it fair.

Make up a game of chance of your own.
Write the rules so that each player has an equal chance of winning.
Now change the rules so one player has a better chance of winning.

Consolidation

Example 1

Choose a number between 0 and 1 to describe the probability that:

a you will win the toss at a cricket match
b the sun will come out tomorrow
c you will weigh 2 kg more tomorrow.

....

a neither likely nor unlikely, 0.5
b likely, 0.7 or above (depending on where you live and the time of year)
c very unlikely, 0.1 or below

Example 2

A bag has 3 blue beads, 2 green beads and 8 red beads. What is the probability that the first bead picked at random is:

a blue?
P(bead is blue)

$$= \frac{\text{number of blue beads}}{\text{number of beads}}$$
$$= \frac{3}{13}$$

b green or red?
P(green or red)

$$= \frac{\text{number of green or red beads}}{\text{number of beads}}$$
$$= \frac{2 + 8}{13}$$
$$= \frac{10}{13}$$

c not blue?
P(not blue) = 1 − P(blue)

$$= 1 - \frac{3}{13} = \frac{10}{13}$$

Exercise 18

1 Choose a number between 0 and 1 that describes the probability that:
  a it will rain tomorrow
  b you will throw a 5 when you roll a dice
  c there will be no school tomorrow.

2 What is the probability of these events?
  a Rolling a dice and getting an even number
  b Picking a single digit at random from the digits 1 to 9 and it being less than 3
  c Picking a yellow ball from a bag containing 3 blue, 4 red and 8 yellow balls

  d Not picking a green ball from a bag containing 7 blue, 2 red, 3 yellow and 8 green balls
  e Picking a letter from the alphabet at random and it is a vowel (a, e, i, o or u)
  f Picking a letter from the alphabet at random and it is not a vowel

3 Jason rolls two dice. Here are his results:

Dice A

Score	1	2	3	4	5	6
Frequency	61	69	58	56	65	64

Dice B

Score	1	2	3	4	5	6
Frequency	103	81	113	90	68	75

  a How many times did he roll:
    i dice A   **ii** dice B?
  b What is the probability of throwing a 1 on:
    i dice A   **ii** dice B?
  c Can you tell if the dice are fair? Explain.

4 In the last year, 129 days were recorded as 'wet' days in Cambridge, UK.

  What is the probability that it will be dry in Cambridge today?

5 Mr Masood decided to give his class double homework if both the spinners below were the same colour when spun.

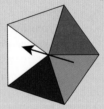 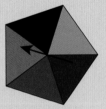

List all possible outcomes.
What is the probability of Mr Masood's class getting double homework?

Summary

You should know ...

1 Some events are more likely to happen than others. In mathematics, the likelihood of an event happening is described by a number between 0 and 1.

This number is the probability of the event.

Impossible	Unlikely	As likely as not	Very likely	Certain
0	0.2	0.5	0.9	1

2 How to find the probability of an event by performing an experiment.
The occurrence of a particular event is called a successful outcome.

This means the probability of a success

$$P(success) = \frac{\text{number of successful outcomes}}{\text{total number of outcomes}}$$

For example:
A coin was thrown 100 times and 52 heads occurred, so
$$P(head) = \frac{52}{100} = \frac{13}{25}$$
is the experimental probability.

3 How to find the probability of simple events without performing experiments.

For example:
If a coin is tossed, the theoretical probability of getting a head is

$$P(head) = \frac{\text{number of successful outcomes}}{\text{total number of outcomes}}$$
$$= \frac{1}{2}$$

4 If the probability of an event occurring is p, then the probability of it not occurring is $1 - p$.

For example:
The probability of picking a red sweet is $\frac{3}{8}$ then the probability of not picking a red sweet is $1 - \frac{3}{8} = \frac{5}{8}$.

Check out

1 Write down two events that are:
a impossible
b certain
c very likely to happen
d unlikely to happen.

2 In a survey of 40 people, 6 said they wore small t-shirts, 15 wore medium and 19 wore large t-shirts.
a What is the probability that a person chosen at random from the sample wears a small t-shirt?
b What is the probability that a person wears a medium or large t-shirt?

3 From the letters in the word ADDING, one letter is chosen. What is the probability that it is:
a a D
b not a D
c a vowel?

4 a If the probability of picking a yellow sweet is $\frac{2}{5}$, what is the probability of not picking a yellow sweet?
b If the probability of it raining tomorrow is 0.7, what is the probability of it not raining tomorrow?
c If the probability of winning a football match is 45%, what is the probability of not winning the football match?

5 When comparing estimated, experimental probabilities with theoretical probabilities, it is important to remember that
  • when experiments are repeated different outcomes may result
  • increasing the number of times an experiment is repeated generally leads to better estimates of probability.

5 Three groups of students conducted an experiment using the same biased coin. Here are their results:

Group 1 results

Outcome	Heads	Tails
Frequency	1	9

Group 2 results

Outcome	Heads	Tails
Frequency	31	69

Group 3 results

Outcome	Heads	Tails
Frequency	256	744

a For each of the three groups, work out the experimental probability for getting heads.

b Which of the three groups' results will give the best estimate of probability? Why?

c The theoretical probability for getting heads with this biased coin is actually 0.25. Why do you think that none of the groups' results, from part **a**, were 0.25?

6 How to list systematically all possible mutually exclusive outcomes for two successive events.

For example:
A 6-sided dice and a tetrahedral dice are both rolled.

The possible outcomes can be listed or shown in a two-way table like this:

		Tetrahedral dice			
		1	**2**	**3**	**4**
	1	1, 1	2, 1	3, 1	4, 1
	2	1, 2	2, 2	3, 2	4, 2
6-sided dice	**3**	1, 3	2, 3	3, 3	4, 3
	4	1, 4	2, 4	3, 4	4, 4
	5	1, 5	2, 5	3, 5	4, 5
	6	1, 6	2, 6	3, 6	4, 6

6 List all the possible outcomes when a letter from A, B or C is chosen at random and a 6-sided dice is rolled. The list has been started below for you.

A1, A2, …

Review C

1 Write each ratio in its simplest form.
  a 20 : 60   **b** 75 : 35
  c 20 : 80 : 25   **d** 27 : 45
  e 91 : 26   **f** 36 : 156 : 72
  g 34 : 119

2 Find
  a the circumference   **b** the area of these circles.

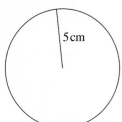
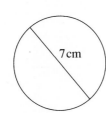

3 Find the first 5 terms of these sequences.
  a The position-to-term rule is *multiply by 3*.
  b The position-to-term rule is *multiply by 4 then subtract 1*.

4 Which is the largest, $\frac{9}{10}$, 0.088, 1.02 or 98%?

5 The probability that a school's netball team wins the next match is $\frac{1}{5}$. What is the probability that it will not win?

6 Copy each diagram onto squared paper. Draw their reflections in the dashed line.

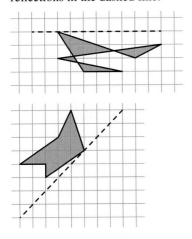

7 Express these fractions as
  i decimals   **ii** percentages

  a $\frac{7}{10}$ **b** $\frac{9}{100}$ **c** $\frac{3}{25}$

  d $4\frac{3}{4}$ **e** $1\frac{2}{5}$ **f** $\frac{3}{8}$

  g $\frac{13}{5}$ **h** $\frac{27}{50}$

8 Look at the diagram below.

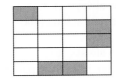

How many more rectangles need to be shaded so that the ratio of shaded to unshaded is 2 : 3?

9 Using a scale of 1 cm to 2 m, make a scale drawing of a room measuring 12 m by 5 m. By measuring your scale drawing, find the distance between opposite corners of the room.

10 Copy and complete this mapping diagram using the function shown.

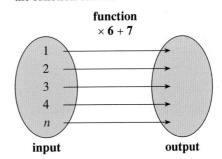

11 On the circle below three lines – black, red and purple – and a blue curve are shown. What are the names of these parts of the circle?

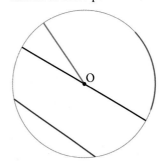

12 A box contains a set of tennis balls. Six are red, four are white and five are green.
They are mixed and one is chosen at random.
Find the probability of each of the following events:
  a a red ball is chosen
  b a white ball is chosen
  c a red or green ball is chosen
  d a red ball is not chosen.

13 a Write the equation that goes with these sentences
    i To find the y-coordinate multiply the x-coordinate by 2 then add 5.
    ii The x-coordinate is always ⁻4.
    iii To find the y-coordinate subtract the x-coordinate from 15.
    iv The y-coordinate is always 3.
 b Draw the graphs of all the lines from part **a**.

14 Change each of these test scores to
  i fractions in their lowest terms
  ii percentages
    a 21 out of 50
    b 105 out of 150
    c 45 out of 75
    d 118 out of 200

15 Copy this diagram.

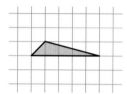

Translate the triangle through these vectors.

a $\begin{pmatrix} 6 \\ 3 \end{pmatrix}$ **b** $\begin{pmatrix} 0 \\ 4 \end{pmatrix}$

c $\begin{pmatrix} ^-5 \\ 0 \end{pmatrix}$ **d** $\begin{pmatrix} 4 \\ ^-8 \end{pmatrix}$

16 Copy and complete:
 a $420\,\text{m} = \square\,\text{km}$
 b $3\,\text{days} = \square\,\text{hours}$
 c $2.7\,\ell = \square\,\text{ml}$
 d $79\,100\,\text{m} = \square\,\text{km}$
 e $600\,\text{kg} = \square\,\text{t}$
 f $420\,\text{minutes} = \square\,\text{hours}$
 g $20\,\text{cm} = \square\,\text{mm}$
 h $320\,\text{ml} = \square\,\ell$
 i $0.4\,\text{kg} = \square\,\text{g}$

17 For the equation $y = 3x + 8$, copy and complete these coordinate pairs by calculating the missing y-coordinate using the given x-coordinate.
 a $(7,\square)$ **b** $(5,\square)$
 c $(^-2,\square)$ **d** $(^-1,\square)$

18 To increase something by 15% the multiplier you should use is 1.15. What multiplier should you use for each of these?
 a An increase of 32%   **b** A decrease of 7%
 c A decrease of 40%   **d** An increase of 35%

19 Use a ratio to compare the quantities:
 a $1\,\text{m}; 20\,\text{cm}$
 b $20\,\text{mm}; 1\,\text{cm}$
 c 3 minutes; 30 seconds
 d 45 minutes; 2 hours
 e $25\,\text{g}; 0.75\,\text{kg}$
 f $3.6\,\text{kg}; 90\,\text{g}$

20 Copy and complete this table.

Fraction	Decimal	Percentage
$\frac{3}{4}$		
	0.1	
		$66\frac{2}{3}\%$
		$12\frac{1}{2}\%$
$1\frac{1}{5}$		

21 On squared paper, make a copy of this L-shape made from 4 squares.

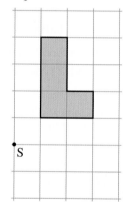

Draw an enlargement of the shape using S as the centre of enlargement and a scale factor of 2. How many squares are needed to make the enlarged L-shape?

22 Calculate:
 a 10% of $65   **b** 25% of $170
 c $12\frac{1}{2}\%$ of $1600   **d** 125% of $60
 e $\frac{1}{2}\%$ of $50

23 Using the numbers 1, 2, 3 and 4 for x, draw a mapping diagram to show:
 a $x \rightarrow x + 5$ **b** $x \rightarrow 2x + 1$

24 Using 3.14 for π, find the circumference of a circle with radius:
 a $12.56\,\text{cm}$ **b** $37.68\,\text{cm}$
 c $50\,\text{cm}$ **d** $157\,\text{cm}$

25 Look at the diagrams below:

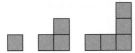

Each time new squares are added around the outside of the previous shape.

a Draw the next few shapes.

b Copy and complete the table.

Shape	Number of squares
1	1
2	3
3	
4	
5	
	85
13	
n	

c Looking at the diagrams, explain why the nth term works.

26 Find the area of each shape:

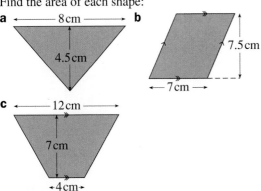

a 8 cm, 4.5 cm **b** 7.5 cm, 7 cm **c** 12 cm, 7 cm, 4 cm

27 It takes Mr Speedy 32 minutes to cycle to work, or 1 hour 48 minutes to walk to work. Write as a ratio the cycling time to the walking time. Simplify your ratio.

28 For the mapping below
a draw the function machine
b write down the ordered pairs
c draw an X, Y graph on a grid

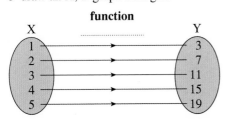

function

X		Y
1	→	3
2	→	7
3	→	11
4	→	15
5	→	19

29 Find the areas of these shapes.

a 5 cm, 16 cm **b** 13.3 cm

30 Write down the first five terms of the sequences described.

a The first term is 80, the term-to-term rule is *subtract 11*

b The first term is 0.5, the term-to-term rule is *multiply by 2*

c The first term is 11, the term-to-term rule is *add 8*

d The *third* term is 200, the term-to-term rule is *divide by 10*

31 You may use a calculator for this question. Find:
a 3.4% of 5700 km **b** 17.5% of $36
c 17% of 36ℓ **d** 12% of $448.50

32 Two cards are drawn at random from a group of ten cards numbered from 1 to 10.
Once a card is selected, it is not replaced.
a List all the possible outcomes.
b Find the probability of drawing each of these outcomes.
   i A 5 and then a 3
   ii Two even numbers
   iii Two numbers greater than 4
   iv A 6 and then an odd number
   v A number greater than 7 and then a number less than 6

33 If 12 cents out of each dollar is paid in tax:
a What fraction is paid in tax?
b What percentage is paid in tax?

34 Copy this diagram.

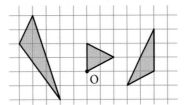

Rotate each triangle 90° anticlockwise about the point O.

35 A green paint mixes yellow and blue paint in the ratio 7 : 3.
  a If 4.8 litres of blue paint are used, how many litres of yellow paint are needed to make the green paint?
  b If 9.1 litres of yellow paint are used, how many litres of blue paint are needed to make the green paint?

36 90 kg of animal feed is delivered to a zoo. After the elephants were given their lunch, this mass went down by 11%. What mass remains?

37

x	-2	-1	0	1	2
y					

Copy and complete the table for each of these equations. Use the table to draw a graph.
  a $y = 3x + 4$ **b** $y = 2x - 1$
  c $y = 7 - x$

38 Copy the diagram.

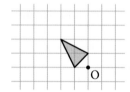

Using the centre of enlargement O and a scale factor of 4, draw the enlargement of the triangle.

39 Neema and Isis each have $180.
Neema spends $\frac{3}{8}$ of her money.
Isis spends 40% of her money.
Who has spent the most money?
Show your working.

40 a Find the area of a triangle which has a height of 7.2 cm and a base of 2.5 cm.
  b The area of a triangle is 40 cm². If the base of the triangle is 2.5 cm find the height of the triangle.

41 A farm has hens, goats and sheep in the proportion 1 : 2 : 5.
  a Find how many goats and sheep there are if there are 8 hens.
  b Find how many hens and goats there are if there are 35 sheep.
  c Find how many hens and sheep there are if there are 12 goats.
  d Find how many of each type of animal there are, if there are altogether
    i   16 animals   **ii** 48 animals
    iii 72 animals   **iv** 120 animals.

42 Look at the graph below.
  a List the ordered pairs for the 7 points.
  b Write down the rule as: $x \rightarrow \square$

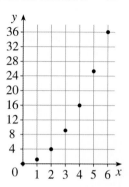

43 Each of the shapes below is made from part of a circle. Use $\pi = 3.14$ and find the area and perimeter of each shape.

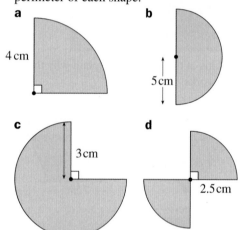

44 What is the volume of these cuboids?

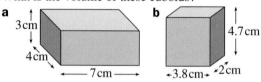

45 What is the surface area of each of the cuboids in Question **44**?

46 It is intended to make a 400 m running track from two 120 m straights and two semicircular ends, as shown:

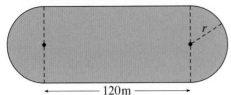

← 120m →

Using $\pi = 3.14$:

a What is the length of the curved part of the track?

b What is the radius of each of the semicircular ends?

47 a Make an accurate scale drawing of the running track in Question **46** using a scale in which 1 cm represents 10 m.

b On your scale drawing measure the length of the diagonal of the rectangle formed by the two straights.

c Use your answer to part (b) to write down the actual length of the diagonal on the running track.

48 Write each of these as a ratio in its simplest whole number form:

a $\frac{1}{2} : 3$ **b** $3.5 : 2.8 : \frac{7}{10}$

c $1.4 : 8.2 : 6.3$ **d** $\frac{2}{25} : 6.4$

49 Find the nth term for the multiples of 12.

50 a Given a set of numbers 1, 20, 2, 43, 6, 7, 3, 106, what is the probability of choosing at random from the set

i   a number greater than 7

ii the number 7

iii a number less than 7?

b What is the sum of the probabilities in part **a**?

51 The shape A goes through two transformations. The first transformation maps A onto A′ by reflecting A in the dashed mirror line. The second transformation maps A′ onto A″ by rotating A′ 90° anticlockwise about the point O.

Copy the diagram onto squared paper. Draw each of the images A′ and A″.

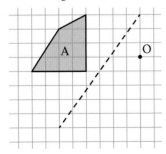

52 Work out the new values.

a Decrease 300 ml by 18%

b Increase $340 by 15%

c Increase 325 km by 16%

d Decrease 4800 cm by 34%

53 If ribbon costs $1.53 for 4.5 m, how much does it cost for 17 m?

54 Which of these points lie on the line $y = 8 - 2x$?
(3, 18)   (0, 8)   (⁻1, 6)   (⁻2, 12)

55 What are **i** the next two terms **ii** the term-to-term rule **iii** the nth term for these sequences?

a 12, 16, 20, 24, 28, …

b 7, 18, 29, 40, 51, …

c 3.5, 4, 4.5, 5, 5.5, 6, …

d 550, 500, 450, 400, 350, …

56 Find the area of each shape:

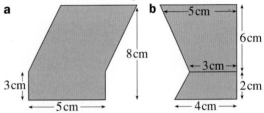

57 Divide:

a 240 m in the ratio 7 : 8

b 540 kg in the ratio 3 : 2 : 4

c $216 in the ratio 3 : 5

d 1400 ml in the ratio 3 : 4

e 0.91 kg in the ratio 7 : 5 : 1

f 0.35 m in the ratio 2 : 3

g 0.119 ℓ in the ratio 5 : 2

h 1 hour in the ratio 1 : 9 : 10

58 Write down the first five terms of the sequence with nth term $3n - 2$

59 A letter from the word STATISTICS is picked at random. Find the probability that the letter is:

a a vowel

b an S

c a consonant

d an I or a C

e also a letter from the word PROBABILITY.

60 If $8200 \times \frac{3}{10} = 2460$, what is

a 15% of 8200   **b** 60 % of 820?

61 6 pens cost $5.28.

a How much will 11 pens cost?

b How many pens can you buy for $22?

62 Sort these lines into one of the three straight line categories shown below

$y = 2x$ ⠀⠀⠀⠀⠀⠀ $x = {}^-4$
$y = {}^-x$ ⠀⠀⠀⠀⠀⠀ $y = 7$
$y = 3x + 0.5$ ⠀⠀ $y = \frac{1}{2}x$
$y = 6x - 2$

Categories:

Horizontal line ⠀⠀ Vertical line ⠀⠀ Diagonal line

63 In a restaurant, a service charge of 15% is added to the price of the meal. Including the service charge, what will be the bill for a meal costing $140 without the service charge?

64 The table shows the number of customers and skate rental at a rollerskating rink during one week in summer.

Day	Customers	Pairs of skates rented
Monday	192	130
Tuesday	328	212
Wednesday	296	222
Thursday	325	195
Friday	456	292

a What is the probability that a customer will rent skates on Wednesday?
b What is the probability that a customer will rent skates on Thursday?
c What is the probability that a customer has their own skates (and doesn't rent any) on Thursday?
d During all five days, what is the probability that a customer will rent skates?

65 Write the first amount as
i a fraction ⠀⠀ **ii** a percentage of the second.
a 112 cm out of 8 m
b 400 g out of 5 kg
c 350 ml out of 7 ℓ
d 15 minutes out of 2 hours
e an angle of 288° out of a full turn
In part **i**, write the fraction in its simplest form.

66 Copy the diagram below.

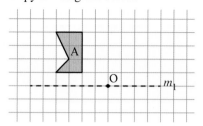

Rotate A 90° clockwise about centre O. Label the image A′. Reflect A′ in the mirror line m_1. Label the image A″.

Finally, translate A″ through $\begin{pmatrix} {}^-6 \\ 2 \end{pmatrix}$. Label the image A‴.

67 What are the missing side lengths in these cuboids?

a

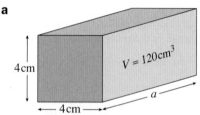

b

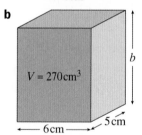

68 a Find the area of a rectangle which is 8 cm wide and 15 cm long.
b A rectangle has an area of 96 cm² and a length of 16 cm. What is the width of the rectangle?

Vectors and matrices

Objectives

- Describe vectors as 2×1 column vectors.
- Add and subtract vectors.
- Associate position vectors with points.
- Use vectors to solve geometry problems.
- Define a matrix as an array of numbers.
- Add, subtract and multiply simple matrices.

The work in this chapter is not in the Cambridge Secondary 1 Mathematics curriculum framework. Vectors and matrices are not in the Checkpoint tests. They are in the Cambridge IGCSE® maths curriculum and are here for you to try if you have completed the work from the other chapters.

What's the point?

We use vectors simply for getting from A to B. They are a great way of describing the space around us and even within us. Vectors are quantities that have both magnitude and direction. They have many applications. A sail-boat captain, for example, has to take account of both the wind and the current in order to steer his course correctly. Both of these are vectors.

Before you start

You should know ...

How to find the midpoint of a line segment AB, given the coordinates of points A and B.

For example:

A (2,3) • ———————— • B (10,7)

The midpoint, M, is at $\left(\dfrac{10 + 2}{2}, \dfrac{7 + 3}{2} \right)$

M (6,5)

Check in

Find the midpoint of the line segment AB if

a A is at $(4, 3)$ and B is at $(10, 5)$

b A is at $(17, 1)$ and B is at $(11, 7)$

19.1 Vectors

You will need squared paper.

A **vector** is a quantity that has both **size** (magnitude) and **direction**.

A vector is distinguished from a **scalar** quantity, which possesses size only.

For example:

Scalar	Vector
Mass	Weight
Speed	Velocity
Distance	Displacement

Note: The **velocity** of a car is its **speed** in a particular direction.

Vectors are represented geometrically by arrows. The length of the arrow represents the size of the vector and the direction of the arrow gives the direction of the vector. Such vectors are usually described in terms of their components as **column vectors**.

EXAMPLE 1

Write the vectors as column vectors.

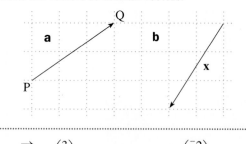

a $\overrightarrow{PQ} = \begin{pmatrix} 3 \\ 2 \end{pmatrix}$

b $x = \begin{pmatrix} ^-2 \\ ^-3 \end{pmatrix}$

Parallel vectors have the same or opposite directions but may differ in magnitude.

EXAMPLE 2

Look at the vectors below.

Explain how **b** and **c** are related to **a**.

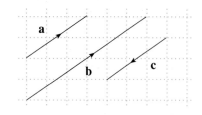

$a = \begin{pmatrix} 3 \\ 2 \end{pmatrix}, b = \begin{pmatrix} 6 \\ 4 \end{pmatrix}, c = \begin{pmatrix} ^-3 \\ ^-2 \end{pmatrix}$

$b = \begin{pmatrix} 6 \\ 4 \end{pmatrix} = 2 \times \begin{pmatrix} 3 \\ 2 \end{pmatrix} = 2a$

b is parallel to **a** and twice its length.

$c = \begin{pmatrix} ^-3 \\ ^-2 \end{pmatrix} = {}^-\begin{pmatrix} 3 \\ 2 \end{pmatrix} = {}^-a$

c is parallel to **a** and has the same length but goes in the opposite direction.

Exercise 19A

1   Write the vectors in the diagram in the form
$\overrightarrow{AB} = \begin{pmatrix} a \\ b \end{pmatrix}$.

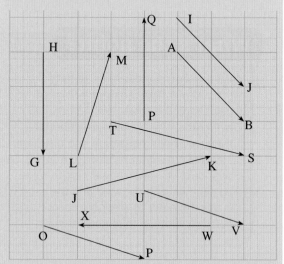

2   Draw these column vectors.

a $a = \begin{pmatrix} 1 \\ 2 \end{pmatrix}$ **b** $b = \begin{pmatrix} 1 \\ ^-2 \end{pmatrix}$

c $c = \begin{pmatrix} ^-1 \\ 2 \end{pmatrix}$ **d** $d = \begin{pmatrix} 2 \\ ^-4 \end{pmatrix}$

e $e = \begin{pmatrix} 2 \\ 4 \end{pmatrix}$ **f** $f = \begin{pmatrix} ^-1 \\ ^-2 \end{pmatrix}$

g $g = \begin{pmatrix} 3 \\ ^-6 \end{pmatrix}$ **h** $h = \begin{pmatrix} 3 \\ 0 \end{pmatrix}$

i $i = \begin{pmatrix} 0 \\ ^-2 \end{pmatrix}$ **j** $j = \begin{pmatrix} 0 \\ 1 \end{pmatrix}$

3 **a** In Question **2**, which pairs of vectors are parallel?

   b Of the parallel pairs, **e** = 2**a**. What other relationships can you find?

4 Using the vectors **a** and **b** given below, show on squared paper the vectors
a + **b**, **a** + 2**b**, **a** − **b**, 2**a** + **b**, 2**a** + 2**b**.

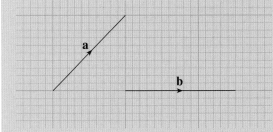

5 Write these vectors in terms of **a**.

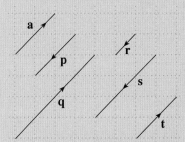

6 If **a** = $\begin{pmatrix} 1 \\ 2 \end{pmatrix}$, draw the vectors

   a 2**a**   **b** 3**a**
   c ⁻**a**   **d** ⁻4**a**

7 Draw $\frac{1}{2}$**a**, $\frac{5}{4}$**a**, 3**a**, $\frac{-2}{3}$**a**, ⁻3**a** for each of these vectors.

   a   **b**

   c   **d**

Vectors as translations

A vector is often used to represent a translation.

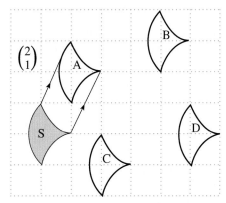

Look at the diagram above. The shaded shape S is mapped on to shape A by a translation described by the column vector $\begin{pmatrix} 1 \\ 2 \end{pmatrix}$.

This vector means move 1 square right and 2 squares up.

Use this diagram for Exercise 19B, which follows.

Exercise 19B

1 Use a column vector to describe the translation that maps shape A in the diagram onto:
   a shape B   **b** shape C   **c** shape D.

2 Which shape does shape C map onto if the translation is described by:

   a $\begin{pmatrix} 3 \\ 1 \end{pmatrix}$ **b** $\begin{pmatrix} 2 \\ 4 \end{pmatrix}$

   c $\begin{pmatrix} ^-2 \\ 1 \end{pmatrix}$ **d** $\begin{pmatrix} ^-1 \\ 3 \end{pmatrix}$?

3 What can you say about the translation that maps shape A onto shape B and the translation that maps shape C onto shape D?

4 Copy and complete this table.

	Translation	Vector	Translation	Vector
a	A → B	$\begin{pmatrix} \circ \\ \circ \end{pmatrix}$	B → A	$\begin{pmatrix} \circ \\ \circ \end{pmatrix}$
b	B → D	$\begin{pmatrix} \circ \\ \circ \end{pmatrix}$	D → B	$\begin{pmatrix} \circ \\ \circ \end{pmatrix}$
c		$\begin{pmatrix} 2 \\ 4 \end{pmatrix}$		$\begin{pmatrix} ^-2 \\ ^-4 \end{pmatrix}$

5 What is the connection between the vectors in each part of Question **4**?

Adding vectors

In geometrical applications of vectors, it is useful to think of vectors as representing translations.

The addition of two vectors can then be thought of as the single translation that is equivalent to the two successive translations.

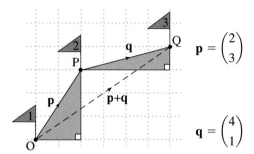

$$\mathbf{p} = \begin{pmatrix} 2 \\ 3 \end{pmatrix}$$

$$\mathbf{q} = \begin{pmatrix} 4 \\ 1 \end{pmatrix}$$

In the diagram, the translation of Flag 1 to Flag 2 is shown by the arrow from O to P.

The translation of Flag 2 to Flag 3 is shown by the arrow from P to Q.

The translation of Flag 1 to Flag 3 could be shown by an arrow from O to Q. It is equivalent to translation **p** followed by translation **q**, that is:

$$\begin{pmatrix} 2 \\ 3 \end{pmatrix} + \begin{pmatrix} 4 \\ 1 \end{pmatrix} = \begin{pmatrix} 6 \\ 4 \end{pmatrix}$$

Vectors are added by placing the arrows which represent them tip to tail:

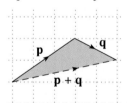

$$\begin{pmatrix} 3 \\ 2 \end{pmatrix} + \begin{pmatrix} 2 \\ -1 \end{pmatrix} = \begin{pmatrix} 5 \\ 1 \end{pmatrix}$$

EXAMPLE 3

If $\mathbf{a} = \begin{pmatrix} -3 \\ -2 \end{pmatrix}$ and $\mathbf{b} = \begin{pmatrix} 2 \\ -1 \end{pmatrix}$, draw a diagram to show $\mathbf{a} + \mathbf{b}$.

....

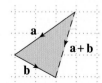

$$\mathbf{a} + \mathbf{b} = \begin{pmatrix} -3 \\ -2 \end{pmatrix} + \begin{pmatrix} 2 \\ -1 \end{pmatrix} = \begin{pmatrix} -1 \\ -3 \end{pmatrix}$$

You should notice that you can add vectors by adding their components:

$$\begin{pmatrix} x_1 \\ y_1 \end{pmatrix} + \begin{pmatrix} x_2 \\ y_2 \end{pmatrix} = \begin{pmatrix} x_1 + x_2 \\ y_1 + y_2 \end{pmatrix}$$

You can subtract vectors in the same way too.

For example,

if $\mathbf{a} = \begin{pmatrix} 2 \\ -1 \end{pmatrix}$ and $\mathbf{c} = \begin{pmatrix} -1 \\ 4 \end{pmatrix}$ then

$$\mathbf{a} - \mathbf{c} = \begin{pmatrix} 2 \\ -1 \end{pmatrix} - \begin{pmatrix} -1 \\ 4 \end{pmatrix}$$

$$= \begin{pmatrix} 2 \\ -1 \end{pmatrix} + \begin{pmatrix} 1 \\ -4 \end{pmatrix} = \begin{pmatrix} 3 \\ -5 \end{pmatrix}$$

Pictorially, this is

$$\mathbf{a} - \mathbf{c} = \mathbf{a} + (^-\mathbf{c})$$

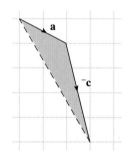

$$\mathbf{c} = \begin{pmatrix} -1 \\ 4 \end{pmatrix}$$

$$^-\mathbf{c} = \begin{pmatrix} 1 \\ -4 \end{pmatrix}$$

Exercise 19C

1 If $\mathbf{a} = \begin{pmatrix} 2 \\ -1 \end{pmatrix}, \mathbf{b} = \begin{pmatrix} -3 \\ 2 \end{pmatrix}, \mathbf{c} = \begin{pmatrix} -1 \\ 4 \end{pmatrix}$

and $\mathbf{d} = \begin{pmatrix} -2 \\ -3 \end{pmatrix}$, draw diagrams to find:

a	$\mathbf{a} + \mathbf{b}$	**b**	$\mathbf{b} + \mathbf{a}$
c	$\mathbf{a} + \mathbf{c}$	**d**	$\mathbf{c} + \mathbf{a}$
e	$\mathbf{b} + \mathbf{d}$	**f**	$\mathbf{c} + \mathbf{d}$
g	$\mathbf{a} + \mathbf{d}$	**h**	$\mathbf{a} - \mathbf{c}$

2 Draw diagrams to show $\mathbf{p} - \mathbf{q}$ when

a $\mathbf{p} = \begin{pmatrix} 4 \\ 3 \end{pmatrix}, \mathbf{q} = \begin{pmatrix} 1 \\ 2 \end{pmatrix}$

b $\mathbf{p} = \begin{pmatrix} 1 \\ 3 \end{pmatrix}, \mathbf{q} = \begin{pmatrix} 2 \\ 1 \end{pmatrix}$

Check your answers by subtracting the column vectors.

3 Draw diagrams to illustrate these vector additions.

a $\begin{pmatrix} 1 \\ 3 \end{pmatrix} + \begin{pmatrix} 2 \\ 4 \end{pmatrix} = \begin{pmatrix} 3 \\ 7 \end{pmatrix}$

b $\begin{pmatrix} 1 \\ 3 \end{pmatrix} + \begin{pmatrix} ^-2 \\ ^-4 \end{pmatrix} = \begin{pmatrix} ^-1 \\ ^-1 \end{pmatrix}$

c $\begin{pmatrix} 2 \\ 0 \end{pmatrix} + \begin{pmatrix} 0 \\ 1 \end{pmatrix} + \begin{pmatrix} ^-2 \\ ^-1 \end{pmatrix} = \begin{pmatrix} 0 \\ 0 \end{pmatrix}$

4 Use the diagram to show that
$\mathbf{p} + \mathbf{q} = \mathbf{q} + \mathbf{p}$

5 Use the diagram to show that
a $\mathbf{p} - \mathbf{q} = \mathbf{p} + (^-\mathbf{q})$
b $\mathbf{q} + \mathbf{p} + (^-\mathbf{q}) = \mathbf{p}$

6 By drawing a sketch find:

a $\begin{pmatrix} 3 \\ 8 \end{pmatrix} + \begin{pmatrix} 1 \\ ^-2 \end{pmatrix}$ **b** $\begin{pmatrix} ^-7 \\ 2 \end{pmatrix} + \begin{pmatrix} 2 \\ ^-5 \end{pmatrix}$

c $\begin{pmatrix} 6 \\ 5 \end{pmatrix} + \begin{pmatrix} 6 \\ 7 \end{pmatrix}$ **d** $\begin{pmatrix} 5 \\ ^-5 \end{pmatrix} + \begin{pmatrix} 6 \\ ^-2 \end{pmatrix}$

e $\begin{pmatrix} 7 \\ 4 \end{pmatrix} + \begin{pmatrix} 2 \\ ^-3 \end{pmatrix}$ **f** $\begin{pmatrix} ^-3 \\ 0 \end{pmatrix} + \begin{pmatrix} 4 \\ 1 \end{pmatrix}$

Check your answers by vector addition.

7 Calculate:

a $\begin{pmatrix} 1 \\ 3 \end{pmatrix} + \begin{pmatrix} ^-2 \\ 1 \end{pmatrix} + \begin{pmatrix} 4 \\ 2 \end{pmatrix}$

b $\begin{pmatrix} 3 \\ 4 \end{pmatrix} + \begin{pmatrix} ^-4 \\ 2 \end{pmatrix} + \begin{pmatrix} ^-6 \\ ^-1 \end{pmatrix}$

c $\begin{pmatrix} 13 \\ 15 \end{pmatrix} - \begin{pmatrix} 20 \\ 25 \end{pmatrix} + \begin{pmatrix} 3 \\ 8 \end{pmatrix}$

d $\begin{pmatrix} 0 \\ ^-1 \end{pmatrix} + \begin{pmatrix} 1 \\ 2 \end{pmatrix} - \begin{pmatrix} ^-1 \\ ^-1 \end{pmatrix}$

Use sketches to check your answers.

Vector algebra

All vectors have the following properties:

1 Two vectors $\mathbf{a}$ and $\mathbf{b}$ can be added.

For example, if $\mathbf{a} = \begin{pmatrix} 2 \\ 3 \end{pmatrix}$ and $\mathbf{b} = \begin{pmatrix} 4 \\ 1 \end{pmatrix}$

then $\mathbf{a} + \mathbf{b} = \begin{pmatrix} 2 + 4 \\ 3 + 1 \end{pmatrix} = \begin{pmatrix} 6 \\ 4 \end{pmatrix}$

2 Vector addition is commutative.

That is, $\mathbf{a} + \mathbf{b} = \mathbf{b} + \mathbf{a}$.

3 Vector $\mathbf{a}$ can be subtracted from vector $\mathbf{b}$ to give vector $\mathbf{b} - \mathbf{a}$.

For example, if $\mathbf{b} = \begin{pmatrix} 4 \\ 1 \end{pmatrix}$ and $\mathbf{a} = \begin{pmatrix} 2 \\ 3 \end{pmatrix}$

then $\mathbf{b} - \mathbf{a} = \begin{pmatrix} 4 - 2 \\ 1 - 3 \end{pmatrix} = \begin{pmatrix} 2 \\ ^-2 \end{pmatrix}$

4 A vector $\mathbf{a}$ can be multiplied by a number k. The result is another vector, $k\mathbf{a}$. When the vector is written as a column vector, each component is multiplied by k.

For example if $\mathbf{a} = \begin{pmatrix} 2 \\ 3 \end{pmatrix}$

then $5\mathbf{a} = \begin{pmatrix} 5 \times 2 \\ 5 \times 3 \end{pmatrix} = \begin{pmatrix} 10 \\ 15 \end{pmatrix}$

5 For each vector $\mathbf{a}$ there is an opposite or **inverse vector** $(^-\mathbf{a})$, and $\mathbf{a} + (^-\mathbf{a}) = \mathbf{0}$.

The result $\mathbf{0}$ is called the **zero vector** or **null vector**.

For example, if $\mathbf{a} = \begin{pmatrix} 3 \\ ^-2 \end{pmatrix}$ then $^-\mathbf{a} = \begin{pmatrix} ^-3 \\ 2 \end{pmatrix}$

and $\mathbf{a} + (^-\mathbf{a}) = \begin{pmatrix} 0 \\ 0 \end{pmatrix}$.

Exercise 19D

1 $\mathbf{r} = \begin{pmatrix} ^-1 \\ 5 \end{pmatrix}, \mathbf{s} = \begin{pmatrix} 4 \\ ^-3 \end{pmatrix}, \mathbf{t} = \begin{pmatrix} 3 \\ ^-15 \end{pmatrix}$

a Write down: **i** $^-\mathbf{r}$ **ii** $^-\mathbf{s}$ **iii** $^-\mathbf{t}$

b Show that:
i $\mathbf{r} + \mathbf{s} = \mathbf{s} + \mathbf{r}$ **ii** $\mathbf{t} + (^-\mathbf{t}) = \mathbf{0}$
iii $\mathbf{t} = ^-3\mathbf{r}$ **iv** $\mathbf{t} + \mathbf{r} = ^-2\mathbf{r}$

c Find:
i $\mathbf{r} + \mathbf{s} + \mathbf{t}$ **ii** $3\mathbf{s} + 2\mathbf{r} + \mathbf{t}$
iii $\frac{1}{3}\mathbf{t}$ **iv** $\frac{2}{3}\mathbf{t}$
v $3\mathbf{t} - 2\mathbf{s}$

2 If $\mathbf{r} = \begin{pmatrix} 5 \\ 6 \end{pmatrix}$ and $\mathbf{s} = \begin{pmatrix} 3 \\ 4 \end{pmatrix}$, find:

a $\mathbf{r} - \mathbf{s}$ **b** $\mathbf{s} - \mathbf{r}$
c $2\mathbf{r} - \mathbf{s}$ **d** $3\mathbf{r} - 2\mathbf{s}$

Show also that $3(\mathbf{r} + \mathbf{s}) = 3\mathbf{r} + 3\mathbf{s}$

3 Find the values of x and y if

a $\begin{pmatrix} x \\ y \end{pmatrix} + \begin{pmatrix} 2 \\ 3 \end{pmatrix} = \begin{pmatrix} 4 \\ 5 \end{pmatrix}$

b $\begin{pmatrix} 3 \\ x \end{pmatrix} + \begin{pmatrix} y \\ 2 \end{pmatrix} = \begin{pmatrix} 6 \\ {}^-3 \end{pmatrix}$

c $3\begin{pmatrix} x \\ 2 \end{pmatrix} - 2\begin{pmatrix} 1 \\ y \end{pmatrix} = \begin{pmatrix} 7 \\ 4 \end{pmatrix}$

4 Given $\mathbf{a} = \begin{pmatrix} 4 \\ {}^-3 \end{pmatrix}$, $\mathbf{b} = \begin{pmatrix} {}^-6 \\ {}^-5 \end{pmatrix}$ and $\mathbf{c} = \begin{pmatrix} {}^-8 \\ 4 \end{pmatrix}$, find:

a $3\mathbf{a} + 2\mathbf{c}$
b $^-3\mathbf{c} + 4\mathbf{b}$
c $2\mathbf{a} - 3\mathbf{b} - \mathbf{c}$

5 If $\mathbf{a} = \begin{pmatrix} 4 \\ 3 \end{pmatrix}$, $\mathbf{b} = \begin{pmatrix} {}^-4 \\ 2 \end{pmatrix}$,

$\mathbf{c} = \begin{pmatrix} {}^-3 \\ {}^-2 \end{pmatrix}$ and $\mathbf{d} = \begin{pmatrix} {}^-6 \\ {}^-5 \end{pmatrix}$,

find:

a $\mathbf{a} - \mathbf{b}$	**b** $2\mathbf{a} + \mathbf{b}$
c $3\mathbf{a} + 2\mathbf{b}$	**d** $\mathbf{a} + 5\mathbf{b}$
e $^-2\mathbf{c} - 3\mathbf{d}$	**f** $2\mathbf{c} - 5\mathbf{d}$
g $\mathbf{a} - 2\mathbf{b} + \mathbf{c}$	**h** $\mathbf{a} - \mathbf{b} - 3\mathbf{c}$
i $^-\mathbf{a} - 4\mathbf{d} - 2\mathbf{c}$	**j** $\mathbf{a} - 2\mathbf{b} - 3\mathbf{c}$
k $^-2\mathbf{c} - \mathbf{d} - \mathbf{a}$	**l** $^-3\mathbf{d} - \mathbf{c} - \mathbf{b}$

19.2 Using vectors in geometry

You can use vectors to prove simple geometrical properties of shapes.

There are two basic ideas:

1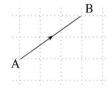

$$\overrightarrow{AB} = {}^-\overrightarrow{BA}$$

2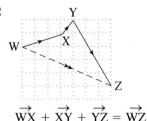

$$\overrightarrow{WX} + \overrightarrow{XY} + \overrightarrow{YZ} = \overrightarrow{WZ}$$

EXAMPLE 4

Two lines, AC and DB, intersect at M. M is the midpoint of both AC and DB.

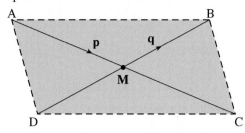

If $\overrightarrow{AM} = \mathbf{p}$ and $\overrightarrow{MB} = \mathbf{q}$, show that AB = DC and AB is parallel to DC.

··

$\overrightarrow{AB} = \overrightarrow{AM} + \overrightarrow{MB} = \mathbf{p} + \mathbf{q}$

As M is the midpoint of AC and DB,
$\overrightarrow{DM} = \overrightarrow{MB} = \mathbf{q}$ and $\overrightarrow{MC} = \overrightarrow{AM} = \mathbf{p}$

So, $\overrightarrow{DC} = \overrightarrow{DM} + \overrightarrow{MC} = \mathbf{q} + \mathbf{p}$

Hence $\overrightarrow{AB} = \overrightarrow{DC}$.

Since the vectors $\overrightarrow{AB}$ and $\overrightarrow{DC}$ are equal, they have the same length (AB = DC) and the same direction, so they are parallel.

Exercise 19E

1

Copy and complete:

a $\overrightarrow{AB} + \overrightarrow{BC} = \square$

b $\overrightarrow{BC} + \overrightarrow{CD} = \square$

c $\overrightarrow{AB} + \overrightarrow{BC} + \overrightarrow{CD} = \square$

2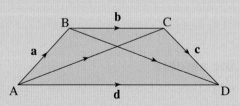

Copy and complete:

a $\overrightarrow{AB} + \overrightarrow{BD} = \square$

b $\overrightarrow{AC} + \overrightarrow{CD} = \square$

c $\overrightarrow{BA} + \overrightarrow{AD} + \overrightarrow{DC} = \square$

3 X and Y are the midpoints of AB and DC, which are sides of the parallelogram ABCD.

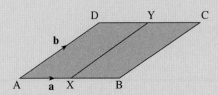

If $\overrightarrow{AX} = \mathbf{a}$ and $\overrightarrow{AD} = \mathbf{b}$, write in terms of $\mathbf{a}$ and $\mathbf{b}$:

a $\overrightarrow{AB}$ **b** $\overrightarrow{DY}$ **c** $\overrightarrow{AY}$

d $\overrightarrow{AC}$ **e** $\overrightarrow{YB}$ **f** $\overrightarrow{CA}$

4 Use the figure to complete:

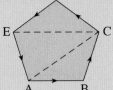

a $\overrightarrow{AB} + \overrightarrow{BC} = \square$

b $\overrightarrow{CD} + \overrightarrow{DE} = \square$

c $\overrightarrow{AC} + \overrightarrow{CD} = \square$

Now explain why $\overrightarrow{AB} + \overrightarrow{BC} + \overrightarrow{CD} + \overrightarrow{DE} + \overrightarrow{EA}$ is the zero vector.

5 Draw any hexagon ABCDEF. Copy and complete:

a $\overrightarrow{AB} + \overrightarrow{BC} = \square$ **b** $\overrightarrow{AC} + \overrightarrow{CD} = \square$

c $\overrightarrow{AD} + \overrightarrow{DE} = \square$ **d** $\overrightarrow{AE} + \overrightarrow{EF} = \square$

Use your results to show that:
$\overrightarrow{AB} + \overrightarrow{BC} + \overrightarrow{CD} + \overrightarrow{DE} + \overrightarrow{EF} + \overrightarrow{FA} = \mathbf{0}$

6 Look at your answers for Question **4** and **5**. Explain why the sum of the vectors representing the sides of any closed polygon, taken in order, is always the zero vector.

7 $\overrightarrow{OA} = \mathbf{a}$ and $\overrightarrow{OB} = \mathbf{b}$

Write in terms of $\mathbf{a}$ and $\mathbf{b}$:

a $\overrightarrow{AO}$

b $\overrightarrow{AB}$

c $\overrightarrow{BA}$

8 In the parallelogram ABCD, $\overrightarrow{AB} = \mathbf{u}$ and $\overrightarrow{BC} = \mathbf{v}$. Write in terms of $\mathbf{u}$ and $\mathbf{v}$:

a $\overrightarrow{AD}$

b $\overrightarrow{CD}$

c $\overrightarrow{AC}$

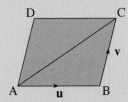

9 In the diagram, M and N are the midpoints of AB and AC respectively. Copy and complete:

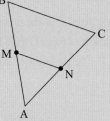

a $\overrightarrow{AM} = \frac{1}{2}(\quad)$

b $\overrightarrow{AN} = \frac{1}{2}(\quad)$

c $\overrightarrow{AB} + \square = \overrightarrow{AC}$

d $\overrightarrow{AM} + \square = \overrightarrow{AN}$

10 In the diagram in Question **9**, if $\overrightarrow{AB} = \mathbf{p}$ and $\overrightarrow{AC} = \mathbf{q}$, write expressions in terms of $\mathbf{p}$ and $\mathbf{q}$ for:

a $\overrightarrow{AM}$ **b** $\overrightarrow{AN}$ **c** $\overrightarrow{BC}$ **d** $\overrightarrow{MN}$

Use your answers to show that $\overrightarrow{MN} = \frac{1}{2}\overrightarrow{BC}$.

11 In Question **9**, the vector representing MN is half the vector representing BC. Explain why this shows that MN is parallel to BC and why the length of MN is half that of BC.

12 OACB is a square with $\overrightarrow{OA} = 2\mathbf{a}$ and $\overrightarrow{OB} = 3\mathbf{b}$. M is the midpoint of BC and N is a point one third of the way along AC.

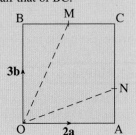

Write down in terms of $\mathbf{a}$ and $\mathbf{b}$ the vectors for:

a $\overrightarrow{BM}$ **b** $\overrightarrow{MC}$ **c** $\overrightarrow{AN}$ **d** $\overrightarrow{CN}$

Use your results to write down the vectors for:

e $\overrightarrow{OM}$ **f** $\overrightarrow{ON}$ **g** $\overrightarrow{BN}$ **h** $\overrightarrow{MN}$

13 In the triangle OAB, C is the midpoint of AB, $\overrightarrow{OA} = \mathbf{a}$ and $\overrightarrow{OB} = \mathbf{b}$.

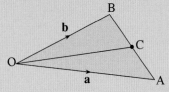

Write in terms of $\mathbf{a}$ and $\mathbf{b}$:

a $\overrightarrow{AO}$ **b** $\overrightarrow{AB}$

c $\overrightarrow{AC}$ **d** $\overrightarrow{OC}$

Position vectors

Instead of using coordinates to describe the position of a point on a graph, you could use a **vector**.

Any point P in the plane defines the vector $\overrightarrow{OP}$ which joins the origin O to P. $\overrightarrow{OP}$ is called the **position vector** of P. The position vector for another point Q is $\overrightarrow{OQ}$.

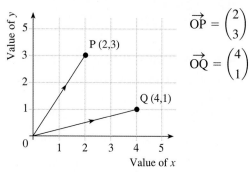

$$\overrightarrow{OP} = \begin{pmatrix} 2 \\ 3 \end{pmatrix}$$

$$\overrightarrow{OQ} = \begin{pmatrix} 4 \\ 1 \end{pmatrix}$$

The coordinates of P are (2,3).

The position vector of P $= \overrightarrow{OP} = \begin{pmatrix} 2 \\ 3 \end{pmatrix}$

A column vector is used for $\overrightarrow{OP}$ to make sure you do not confuse it with the coordinates of P.

EXAMPLE 6

XY is a line with midpoint M, where X is the point (2, 3) and Y is the point $(4, {}^-2)$.

a Write down the position vectors of X, Y and M.

b If N is the point (6, 4), find the components of $\overrightarrow{MN}$.

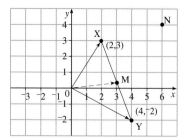

a $\overrightarrow{OX} = \begin{pmatrix} 2 \\ 3 \end{pmatrix}, \overrightarrow{OY} = \begin{pmatrix} 4 \\ -2 \end{pmatrix}$

M has coordinates $\left(\dfrac{2+4}{2}, \dfrac{3+({}^-2)}{2} \right)$

$= (3, 0.5)$

So $\overrightarrow{OM} = \begin{pmatrix} 3 \\ 0.5 \end{pmatrix}$

b $\overrightarrow{ON} = \begin{pmatrix} 6 \\ 4 \end{pmatrix}$

$\overrightarrow{MN} = \overrightarrow{MO} + \overrightarrow{ON} = {}^-\overrightarrow{OM} + \overrightarrow{ON}$

$= {}^-\begin{pmatrix} 3 \\ 0.5 \end{pmatrix} + \begin{pmatrix} 6 \\ 4 \end{pmatrix} = \begin{pmatrix} 3 \\ 3.5 \end{pmatrix}$

The next example shows how position vectors can assist you when a line is divided in a ratio.

EXAMPLE 7

Find the coordinates of the point K which divides the line PQ in the ratio $2:3$.

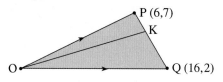

In the diagram $PK:KQ = 2:3$. So if PQ is divided into five parts, PK is two parts and KQ is three parts. In other words $PK = \frac{2}{5}PQ$.

$\overrightarrow{OP} = \begin{pmatrix} 6 \\ 7 \end{pmatrix}, \overrightarrow{OQ} = \begin{pmatrix} 16 \\ 2 \end{pmatrix}$

$\overrightarrow{PQ} = \overrightarrow{PO} + \overrightarrow{OQ} = {}^-\overrightarrow{OP} + \overrightarrow{OQ}$

$= {}^-\begin{pmatrix} 6 \\ 7 \end{pmatrix} + \begin{pmatrix} 16 \\ 2 \end{pmatrix} = \begin{pmatrix} 10 \\ -5 \end{pmatrix}$

$\overrightarrow{OK} = \overrightarrow{OP} + \overrightarrow{PK} = \overrightarrow{OP} + \frac{2}{5}\overrightarrow{PQ}$

So $\overrightarrow{OK} = \begin{pmatrix} 6 \\ 7 \end{pmatrix} + \frac{2}{5}\begin{pmatrix} 10 \\ -5 \end{pmatrix}$

$= \begin{pmatrix} 6 \\ 7 \end{pmatrix} + \begin{pmatrix} 4 \\ -2 \end{pmatrix}$

$= \begin{pmatrix} 10 \\ 5 \end{pmatrix}$

The coordinates of K are therefore (10, 5).

Exercise 19F

1 A has coordinates $(1,3)$ and B has coordinates $(^-1,4)$. What is:
  a the position vector of A
  b the position vector of B
  c the vector $\overrightarrow{AB}$?

2 Given $A(3,^-2)$ and $\overrightarrow{AB} = \begin{pmatrix} 2 \\ 4 \end{pmatrix}$, find:
  a the position vector of A
  b the position vector of B
  c the coordinates of B.

3 Given $B(^-3,5)$ and $\overrightarrow{AB} = \begin{pmatrix} 7 \\ 3 \end{pmatrix}$, find:
  a the position vector of A
  b the position vector of B
  c the coordinates of A.

4 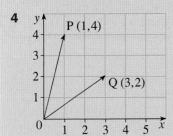 $\overrightarrow{OP} = \begin{pmatrix} 1 \\ 4 \end{pmatrix}$

$\overrightarrow{OQ} = \begin{pmatrix} 3 \\ 2 \end{pmatrix}$

  a For the diagram above, complete the statement: $\overrightarrow{PQ} = \overrightarrow{PO} + \square = {}^-\overrightarrow{OP} + \square$
  b Find the column vector for PQ.
  c If M is the midpoint of PQ, write down the column vector for $\overrightarrow{PM}$.
  d Complete the statement $\overrightarrow{OM} = \overrightarrow{OP} + \square$.
  e Write down the position vector of M.

5 Compare the column vectors for $\overrightarrow{OP}$, $\overrightarrow{OQ}$ and $\overrightarrow{OM}$ in Question **4**.
  Can you find a quick way of obtaining $\overrightarrow{OM}$ from $\overrightarrow{OP}$ and $\overrightarrow{OQ}$?

6 In the diagram, O is the origin. R and S are points with position vectors **r** and **s**. M is the midpoint of RS.

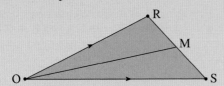

  a Write $\overrightarrow{RS}$ in terms of **r** and **s**.
  b By first finding $\overrightarrow{RM}$ write the position vector of M in terms of **r** and **s**.
  c Use this result to check your answer for Question **4**.

19.3 Matrices

Number tables are matrices

The number of runs made by the five best batsmen in High Town School in the first innings of four cricket matches are given in the table. The opposing schools are called P, Q, R and S.

	P	Q	R	S
Benhoe	8	10	21	9
Lee	10	3	30	15
Singh	7	15	2	23
Marshall	18	6	14	12
Williams	3	12	0	22

The runs made by these same batsmen in their second innings were:

	P	Q	R	S
Benhoe	17	3	1	7
Lee	3	8	5	15
Singh	7	16	21	4
Marshall	0	27	3	6
Williams	18	4	1	16

A **matrix** is an array of numbers arranged in rows and columns. Most numerical tables, like the ones above, can be written as matrices.

A matrix is written in brackets. It can also be named using one of the letters of the alphabet.
For example, the two tables could be written as:

$$\mathbf{A} = \begin{pmatrix} 8 & 10 & 21 & 9 \\ 10 & 3 & 30 & 15 \\ 7 & 15 & 2 & 23 \\ 18 & 6 & 14 & 12 \\ 3 & 12 & 0 & 22 \end{pmatrix} \quad \text{and}$$

$$\mathbf{B} = \begin{pmatrix} 17 & 3 & 1 & 7 \\ 3 & 8 & 5 & 15 \\ 7 & 16 & 21 & 4 \\ 0 & 27 & 3 & 6 \\ 18 & 4 & 1 & 16 \end{pmatrix}$$

The first **row** of **A** is (8 10 21 9)

The first **column** of **B** is $\begin{pmatrix} 17 \\ 3 \\ 7 \\ 0 \\ 18 \end{pmatrix}$

A has 5 rows and 4 columns. It is called a 5×4 matrix.

(This is read as a 'five by four matrix'.)

B is also a 5×4 matrix.

Exercise 19G

1   Use **A** to find the total number of runs made by Benhoe in the first innings of the four matches.

2   Use **B** to find the total number of runs scored by the five batsmen in the second innings against school Q.

3   Find:
   a   the total number of runs scored by Benhoe against school P
   b   the total number of runs scored by Singh against school R.

4   The total scores by the five batsmen for both innings will be given by the matrix **A** + **B**. Copy and complete the matrix **A** + **B** by adding the components of **A** and **B**.

$$\mathbf{A} + \mathbf{B} = \begin{pmatrix} 25 & 13 & 22 & 16 \\ & & & \\ & & & \\ & & & \\ & & & \end{pmatrix}$$

5   What type of matrix (how many rows and columns) is each of the following?

   a $(1\ 2\ 3)$
   b $\begin{pmatrix} 2 & ^-1 \\ 1 & 6 \end{pmatrix}$
   c $\begin{pmatrix} 2 \\ 3 \end{pmatrix}$
   d $\begin{pmatrix} 1 & ^-2 & 4 \\ 2 & 3 & ^-1 \end{pmatrix}$
   e $\begin{pmatrix} 3 & ^-2 \\ 4 & 1 \\ 2 & 6 \end{pmatrix}$
   f $\begin{pmatrix} 1 & 2 & 4 \\ 6 & ^-2 & 3 \\ 1 & 7 & 8 \end{pmatrix}$

Adding and subtracting matrices

Only matrices of the same type can be added.

EXAMPLE 8

If possible, work out:

a $\begin{pmatrix} 2 & 3 \\ 4 & 1 \end{pmatrix} + \begin{pmatrix} 1 & ^-2 \\ 3 & ^-3 \end{pmatrix}$
b $\begin{pmatrix} 2 & 3 \\ 4 & 1 \end{pmatrix} + \begin{pmatrix} 1 \\ 3 \end{pmatrix}$

. . . .

a $\begin{pmatrix} 2 & 3 \\ 4 & 1 \end{pmatrix} + \begin{pmatrix} 1 & ^-2 \\ 3 & ^-3 \end{pmatrix} = \begin{pmatrix} 2 + 1 & 3 + ^-2 \\ 4 + 3 & 1 + ^-3 \end{pmatrix}$

$= \begin{pmatrix} 3 & 1 \\ 7 & ^-2 \end{pmatrix}$

b $\begin{pmatrix} 2 & 3 \\ 4 & 1 \end{pmatrix} + \begin{pmatrix} 1 \\ 3 \end{pmatrix}$

This cannot be done as the first matrix is 2×2 and the second is a 2×1 matrix.

Exercise 19H

1   Here are some matrices:

$$\mathbf{A} = \begin{pmatrix} 2 & 4 \\ 3 & 9 \end{pmatrix} \qquad \mathbf{B} = \begin{pmatrix} 3 & ^-4 \\ 11 & 6 \end{pmatrix}$$

$$\mathbf{C} = \begin{pmatrix} ^-3 & 4 \\ ^-11 & ^-6 \end{pmatrix} \qquad \mathbf{D} = \begin{pmatrix} 0 & 0 \\ 0 & 0 \end{pmatrix}$$

$$\mathbf{E} = \begin{pmatrix} ^-3 \\ 1 \end{pmatrix} \qquad \mathbf{F} = (2\ 6)$$

Where possible, work out:
   a   **A** + **B**   b   **A** + **D**
   c   **B** + **A**   d   **E** + **C**
   e   **F** + **E**   f   **B** + **C**

2   a   Which of the matrices in Question **1** are 2×2?
   b   What is the effect of adding matrix **D** to another matrix?
   c   What can you say about **B** and **C**?
   d   Can you add **A** to **E**? Why not?

3   For the matrices in Question **1**, work out:
   a $(\mathbf{A} + \mathbf{B}) + \mathbf{C}$ b $\mathbf{A} + (\mathbf{B} + \mathbf{C})$

   Are the results the same?

4 **X** and **Y** are matrices:

$$\mathbf{X} = \begin{pmatrix} 2 & ^-3 \\ ^-1 & 0 \end{pmatrix}, \quad \mathbf{Y} = \begin{pmatrix} 1 & ^-2 \\ 2 & 3 \end{pmatrix}$$

The matrix $\mathbf{X} - \mathbf{Y} = \begin{pmatrix} 1 & ^-1 \\ ^-3 & ^-3 \end{pmatrix}$

a Explain the rule for subtracting a matrix.
b For the matrices in Question **1**, where possible find:

    i $\mathbf{A} - \mathbf{B}$ **ii** $\mathbf{D} - \mathbf{A}$
    iii $\mathbf{E} - \mathbf{F}$ **iv** $\mathbf{F} - \mathbf{E}$
    v $\mathbf{A} - \mathbf{D}$ **vi** $\mathbf{C} - \mathbf{B}$

5 If $\mathbf{A} = \begin{pmatrix} 3 & 1 \\ 5 & 4 \end{pmatrix}$, $\mathbf{B} = \begin{pmatrix} 5 & 0 \\ ^-2 & 7 \end{pmatrix}$ and

$$\mathbf{C} = \begin{pmatrix} ^-1 & 3 \\ ^-4 & 0 \end{pmatrix}$$

find:
a $\mathbf{A} - \mathbf{B}$ **b** $\mathbf{B} - \mathbf{A}$
c $\mathbf{A} + \mathbf{C} - \mathbf{B}$ **d** $\mathbf{C} - \mathbf{A} + \mathbf{B}$

Multiplying by a scalar

You can multiply a matrix by a number (or scalar) by multiplying every component (or element) in the matrix by that number.
For example:

$$5 \times \begin{pmatrix} 4 & 3 \\ 1 & 6 \end{pmatrix} = \begin{pmatrix} 5 \times 4 & 5 \times 3 \\ 5 \times 1 & 5 \times 6 \end{pmatrix}$$

$$= \begin{pmatrix} 20 & 15 \\ 5 & 30 \end{pmatrix}$$

Exercise 19I

1 A manufacturer of candies produces two flavours (chocolate and strawberry), and supplies them in two sizes (large and small). The manager of Green Valley Stores finds he has the following stock:

	Size	
	Large	**Small**
Chocolate	10	22
Strawberry	15	36

a The manager wants twice as many of each in stock.
Write down the table for this.
b In matrix form you can show the multiplication by 2 as: $2\begin{pmatrix} 10 & 22 \\ 15 & 36 \end{pmatrix}$

Write down the answer as a matrix.
c In the multiplication, 2 is a **scalar**.
Explain how you would multiply a 2 × 2 matrix by the scalar 3.

2 Let **a** be the vector $\begin{pmatrix} 5 \\ 2 \end{pmatrix}$. Write down the column vector 4**a**. Is the rule for multiplying by a scalar the same for both vectors and matrices?

3 Write as a single matrix:

a $2\begin{pmatrix} 3 & 0 \\ 0 & 1 \end{pmatrix}$ **b** $5\begin{pmatrix} 3 \\ 4 \end{pmatrix}$

c $3\begin{pmatrix} 2 & 1 \\ ^-1 & 0 \end{pmatrix} + \frac{1}{2}\begin{pmatrix} 2 & 4 \\ 1 & 6 \end{pmatrix}$

d $3\begin{pmatrix} 1 \\ 2 \end{pmatrix} - 2\begin{pmatrix} 3 \\ ^-1 \end{pmatrix}$

4 Find the value of each letter in these matrices:

a $2\begin{pmatrix} 1 & 2 \\ 3 & 4 \end{pmatrix} = \begin{pmatrix} a & 4 \\ 6 & b \end{pmatrix}$

b $3\begin{pmatrix} 1 & ^-2 \\ c & d \end{pmatrix} = \begin{pmatrix} 3 & ^-6 \\ d & 12 \end{pmatrix}$

c $\begin{pmatrix} e & 2 \\ f & ^-3 \end{pmatrix} + 2\begin{pmatrix} 3 & ^-2 \\ f & g \end{pmatrix} = \begin{pmatrix} 10 & ^-2 \\ ^-12 & 6 \end{pmatrix}$

d $h\begin{pmatrix} 0 & h \\ i & ^-3 \end{pmatrix} - \begin{pmatrix} 1 & 2 \\ 3 & j \end{pmatrix} = \begin{pmatrix} k & 14 \\ 5 & ^-12 \end{pmatrix}$

5 $\mathbf{A} = \begin{pmatrix} 3 & ^-5 \\ 8 & 4 \end{pmatrix}$ $\mathbf{B} = \begin{pmatrix} ^-2 & 3 \\ 1 & 5 \end{pmatrix}$ $\mathbf{C} = \begin{pmatrix} ^-4 & 0 \\ 5 & ^-1 \end{pmatrix}$

Evaluate **a** $2\mathbf{B}$ **b** $2\mathbf{A} + 2\mathbf{B}$ **c** $3\mathbf{C} - 2\mathbf{A}$
       d $\frac{1}{2}\mathbf{A} - \frac{1}{2}\mathbf{B}$ **e** $\frac{1}{3}\mathbf{B} + \frac{3}{4}\mathbf{C}$

19.4 Multiplying matrices

To multiply matrices, you have to multiply the **rows of the first matrix** by the **columns of the second matrix**.

EXAMPLE 9

Work out $\begin{pmatrix} 1 & 2 \\ 3 & 4 \end{pmatrix}\begin{pmatrix} 5 \\ 6 \end{pmatrix}$

....

Divide the first matrix into rows, R1 and R2 and the second matrix into columns, C1:

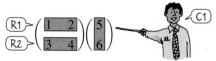

Now multiply each element in R1 by the corresponding element in C1 and add:

$1 \times 5 = 5$
$2 \times 6 = \underline{12 +}$
17

Similarly, multiply the elements of R2 and C1:

$3 \times 5 = 15$
$4 \times 6 = \underline{24 +}$
39

This gives the elements of the product of the two matrices:

$\begin{pmatrix} 1 & 2 \\ 3 & 4 \end{pmatrix}\begin{pmatrix} 5 \\ 6 \end{pmatrix} = \begin{pmatrix} 17 \\ 39 \end{pmatrix}$

Notice that the size of the resulting 2×1 matrix is determined by the number of *rows* in the first matrix (2) and the number of *columns* in the second matrix (1).

The number of boys and girls in the first and second years of Green Trees High School is as follows:

	Boys	**Girls**
Year 1	10	17
Year 2	14	9

This can be written as the matrix $\mathbf{X} = \begin{pmatrix} 10 & 17 \\ 14 & 9 \end{pmatrix}$

In the first and second years, the boys and girls all do pottery and art.
During each year, they each have to complete pieces of pottery as shown on the right.

	Pottery
Boys	3
Girls	1

This can be written as the matrix $\mathbf{Y} = \begin{pmatrix} 3 \\ 1 \end{pmatrix}$

Exercise 19J

1 What is the total number of pieces of pottery to be completed by:
   a all the boys in Year 1
   b all the boys in Year 2
   c all the girls in Year 1?

2 What is the total number of pieces of pottery to be completed:
   a in Year 1   **b** in Year 2?

3 Write the answer to Question **2** in a table:

Pottery

Year 1	
Year 2	

4 Write the answer to Question **3** as a matrix **Z**.

5 The matrix **Z** could be obtained from the matrix multiplication $\mathbf{X} \times \mathbf{Y}$, as shown below:

$$\begin{array}{c} \; B \quad\; G \\ \begin{array}{c} Y_1 \\ Y_2 \end{array}\!\begin{pmatrix} 10 & 17 \\ 14 & 9 \end{pmatrix} \end{array} \times \begin{array}{c} P \\ \begin{array}{c} B \\ G \end{array}\!\begin{pmatrix} 3 \\ 1 \end{pmatrix} \end{array} = \begin{array}{c} P \\ \begin{array}{c} Y_1 \\ Y_2 \end{array}\!\begin{pmatrix} 47 \\ 51 \end{pmatrix} \end{array}$$

For the matrix **Z**:
   a what does the 47 tell you
   b what does the 51 tell you?

6 The number of art pieces that have to be completed is given by the matrix **D** where:

$$\mathbf{D} = \begin{array}{c} A \\ \begin{array}{c} B \\ G \end{array}\!\begin{pmatrix} 2 \\ 5 \end{pmatrix} \end{array}$$

What is the total number of art pieces that must be completed by:
   a all the students in Year 1
   b all the students in Year 2?

7 Set out the multiplication $\mathbf{X} \times \mathbf{D}$ in the same way as $\mathbf{X} \times \mathbf{Y}$ in Question **5**.

8 Work out:

   a $\begin{pmatrix} 1 & 2 \\ 3 & 4 \end{pmatrix}\begin{pmatrix} 7 \\ 8 \end{pmatrix}$ **b** $\begin{pmatrix} 1 & 3 \\ 4 & 2 \end{pmatrix}\begin{pmatrix} 2 \\ 1 \end{pmatrix}$

   c $\begin{pmatrix} 3 & 0 \\ 1 & ^-4 \end{pmatrix}\begin{pmatrix} 1 \\ ^-2 \end{pmatrix}$ **d** $\begin{pmatrix} 4 & ^-1 \\ ^-3 & 0 \end{pmatrix}\begin{pmatrix} 2 \\ ^-1 \end{pmatrix}$

   e $\begin{pmatrix} 1 & 0 \\ 0 & 1 \end{pmatrix}\begin{pmatrix} 3 \\ 4 \end{pmatrix}$ **f** $\begin{pmatrix} 0 & 1 \\ 1 & 0 \end{pmatrix}\begin{pmatrix} 3 \\ 4 \end{pmatrix}$

   g $\begin{pmatrix} ^-2 & ^-1 \\ ^-3 & 0 \end{pmatrix}\begin{pmatrix} 1 \\ 4 \end{pmatrix}$ **h** $\begin{pmatrix} a & b \\ c & d \end{pmatrix}\begin{pmatrix} 1 \\ ^-3 \end{pmatrix}$

The method can be extended to multiply a 2×2 matrix by a 2×2 matrix.

At Green Trees High School, the number of pieces of pottery and art to be done by each boy and girl can be written as a single matrix T:

$$\mathbf{T} = \begin{array}{cc} & \begin{array}{cc} P & A \end{array} \\ \begin{array}{c} B \\ G \end{array} & \begin{pmatrix} 3 & 2 \\ 1 & 5 \end{pmatrix} \end{array}$$

The product of **T** with the matrix **X** can be worked out as follows:

Step 1

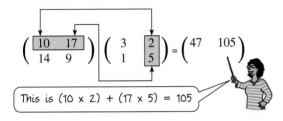

$$\begin{pmatrix} 10 & 17 \\ 14 & 9 \end{pmatrix} \begin{pmatrix} 3 & 5 \\ 1 & 2 \end{pmatrix} = \begin{pmatrix} 30 + 17 \\ \end{pmatrix}$$

10 x 3 = 30
17 x 1 = 17
Now add them.

Step 2

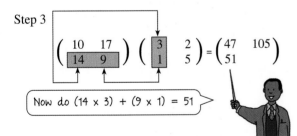

$$\begin{pmatrix} 10 & 17 \\ 14 & 9 \end{pmatrix} \begin{pmatrix} 3 & 2 \\ 1 & 5 \end{pmatrix} = \begin{pmatrix} 47 & 105 \end{pmatrix}$$

This is (10 x 2) + (17 x 5) = 105

Step 3

$$\begin{pmatrix} 10 & 17 \\ 14 & 9 \end{pmatrix} \begin{pmatrix} 3 & 2 \\ 1 & 5 \end{pmatrix} = \begin{pmatrix} 47 & 105 \\ 51 & \end{pmatrix}$$

Now do (14 x 3) + (9 x 1) = 51

Step 4

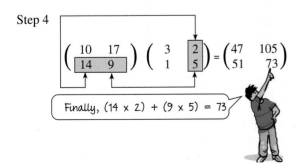

$$\begin{pmatrix} 10 & 17 \\ 14 & 9 \end{pmatrix} \begin{pmatrix} 3 & 2 \\ 1 & 5 \end{pmatrix} = \begin{pmatrix} 47 & 105 \\ 51 & 73 \end{pmatrix}$$

Finally, (14 x 2) + (9 x 5) = 73

You can multiply matrices so long as the number of columns in the first matrix is the same as the number of rows in the second.

EXAMPLE 10

Work out:

$$\begin{pmatrix} 1 & 2 \\ 3 & 4 \end{pmatrix} \begin{pmatrix} 5 & 7 \\ 6 & 8 \end{pmatrix}$$

....

$$\begin{pmatrix} 1 & 2 \\ 3 & 4 \end{pmatrix} \begin{pmatrix} 5 & 7 \\ 6 & 8 \end{pmatrix}$$

$$= \begin{pmatrix} (1\times5) + (2\times6) & (1\times7) + (2\times8) \\ (3\times5) + (4\times6) & (3\times7) + (4\times8) \end{pmatrix}$$

$$= \begin{pmatrix} 17 & 23 \\ 39 & 53 \end{pmatrix}$$

Exercise 19K

1 Copy and complete:

a $\begin{pmatrix} 8 & 1 \\ 2 & 3 \end{pmatrix} \begin{pmatrix} 1 & 4 \\ 2 & 3 \end{pmatrix} = \begin{pmatrix} 10 & \\ & \end{pmatrix}$

b $\begin{pmatrix} 3 & 5 \\ {}^{-}2 & 4 \end{pmatrix} \begin{pmatrix} 2 & {}^{-}1 \\ 3 & 0 \end{pmatrix} = \begin{pmatrix} & \\ & 2 \end{pmatrix}$

c $\begin{pmatrix} {}^{-}2 & 3 \\ 6 & 7 \end{pmatrix} \begin{pmatrix} 2 & 1 \\ {}^{-}1 & 3 \end{pmatrix} = \begin{pmatrix} & 7 \\ & \end{pmatrix}$

d $\begin{pmatrix} 3 & 4 \\ {}^{-}2 & 1 \end{pmatrix} \begin{pmatrix} 5 & {}^{-}1 \\ 2 & 3 \end{pmatrix} = \begin{pmatrix} & \\ {}^{-}8 & \end{pmatrix}$

2 Find the products of these 2 × 2 matrices:

a $\begin{pmatrix} 5 & 2 \\ 4 & 3 \end{pmatrix} \begin{pmatrix} 2 & 8 \\ 7 & 6 \end{pmatrix}$ **b** $\begin{pmatrix} 3 & 8 \\ 2 & {}^{-}1 \end{pmatrix} \begin{pmatrix} 2 & 6 \\ 5 & {}^{-}3 \end{pmatrix}$

c $\begin{pmatrix} 1 & 0 \\ 0 & 1 \end{pmatrix} \begin{pmatrix} 2 & 3 \\ 4 & 2 \end{pmatrix}$ **d** $\begin{pmatrix} 2 & 3 \\ 4 & 2 \end{pmatrix} \begin{pmatrix} 3 & 0 \\ 0 & 6 \end{pmatrix}$

e $\begin{pmatrix} 0 & 1 \\ 1 & 0 \end{pmatrix} \begin{pmatrix} {}^{-}2 & 1 \\ 3 & 1 \end{pmatrix}$ **f** $\begin{pmatrix} {}^{-}2 & 0 \\ 1 & 2 \end{pmatrix} \begin{pmatrix} 2 & 3 \\ 4 & 5 \end{pmatrix}$

g $\begin{pmatrix} 1 & 1 \\ 2 & {}^{-}1 \end{pmatrix} \begin{pmatrix} 2 & 4 \\ 0 & 5 \end{pmatrix}$ **h** $\begin{pmatrix} {}^{-}1 & {}^{-}1 \\ 0 & 2 \end{pmatrix} \begin{pmatrix} {}^{-}3 & {}^{-}1 \\ {}^{-}2 & {}^{-}4 \end{pmatrix}$

Consolidation

Example 1

Draw the column vectors:

a $\begin{pmatrix} 3 \\ -2 \end{pmatrix}$

b $\begin{pmatrix} -2 \\ -5 \end{pmatrix}$

....

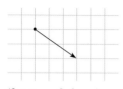 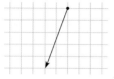

(3 across, 2 down)   (2 left across, 5 down)

Example 2

If $\mathbf{A} = \begin{pmatrix} 3 \\ -2 \end{pmatrix}$, $\mathbf{B} = \begin{pmatrix} 2 \\ 1 \end{pmatrix}$ and $\mathbf{C} = \begin{pmatrix} 0 \\ 2 \end{pmatrix}$, what is:

a $\mathbf{A} + \mathbf{B} + \mathbf{C}$

$$\mathbf{A} + \mathbf{B} + \mathbf{C} = \begin{pmatrix} 3 \\ -2 \end{pmatrix} + \begin{pmatrix} 2 \\ 1 \end{pmatrix} + \begin{pmatrix} 0 \\ 2 \end{pmatrix}$$

$$= \begin{pmatrix} 5 \\ 1 \end{pmatrix}$$

b $3\mathbf{A} - 2\mathbf{B}$?

$$3\mathbf{A} - 2\mathbf{B} = 3\begin{pmatrix} 3 \\ -2 \end{pmatrix} - 2\begin{pmatrix} 2 \\ 1 \end{pmatrix}$$

$$= \begin{pmatrix} 9 \\ -6 \end{pmatrix} - \begin{pmatrix} 4 \\ 2 \end{pmatrix} = \begin{pmatrix} 5 \\ -8 \end{pmatrix}$$

Example 3

In the parallelogram PQRS, $\overrightarrow{PQ} = A$ and $\overrightarrow{PS} = B$.

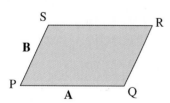

Write **a** $\overrightarrow{PR}$ and **b** $\overrightarrow{QS}$ in terms of **A** and **B**.

....

a $\overrightarrow{PR}$
$\overrightarrow{PR} = \overrightarrow{PQ} + \overrightarrow{QR}$
$= \mathbf{A} + \mathbf{B}$

b $\overrightarrow{QS}$
$\overrightarrow{QS} = \overrightarrow{QP} + \overrightarrow{PS}$
$= ^-\mathbf{A} + \mathbf{B}$
$= \mathbf{B} - \mathbf{A}$

Example 4

Given the matrices

$$\mathbf{A} = \begin{pmatrix} 3 & ^-2 \\ 2 & 4 \end{pmatrix}, \mathbf{B} = \begin{pmatrix} 1 \\ ^-3 \end{pmatrix}, \mathbf{C} = \begin{pmatrix} 4 & 3 \\ ^-1 & 5 \end{pmatrix}, \text{ find}$$

a AB   **b** AC

....

a
$$\mathbf{AB} = \begin{pmatrix} 3 & ^-2 \\ 2 & 4 \end{pmatrix}\begin{pmatrix} 1 \\ ^-3 \end{pmatrix} = \begin{pmatrix} 3 \times 1 + (^-2) \times (^-3) \\ 2 \times 1 + 4 \times (^-3) \end{pmatrix}$$

$$= \begin{pmatrix} 3 + 6 \\ 2 - 12 \end{pmatrix} = \begin{pmatrix} 9 \\ ^-10 \end{pmatrix}$$

b $\mathbf{AC} = \begin{pmatrix} 3 & ^-2 \\ 2 & 4 \end{pmatrix}\begin{pmatrix} 4 & 3 \\ ^-1 & 5 \end{pmatrix}$

$$= \begin{pmatrix} 3 \times 4 + (^-2) \times (^-1) & 3 \times 3 + (^-2) \times 5 \\ 2 \times 4 + 4 \times (^-1) & 2 \times 3 + 4 \times 5 \end{pmatrix}$$

$$= \begin{pmatrix} 12 + 2 & 9 - 10 \\ 8 - 4 & 6 + 20 \end{pmatrix} = \begin{pmatrix} 14 & ^-1 \\ 4 & 26 \end{pmatrix}$$

Exercise 19

1   Draw these column vectors.

a $\begin{pmatrix} 2 \\ 3 \end{pmatrix}$ **b** $\begin{pmatrix} 3 \\ 5 \end{pmatrix}$ **c** $\begin{pmatrix} 2 \\ 0 \end{pmatrix}$

d $\begin{pmatrix} 0 \\ -2 \end{pmatrix}$ **e** $\begin{pmatrix} -2 \\ 1 \end{pmatrix}$ **f** $\begin{pmatrix} -3 \\ -1 \end{pmatrix}$

2   Given $\mathbf{A} = \begin{pmatrix} 1 \\ 3 \end{pmatrix}$, $\mathbf{B} = \begin{pmatrix} 2 \\ -1 \end{pmatrix}$ and $\mathbf{C} = \begin{pmatrix} -3 \\ -5 \end{pmatrix}$, find:

a $\mathbf{A} + \mathbf{B}$ **b** $3\mathbf{A}$
c $2\mathbf{B}$ **d** $\mathbf{A} - \mathbf{C}$
e $3\mathbf{A} + \mathbf{C}$ **f** $3\mathbf{A} + 2\mathbf{B}$
g $\mathbf{A} + \mathbf{B} - \mathbf{C}$ **h** $4\mathbf{C} - \mathbf{A}$
i $5\mathbf{C} + 2\mathbf{B}$

3   Given $\mathbf{A} = \begin{pmatrix} 2 \\ 1 \end{pmatrix}$, $\mathbf{B} = (4 \ 3)$, $\mathbf{C} = \begin{pmatrix} 2 & -1 \\ 1 & 4 \end{pmatrix}$

and $\mathbf{D} = \begin{pmatrix} 3 & 0 \\ -2 & 4 \end{pmatrix}$, find:

a BA   **b** AB   **c** CA
d DA   **e** BC   **f** BD
g CD   **h** DC   **i** CDA

4 In triangle OAB, the midpoints of OA, OB and AB are W, X and Y respectively.

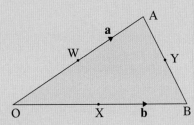

If $\overrightarrow{OA}$ = **a** and $\overrightarrow{OB}$ = **b**

a Write down expressions for:

i $\overrightarrow{OY}$ **ii** $\overrightarrow{AX}$ **iii** $\overrightarrow{BW}$ in terms of **a** and **b**.

b Show that $\overrightarrow{OY}$ + $\overrightarrow{AX}$ + $\overrightarrow{BW}$ = 0.

5 A ship leaves port and sails 5 km south and 4 km west. It then changes course and sails 2 km north and 3 km west.

a Using a scale of 1 cm = 1 km, draw the ship's course on a grid.

b Write down the ship's journey as the sum of two vectors.

c How far is the ship from port?

d The ship is now returning directly to port. Write down a vector that represents its track.

Summary

You should know ...

1 How to use a column vector to describe a translation.

2 How to add or subtract vectors.

For example:

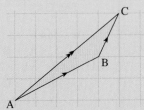

$$\overrightarrow{AB} + \overrightarrow{BC} = \begin{pmatrix} 4 \\ 2 \end{pmatrix} + \begin{pmatrix} 1 \\ 2 \end{pmatrix} = \begin{pmatrix} 5 \\ 4 \end{pmatrix} = \overrightarrow{AC}$$

Check out

1 Describe these translations as column vectors.

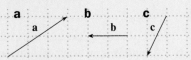

2 If $\mathbf{a} = \begin{pmatrix} 4 \\ 3 \end{pmatrix}$, $\mathbf{b} = \begin{pmatrix} ^-2 \\ 4 \end{pmatrix}$, $\mathbf{c} = \begin{pmatrix} ^-3 \\ ^-2 \end{pmatrix}$

find the value of:

a $\mathbf{a} + \mathbf{b}$

b $\mathbf{a} - \mathbf{b}$

c $2\mathbf{a} - 3\mathbf{c}$

d $3\mathbf{a} + \mathbf{b} - \mathbf{c}$

3 How vectors can be used to solve geometric problems.
For example:

In the diagram $\overrightarrow{AD}$ can be written in terms of **a** and **b**.

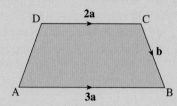

$$\overrightarrow{AD} = \overrightarrow{AB} + \overrightarrow{BC} + \overrightarrow{CD}$$
$$= 3\mathbf{a} - \mathbf{b} - 2\mathbf{a}$$
$$= \mathbf{a} - \mathbf{b}$$

3

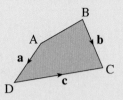

ABCD is a quadrilateral with
$\overrightarrow{AD} = \mathbf{a}, \overrightarrow{BC} = \mathbf{b}$ and $\overrightarrow{DC} = \mathbf{c}$

a Find in terms of **a**, **b** and **c**:
   i $\overrightarrow{AC}$
   ii $\overrightarrow{AB}$

b If M is the midpoint of DC,
find $\overrightarrow{AM}$.

---

4 Any point can be represented by a position vector.
For example:

the point, P (a, b) has position vector $\overrightarrow{OP} = \begin{pmatrix} a \\ b \end{pmatrix}$

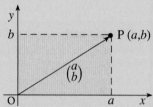

4 **a** Write down the position vectors
of the points A(2,3) and B(6,1).

b Hence find the position vector
of the midpoint, M, of AB.

---

5 A **matrix** is an array of numbers arranged in rows
and columns.
A **2 × 3 matrix** has **2 rows** and **3 columns**.
a To add matrices, you add the corresponding
components.
For example:

$$\begin{pmatrix} 1 & ^-2 \\ 3 & 4 \end{pmatrix} + \begin{pmatrix} ^-1 & 3 \\ 1 & 0 \end{pmatrix} = \begin{pmatrix} 0 & 1 \\ 4 & 4 \end{pmatrix}$$

b To multiply matrices, divide the first matrix into rows
and the second into columns, then multiply each row
by each column.
For example:

$$\begin{array}{cc} & \text{C1} \quad \text{C2} \\ \begin{array}{c} \text{R1} \\ \text{R2} \end{array} & \begin{pmatrix} 3 & ^-1 \\ 4 & 3 \end{pmatrix} \end{array} \begin{pmatrix} 2 & 0 \\ ^-1 & 3 \end{pmatrix} = \begin{pmatrix} 7 & ^-3 \\ 5 & 9 \end{pmatrix}$$

R1 × C1 = 3 × 2 + ¯1 × ¯1 = 6 + 1 = 7
R1 × C2 = 3 × 0 + ¯1 × 3 = 0 + ¯3 = ¯3
R2 × C1 = 4 × 2 + 3 × ¯1 = 8 − 3 = 5
R2 × C2 = 4 × 0 + 3 × 3 = 0 + 9 = 9

5 $\mathbf{A} = \begin{pmatrix} 3 & 2 \\ ^-1 & 4 \end{pmatrix}, \quad \mathbf{B} = \begin{pmatrix} ^-1 & 2 \\ 3 & 5 \end{pmatrix}$

and $\mathbf{C} = \begin{pmatrix} 4 & 0 \\ ^-5 & 2 \end{pmatrix}$

a Find:
   i $\mathbf{A} + \mathbf{B}$
   ii $3\mathbf{A} + 2\mathbf{C}$
   iii $\mathbf{B} - \mathbf{C}$
   iv $5\mathbf{B} - 2\mathbf{A}$
   v $2\mathbf{A} - \mathbf{C}$
   vi $\mathbf{AB}$
   vii $\mathbf{BA}$

b Does $\mathbf{AB} = \mathbf{BA}$?

Index